Economic Analysis of Solar Thermal Energy Systems

Solar Heat Technologies: Fundamentals and Applications
Charles A. Bankston, editor-in-chief

1. *History and Overview of Solar Heat Technologies*
Donald A. Beattie, editor

2. *Solar Resources*
Roland L. Hulstrom, editor

3. *Economic Analysis of Solar Thermal Energy Systems*
Ronald E. West and Frank Kreith, editors

4. *Fundamentals of Building Energy Dynamics*
Bruce Hunn, editor

5. *Solar Collectors, Energy Storage, and Materials*
Francis de Winter, editor

6. *Active Solar Systems*
George Löf, editor

7. *Passive Solar Buildings*
J. Douglas Balcomb and Bruce Wilcox, editors

8. *Passive Building Cooling*
Jeffrey Cook, editor

9. *Solar Building Architecture*
Bruce Anderson, editor

10. *Fundamentals of Concentrating Systems*
Lorin Vant-Hull, editor

11. *Distributed and Central Receiver Systems*
Lorin Vant-Hull, editor

12. *Implementation of Solar Thermal Technology*
Ronal Larson and Ronald E. West, editors

Economic Analysis of Solar Thermal Energy Systems

edited by Ronald E. West and Frank Kreith

The MIT Press
Cambridge, Massachusetts
London, England

This book was set in Times New Roman by Asco Trade Typesetting Ltd. in Hong Kong and printed and bound by Halliday Lithograph in the United States of America.

Library of Congress Cataloging-in-Publication Data

Economic analysis of solar thermal energy systems.

(Solar heat technologies: fundamentals and applications; 3)
Bibliography: p.
Includes index.
1. Solar energy—Economic aspects—United States. 2. Solar heating—Economic aspects—United States. I. West, Ronald E. II. Kreith, Frank. III. Series: Solar heat technologies; 3.
TJ809.95.S68   vol. 3   697'.78s   87-37864
ISBN 0-262-23140-9 (v. 3) [338.4'769778]

Contents

Series Foreword

Charles A. Bankston

This series of twelve volumes summarizes research, development, and implementation of solar thermal energy conversion technologies carried out under federal sponsorship during the last eleven years of the National Solar Energy Program. During the period from 1975 to 1986 the U.S. Department of Energy's Office of Solar Heat Technologies spent more than $1.1 billion on research development, demonstration, and technology support projects, and the National Technical Information Center added more than 30,000 titles on solar heat technologies to its holdings. So much work was done in such a short period of time that little attention could be paid to the orderly review, evaluation, and archival reporting of the significant results.

It was in response to the concern that the results of the national program might be lost that this documentation project was conceived. It was initiated in 1982 by Frederick H. Morse, director of the Office of Solar Heat Technologies, Department of Energy, who had served as technical coordinator of the 1972 NSF/NASA study "Solar Energy as a National Resource" that helped start the National Solar Energy Program.

The purpose of the project has been to conduct a thorough, objective technical assessment of the findings of the federal program using leading experts from both the public and private sectors, and to document the most significant advances and findings. The resulting volumes are neither handbooks nor textbooks, but benchmark assessments of the state of technology and compendia of important results. There is a historical flavor to many of the chapters, and volume 1 of the series will offer a comprehensive overview of the programs, but the emphasis throughout is on results rather than history.

The goal of the series is to provide both a starting point for the new researcher and a reference tool for the experienced worker. It should also serve the needs of government and private-sector officials who want to see what programs have already been tried and what impact they have had. And it should be a resource for entrepreneurs whose talents lie in translating research results into practical products.

The scope of the series is broad but not universal. It is limited to solar technologies that convert sunlight to heat in order to provide energy for application in the building, industrial, and power sectors. Thus it explicitly excludes photovoltaic and biological energy conversion and such thermally driven processes as wind, hydro, and ocean thermal power. Even with this limitation, though, the series assembles a daunting amount of information. It represents the collective efforts of more than 200 authors and editors. The volumes are logically divided into those dealing with general topics such as the availability, collection, storage, and economic analysis of solar energy and those dealing with applications.

The present volume covers the economic methods developed to analyze solar energy systems. The common characteristic of such systems is that they require substantial initial investments but have low operating costs relative to conventional energy systems. Analysts have therefore paid a great deal of attention to the evaluation of such investments in a period of highly uncertain energy futures. The volume provides sufficiently detailed information to serve as a sourcebook for anyone interested in the economic evaluation of solar investments. Except in the context of examples, though, it does not present results and does not attempt to compare or rank technologies. The dynamics of the energy and money markets and the level of maturity of the industry supplying solar technology are such that comparisons made prior to the time of an investment are usually inadequate. For further specific information, readers should consult the application-oriented volumes in this series, volumes 6–11, which contain economic evaluations of projects and technologies that have become part of program development and present the context in which these evaluations were made.

Acknowledgments

Ronald E. West and Frank Kreith

The publication of this volume would not have been possible without the belief in and dedication to solar energy by a number of people over several years. Frederick Morse and Charles Bankston energized the project and provided leadership and guidance. Lynda McGovern-Orr was a patient problem solver who helped overcome many hurdles. Paul Notari and Charles Berberich organized assistance through the Solar Energy Research Institute and were helpful in finding ways to get things done. Nancy Reece edited all the manuscripts carefully and well. Judy Hulstrom cheerfully word-processed each chapter—several times. Jack Roberts and Oscar Hillig dealt cooperatively with editors and authors on the financial aspects of the work. Ronal Larson conceived the original outline and structure of the book and solicited many of the authors; he played an important role in shaping the result.

We are also grateful to those professionals who reviewed draft chapters and whose comments were valuable in improving the final material for the book. These reviewers are Tung Au, Carnegie-Mellon University; William A. Beckman, University of Wisconsin; Roger Bezdek, U.S. Department of Treasury; Bruce W. Cone, U.S. Department of Agriculture; the late James Easterling, Pacific Northwest Laboratories; Theresa Flaim, Niagara Mohawk Power Corporation; William R. Gates, Jet Propulsion Laboratory (JPL); Charles Hall, Cornell University; Rodney Hardee, Los Alamos National Laboratory (LANL); Russell Hewett, Solar Energy Research Institute (SERI); J. J. Iannucci, Sandia National Laboratory; Ralph J. Johnson, National Association of Home Builders Research Foundation; David Kearney, Consulting Engineer; Henry C. Kelley, Office of Technology Assessment, U.S. Congress; Thomas A. King, Mueller Associates, Inc.; Ronal W. Larson, EMA, Inc.; Henry Lee, Harvard University; George O. G. Löf, Colorado State University (CSU); E. Kenneth May, Industrial Solar Technology, Inc.; Frederick H. Morse, U.S. Department of Energy (DOE); Stanley Mumma, Pennsylvania State University; L. Marty Murphy, SERI; Patrick J. Pesacreta, DOE; David B. Reister, Institute for Energy Analysis, Oak Ridge; Frederick Roach, LANL; Rosalie Ruegg, National Bureau of Standards; Jefferson Shingleton, Consulting Engineer; Katsuaki L. Terasawa, JPL; Rebecca Vories, Infinite Energy; C. Bryon Winn, CSU; Anthony M. Yezer, George Washington University; and Michael Yokell, Energy and Resources Consultants, Inc.

We express great thanks to those mentioned above and especially to the authors. Others have made contributions to this work in less tangible ways, and their help is also acknowledged. We have tried to incorporate all the constructive suggestions and criticisms to the best of our ability but to err is human. As editors, we accept the responsibility for any errors that may have crept into the final product.

Economic Analysis of Solar Thermal Energy Systems

1 Introduction

Ronald E. West

1.1 The Importance of Economics in Solar Energy Analyses

Economics plays a crucial role in the development and implementation of all types of solar thermal energy systems. Because of its importance, economic considerations are presented at the beginning of this series, which assesses the development and the status of these technologies.

Economic evaluation consists of describing a project in economic terms, that is, expressing the costs and benefits of an activity in monetary units (such as dollars) and then comparing the costs and benefits. Economic evaluation addresses the following questions: Is a project economically efficient? How does a project compare economically with alternative projects? What is the most economically favorable timing of the investment? Answers to these questions are important in making research funding decisions, such as deciding on a potential expenditure and deciding how to allocate funds among various possible expenditures.

The economics of solar systems are defined by the solar resource: sunshine may be free, but collecting it is not! The density of solar radiation striking the earth is low, and large collector areas are necessary to obtain substantial energy outputs. Traditionally energy users have not made the investment necessary to make an energy source, such as a fossil fuel, available, but have purchased the energy source from a supplier. The user of solar thermal energy, though, typically does make the investment that is necessary to make the solar energy usable. Thermal applications of solar energy require large initial investments in construction and in energy before any energy is delivered or any financial savings can be realized. Thus solar systems are characterized by large initial capital investments by the energy user that require careful economic analysis and planning. The cost of operating a solar thermal system, on the other hand, is usually small compared to the cost of buying fossil fuels. Once a solar system has been built, its operating expenses are small, consisting only of modest power costs to drive and control the system plus repair and maintenance costs. Fossil fuels, however, must be supplied and paid for in proportion to the heat requirement. So the user benefit from a solar heat system is the saving in operating expense for fuel, whereas the cost is primarily the initial investment.

The large initial investment characteristic of solar thermal applications makes them candidates for investigation by the classical methods of investment analysis. These methods compare the monetary value of an investment with the monetary value of the earnings or savings produced by the investment over its lifetime. The methods assess the merit of future earnings or savings compared to the investment of capital and are used routinely for business and industrial investment decisions and sometimes for decisions in the public sector. They are less commonly used by individuals, such as a homeowner considering installation of a solar thermal water heater. But any investment decision involves weighing benefits and costs and therefore can gain from an economic analysis. In a broad sense an economic analysis is desirable for every investment decision in order to allocate total available capital better.

Economic analysis of solar thermal systems encompasses more than just investment decisions. Economic modeling by mathematical representation of the economic performance of a system has been applied to issues such as prediction of market penetration, cost requirements for new technologies, applications analysis, and the analysis of effects of governmental taxation and incentive policies on technology development and utilization. Economic analysis techniques have also been used to provide guidance for research and development, as well as commercialization planning. These subjects are important to the government as promoter and user of the technology, to the private sector as supplier and user, and to individuals as users. Hence economic analysis is important because of its pervasive influence on investment decisions, on the research and development climate, and on commercialization of solar thermal technologies.

Investment decisions and economic analyses require prediction of conditions in the future because the benefits of most investments accrue after the investment has been made. Our ability to predict trends in energy economics is not good, even for the short range, as demonstrated by energy market events between 1973 and 1986. General economic conditions influence the energy market, of course. But energy supply and demand, which affect price, are also influenced by government policies, consumer attitudes, and especially international events and the policies of oil-exporting countries. Although over the long term energy prices will eventually rise compared to the prices of other goods and services, estimates of when this will happen are made with far less confidence in 1986 than they were in 1976 or even 1983. The results of economic analyses of solar energy projects are no better than the assumptions about the future trends of parameters such as general inflation rate, discount rate, and fuel price escalation rate. The more promising a project is in the short term, the less it depends on the accuracy of predictions.

This volume is not intended to cover all of the vast literature on solar thermal economics; we have chosen to review only the most important contributions related to the federal solar programs. Our emphasis is on economic methods and modeling and their use in studying policy alternatives. This volume also reviews applications analysis and net energy analysis. Applications analysis lies between economics and the technology, and net energy analysis is a methodology similar to economic analysis, but it uses energy rather than money as the quantity of measurement. The tool of cost requirements, used in program planning, is reviewed for active heating and cooling, for passive heating and cooling, and for electric-power generation and industrial process heat. Because one indicator of the success of the solar thermal programs is how the cost of solar systems changed over time, solar systems' cost histories are also reviewed.

Progress in the use of economics for solar energy decisions made under federally sponsored solar thermal programs is recorded in this volume; it should not be considered a tutorial or a complete review of the literature. We hope that it will assist those in the energy field to build on the experience and techniques in economics that have been developed over the past ten to fourteen years. Methods described here are used throughout this series, especially for the design and optimization analyses in volumes 6, 7, 10, and 11 and for discussions of research and development planning in volume 5.

This volume has two main objectives: to report on the research and development of economic techniques primarily under federal sponsorship, between approximately 1972 and 1985, and to describe applications of these techniques and how they influenced research, development, and demonstration activities in the federally sponsored programs.

1.2   Requirements for an Economic Analysis

A complete economic analysis of a solar thermal system requires not only an economic methodology with economic parameter values but also cost and performance data. Figure 1.1 indicates the information inputs needed to model the performance of a solar thermal system in economic terms:

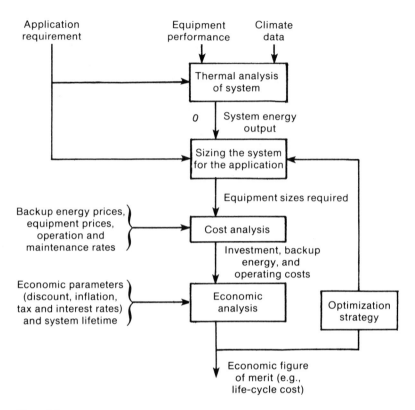

Figure 1.1
Economic performance model of a solar thermal system.

the requirements of the application (heat rate, timing, and temperature), the solar system performance characteristics (optical properties, heat loss rates), and the local climate data (insolation and weather) are all needed to predict the energy output of a solar system. With a knowledge of this output the system can be sized for a particular heat delivery rate; that is, equipment sizes and backup energy requirements may be estimated. To do so requires appropriate system performance models, and several such models are widely available (see chapter 3). Equipment costs, operating and maintenance (O&M) rates, and the price for the alternative conventional energy source can then be combined with the sizing results to compute the capital investment, the backup energy supply cost, and the O&M expense. This information along with an estimate of the useful lifetime of the system and values for the necessary economic analysis parameters (for example, interest, inflation, and discount rates, and tax policies) are the inputs to the economic evaluation itself. This economic evaluation produces a "figure of merit," or "objective criterion," such as the life-cycle cost. Usually the steps from the system sizing through the economic analysis are repeated until an economically optimal system size is found (see chapter 3 for a discussion of optimization).

Solar system performance and thus solar economics depend on location and load. Volume 2 of this series provides the pertinent insolation and climate data. The prediction of the performance of solar systems is a major concern of this series, and volumes 5 through 11 all contain relevant information. These volumes also contain information on the costs of components of solar systems, an essential ingredient in any economic evaluation. A historical perspective on installed costs of solar heat systems is presented in chapter 10.

1.3   Issues in the Economic Analysis of Solar Systems

The investment of capital in a particular project to generate income or savings in the future raises the question, Is it worthwhile to invest capital in this project? Capital can always be invested in alternative ways, so criteria are needed to guide the investment decision. Suitable criteria, such as life-cycle cost, net present worth, or payback time, and their calculation are the subject of chapter 2, where eight different methods are presented. It is useful here to consider some of the fundamental quantities that go into such an analysis. "Interest" is, broadly, the return obtainable from the productive investment of capital; it usually refers to money paid for the use of borrowed money. Because money can always be invested at some interest rate, an amount of money available at a certain time is equivalent to a larger amount in the future by the amount of interest earned. To turn this around, an amount available in the future is equivalent to a smaller sum available at the present time, again because interest could be earned from the present at some rate. This is the concept of "time value of money," or the "present worth" (also called "present value") of an amount of money available in the future. The rate used to compute present worth is usually called the "discount rate" and converting future monies to their present value is called "discounting." The discount rate need not be the same as the interest rate paid on borrowed money, as discussed later.

Present worth is not to be confused with inflation; inflation refers to an increase in price level so that the same amount of money has less purchasing power, that is, will buy less goods and services than at an earlier time. Thus money available at a given time in an inflationary economy is worth more than the same amount of money available in the future because of the combination of both the time value of money and inflation.

Inflation raises an important issue with respect to the rate of interest, namely, the differential between the interest rate and the inflation rate. If the general rate of inflation is, say, 5% and interest is being earned at a rate of 5%, there is no growth in the purchasing power of interest-bearing capital. The actual rate at which interest is being paid is called the "nominal" rate, and the effective rate of interest over and above the rate of inflation is called the "real" rate. The effective or real interest rate is approximately equal to the difference between the nominal interest rate and the rate of inflation (but not exactly, as these values are compounded and the difference must be calculated using the correct compounding formulas). This concept of a real rate applies to any other economic parameter that is a time rate, such as the discount rate. Different inflation rates may be associated with different goods and services. For example, throughout the 1970s and early 1980s the rate of inflation of energy fuels was much greater than that of the general economy. The high energy inflation rate was a major contributing cause to the general inflation. The concept of a real rate applies to the fuel price inflation rate as well.

Considerations other than purely economic ones may weigh heavily in energy policy and economic decisions. From a national perspective, security of supply and independence from foreign suppliers may be overriding considerations. To an individual or firm, public relations, security, and independence may also be important, even overriding. Other factors such as aesthetics, patriotism, and environmental protection are often paramount to an individual. Such factors are considered in volume 12.

1.3.1 Economic Parameter Values

The values used for such parameters as the interest, discount, and inflation rates, project lifetime, and the prices of alternative energy sources are critical in making economic evaluations of solar energy systems. Current and future values of these quantities must be estimated based on the best information available at the time of the evaluation. Factors entering into any economic evaluation will change in unpredictable ways, so it is likely that any economic evaluation will turn out to be inaccurate! We do not recommend values of economic parameters and future fuel prices to use. Instead, chapter 2 includes, in its examples, parameter values used in past economic studies. Solar system lifetimes are discussed in subsequent technology volumes of this series. Future prices for various energy sources, perhaps the most important of all the values used in solar energy economic evaluation, have been the subject of many studies. Chapters 4 and 10 discuss several of these studies and give useful references on this difficult issue. The Energy Information Administration (EIA) of the Department of Energy (DOE) publishes an *Annual Energy Outlook* that includes projections of the prices of the major energy sources to the user. Prices are projected ten years ahead in constant dollars of the year of the report (EIA, 1986).

The discount rate value to use in the evaluation of energy investments is a controversial issue. An excellent discussion is given by Lind et al. (1982). The discount rate reflects the required rate of return on an investment plus an allowance for risk associated with the investment. In the private sector the required rate of return is usually taken as the cost of capital, that is, the market rate of interest or the cost of equity financing. For individuals the required rate of return is the minimum rate at which a consumer is willing to forgo consumption now for consumption in the future. In theory the private sector rate is the before-tax cost of capital and the individual rate is the after-tax cost of capital; in either case the discount rate is increased to account for risk (uncertainty) inherent in predicting the future and in the expected performance of the investment project. The Office of Management and Budget directed in 1972 that a real discount rate of 10% be used in the evaluation of federal projects. The 1980 Energy Security Act (PL96-294) required that a 7% real discount rate be used to evaluate federal solar energy projects (Ruegg, 1986). This value is based on private sector discount rates. Most energy investments are made in the private sector, so it is often argued that private sector financing criteria be used in energy project evaluations. Nonetheless, many different discount rates have been used in energy studies, ranging at least from 0% to 13%.

1.3.2 Inflation Adjustment

This volume discusses the developments in solar energy economics covering the fourteen years of 1972–1985. This was a period of profound change in the energy business worldwide as well as a period of steep inflation, partly because of energy costs. Direct comparisons of costs and prices are difficult because of the rapid changes. Therefore in this volume cost and price values have been converted into dollars for a common year, namely 1985, using the U.S. Department of Commerce Gross National Product (GNP) Implicit Price Deflator Index. We could have used several other indexes, such as the Engineering News-Record Index. Some of these are related to construction or manufacturing price changes and might therefore reflect changes in the cost of solar systems more closely than the GNP index. There is not, however, a "solar equipment cost index" or anything equivalent. The GNP index was chosen, therefore, because it reflects general inflation in the overall economy and because it is readily available.

Table 1.1 includes the GNP deflator index for 1972 through 1985, as used throughout this series. For other years the Survey of Current Business, published by the U.S. Department of Commerce, should be con-

Table 1.1
Gross National Product Implicit Price Deflator Index

Year	Index	Conversion to 1985 dollars
1972	46.5	2.402
1973	49.5	2.257
1974	54.0	2.068
1975	59.3	1.883
1976	63.1	1.770
1977	67.3	1.659
1978	72.2	1.547
1979	78.6	1.421
1980	85.7	1.303
1981	94.0	1.188
1982	100.0	1.117
1983	103.8	1.076
1984	108.1	1.033
1985	111.7	1.000

Source: U.S. Department of Commerce (1986).

sulted. To use the table, one takes a price known for a particular year, say, 1973, and converts that price into 1985 dollars by multiplying by the ratio of the 1985 index (111.7) to the 1973 index (49.5), that is, by 2.257. The result does not mean that an item priced at $1 in 1973 would be priced $2.26 in 1985. Rather, it means that because of the decline in the purchasing power of the dollar, it took $2.26 in 1985 to buy the same "market basket" of goods that $1 purchased in 1973. The 1985 index value was reported in March 1986 and subsequently may have changed slightly as the basic economic data were refined. The procedure in this series has been to report values in the same terms as they were originally reported (for example, 1973 dollars) and then to convert and give the value in 1985 dollars. Sometimes on tables or graphs, especially when data from several different years are included, values are presented in the current year dollars, as they were reported, and the multiplying factors to convert to 1985 dollars are given.

The term "current year dollars" refers to numbers that are expressed as the dollar value when they were stated. The term "constant dollars" is applied to numbers that have been converted into the dollar value in one particular year, 1985 in this series.

1.4   Economic Methods in the Federal Solar Program

The U.S. government's support for research, development, and demonstration (RD&D) on solar energy was about $2 million in fiscal year (FY) 1972 (about $5 million in 1985 dollars) (Herwig, 1981). This had increased to a peak of about $1 billion in FY 1982 (about $1.1 billion in 1985 dollars). The total federal commitment to solar energy, including tax incentives, approached $1.5 billion in FY 1982 (about $1.7 billion in 1985 dollars) (DOE, 1982). This growth in support for solar energy was spurred by the oil boycott in the fall of 1973, which created an awareness of impending energy supply shortages and the dependence of the United States on imported oil, and again by sharp price increases in 1979. Solar budget authorizations have declined since 1982 as the federal government's policies and the world energy market have changed.

A goal of the U.S. government's solar energy program was "to develop and introduce, at an early date, economically competitive . . . solar energy systems" (ERDA, 1975). Thus economic analysis was an essential ingredient in designing, planning, guiding, and evaluating the research and development program. Economic methods needed to be incorporated into the technology research, development, and commercialization programs. Moreover, as the programs progressed, needs were recognized for common methods, assumptions, values, and criteria in order to make appropriate comparisons of projects and technologies. Also recognized was that planning solar programs and evaluating solar projects required extensions of existing economic tools and methods, including combining these methods with techniques and models for predicting the performance of solar systems. For example, predicting market penetration, establishing component cost goals, and estimating the profitability of solar projects required that economics techniques be developed to mesh with the peculiar characteristics of solar energy systems.

Before 1976, in fact, the federal solar energy budget included only research, development, and demonstration programs; economic analyses were not separate programmatic or budget items. Between 1976 and 1982 programs other than RD&D were budgeted. One category in the DOE solar budget was market analysis, which included the evaluation of potential markets for all solar technologies and the development of market strategies for those technologies. Those economic analysis activities that were explicitly budgeted were included under this category. The total amount authorized for market analysis in the solar heat technologies (passive and active heating and cooling, agricultural and industrial process heat, and solar thermal power) was $8.4 million in FY 1980 ($10.9 million in 1985 dollars), for example—about 1.4% of that year's total solar energy budget and 3.3% of that budgeted for solar heat programs (DOE, 1980). These economic analysis activities clearly were never a big portion of the solar heat programs, and they were always controversial. Nonetheless, there was a period when their role was significant.

The activities conducted under the market analysis function are listed in table 1.2, including the percentage of the total DOE solar market analysis budget devoted to each activity in FY 1980. Other economic analyses, such as cost goals studies and specific project evaluations, were conducted as technology program activities and did not appear as items in the overall DOE solar budget, so total expenditures on such economics studies were greater than the numbers would indicate. I could not reconstruct a budget history of economics studies for solar heat technologies from the

Table 1.2
Market analysis activities in the DOE Solar Program, FY 1980[a]

Activity	% of budget allocated to activity
Economic modeling and analysis	20
Economic and financial incentives	8
Public utility/solar energy interface	8
Legal issues	17
Employment and work force assessment	17
Marketing and consumer response	27
International market analysis	3

Source: DOE (1980).
a. Total authorized for market analysis activities in 1980 = $8.4 million.

public record. The information given for 1980 is near the high water mark for the amount explicitly spent for economics studies. By 1983 explicit expenditures on economics had disappeared from the federal solar energy budget. Nonetheless, some types of economic analyses continue to be performed for technology programs because they are needed as guidance and in evaluations of those programs. From 1976 through 1982 economics studies were a small but significant feature of the DOE solar program and have continued to be an important facet of program activities.

1.5   Contents of This Book

Chapter 2, "Economic Methods," is coauthored by Rosalie Ruegg and Walter Short. This chapter gives basic background on the principal methods of economic evaluation, stressing advances in the methodologies and their application to solar systems made as a consequence of the federal solar heat programs. Four major solar areas subject to economic analyses are described: the economic feasibility of solar heat projects, the estimation of market potential and penetration by solar technologies, the economic impacts of their implementation, and the formulation of government policies and plans toward solar energy.

Chapter 2 is not a handbook on how to apply these techniques, although much information about them is given. Rather, it reviews the state of the art of the methods and the impact of federal programs on them. The authors point out that the contributions of solar studies to economic methods generally have been adaptive, but they are nonetheless extensive and important. Eight methods of economic evaluation are presented, and many applications of these methods to solar and economic issues are discussed. Six detailed examples are worked out using the economic evaluation methods; these illustrate the similarities and differences of the methods as well as typical values used for the economic parameters in studies of solar energy investments in the 1970s and early 1980s. The reader is also referred to sources that document economic studies and give instruction on the use of the methods.

Economic models include the methods of evaluation described in chapter 2 and extend beyond these to include criteria and techniques for analysis of such decisions as optimal size and timing solar investments and optimal combinations of solar and conservation technologies. G. Thomas Sav wrote chapter 3, "Economic Models," to present background on the theory of economic modeling as it applies to design decisions, market penetration studies, and analysis of incentives for solar heat systems. This theoretical background is important to understand the fundamental assumptions made in economic modeling of solar heat projects.

Sav reviews the integration of economic models with solar system performance models, one of the more challenging aspects of solar economic analysis because of the unique problems encountered in predicting solar performance. He discusses six integrated, computer-based simulation models. For each of these models he emphasizes the economic modeling and compares optimal design results obtained by using each. In addition, he discusses two simplified deterministic models and one econometric model for solar system optimization. This useful review includes a discussion of the advantages and disadvantages of each of these models.

Whereas chapters 2 and 3 present fundamental concepts and methods of economic analysis, the next six chapters relate to applications of the methods to solar issues: market penetration, incentives, applications and net energy analysis, and cost requirements. In chapter 4, "Assessing Market Potential," Gerry Bennington writes on assessment of the market potential of solar technologies. Market potential refers to the portion of a market, say home heating or cooling, which might be captured by a technology. Market potential assessments are needed for policy and incentives analysis, goal setting, RD&D planning, commercialization strategies, and impact assessments. The expected magnitude and timing of potential markets can determine the allocation of resources among these activities. The anticipated cost of solar systems is a major input into market assessment, and economic evaluation is one of its main tools.

Bennington describes components of market potential assessment such as national energy supply and demand models, national solar models, and market segment and technology models. He reviews eight major solar market potential studies, presenting the main points of their methodologies and comparing projections made with each. Penetration of a technology into a market is a function of its economics relative to that of competing technologies. Thus future cost-trend projections are essential to market potential studies; the timing of these trends is especially difficult to predict. The usefulness of these several market potential studies is weighed by Bennington. Although he indicates that they may provide guidance as to comparative trends for different technologies, he cautions that their results should not be treated as predictions. The variables are too numerous and the uncertainties too great to be able to predict the market penetration by a new technology.

The use of incentives by the federal government to assist in the development of a solar energy alternative has been a major policy issue. Peter Spewak addresses this difficult subject in chapter 5, "Analyzing the Effect of Economic Policy on Solar Markets." He presents the rationale for government incentives in energy resource development in general and for solar energy in particular. Seven categories of government incentives are listed: creation or prohibition of organizations, taxation, disbursements, requirements, traditional and nontraditional government services, and market activity.

Spewak reviews arguments regarding the appropriate type, level of support, and timing of incentives; discusses the potential benefits and costs of incentives; and gives the incentives that were proposed and those enacted for solar energy. Evidence regarding the effectiveness of the incentives programs is presented and discussed. Spewak concludes that the solar energy incentives were not as effective as originally expected. The uses of incentives in solar technology implementation are covered in volume 12 of this series.

Chapter 6, "End-Use Matching and Applications Analysis Methodologies," by Kenneth Brown, deals with applications analysis. Applications analysis uses economic evaluation methods but is primarily concerned with the end-use of the energy. Brown reviews the federal activity for development of an applications analysis methodology to guide the

solar industrial process heat (SIPH) and the solar thermal power programs. Applications analysis tests the compatibility of an energy supply
technology with the usage requirements of the energy. It is not a detailed
design but looks for temporal and temperature matches in order to define
a potential market and to establish technical needs. The technique developed includes requirements analysis—a database on the temperature,
schedule, use rate, and geographical distribution of industrial process
heat users and power generators—and end-use matching—a procedure
for finding the solar collector type that most economically meets a set
of process requirements. Together, requirements analysis and end-use
matching can provide a detailed mapping of applications opportunities.

Brown traces the historical development of requirements and end-use
analysis under the federal program, beginning with several studies in the
early 1970s and leading to the end-use matching work done by the Solar
Energy Research Institute in the early 1980s. By that time much of the
definition and clarification needed to focus on candidate applications
was done and more detailed phases were underway. This program ended
in 1982, but it had been useful in identifying potential markets and in
establishing the end-use matching methodology.

Robert Herendeen introduces net energy analysis in chapter 7, "Net
Energy Considerations." Net energy analysis employs techniques similar
to those of economic evaluation but with energy instead of money as the
quantity of measurement. Herendeen outlines the philosophy and the
procedures of this approach. Clearly a technology that consumes more
energy in putting it in place than it can produce during its useful life is not
desirable as an energy supplier. Conceptual and procedural difficulties
that occur in the application of the method are pointed out. Herendeen
presents the results of many net energy analyses conducted by himself
and others for nonsolar and solar heat technologies; he uses the incremental energy ratio (IER), the total energy supplied to the rest of the
economy divided by the total energy input from the economy. The results
and significance of these studies are discussed. Herendeen points out several insights that net-energy analysis provides, such as the decline in the
IER for fossil fuel resources in recent years. Herendeen claims the
method has too many inherent problems to use it for comparison and
ranking of different technologies.

Cost requirements are the subject of chapters 8, 9, and 10. Each chapter is directed to a specific solar technology—active heating and cooling,

passive heating and cooling, and process heat and thermal power—but each author places different emphasis with regard to the estimation of requirements. Cost and performance requirements (or goals) are estimated values that technologies must reach by a given time in order to become competitive. They are of great value in program planning: for choices between projects, as guideposts to evaluate progress, and in setting the direction of project activities. Establishing these requirements or goals involves economic evaluations and market potential projections.

Mashuri Warren addresses active solar heating and cooling in chapter 8, "Cost Requirements for Active Heating and Cooling." He gives a detailed description of each step in the procedure to establish requirements as practiced in the active solar program. This procedure uses an economic evaluation method, a market penetration study, and a solar system performance model. The market penetration study provides a relationship between the fraction of the market captured and a solar system cost/performance parameter (expressed, for example, as payback time). A desired value of the cost/performance parameter plus the economic and the performance models permits fixing the amount and the time at which to invest. This set of values then becomes the system cost and performance goals at a certain date. The system cost and performance goals must be broken down into component goals to be useful. This is usually done as an allocation based on experience; there is not a unique set of component goals.

Warren shows a detailed example of a solar-driven absorption-cooling system. The economic and performance models and parameter values used, which are crucial to the outcome of the analysis, are given and discussed. He observes that the fuel price escalation rate used affects primarily the timing of the investment as opposed to the savings produced. He shows that a comparison of component goals with current values is useful in identifying opportunities and needs for component improvements. In his example he shows the importance and necessity of reducing electrical parasitics for the absorption-cooling technology.

In chapter 9, "Cost Requirements for Passive Solar Heating and Cooling," Charles Hauer takes a broad view as he discusses the technical and economic factors unique to passive designs. Because passive features are closely allied to the architecture of a building, the cost and performance increments they produce are difficult to isolate and quantify. Cost requirements are based on the realities of marketplace, salability, and price.

Hauer reviews and evaluates modeling methods that have been developed to estimate the performance and cost of passive design features. He emphasizes the importance of making the results of this work easily usable by designers and builders. Some key features of the Denver Metro passive solar program are presented. Hauer summarizes key concepts of several passive solar models. The solar load ratio method of Los Alamos National Laboratory is emphasized. He concludes that uncertainties in the cost and performance of passive designs make tenuous their economic analysis.

In chapter 10, "Cost Requirements for Solar Thermal Electric and Industrial Process Heat," Ronald Edelstein presents goals recently established for the Solar Thermal Program. He emphasizes the basic issues and methodological problems involved in establishing cost goals. These include value versus attainability as the basis for setting the goals (that is, what is necessary in order to be competitive versus what appears to be possible technically), the choice of target markets and alternative fuels, the time frame for the analysis, the financial and technical parameter values to be assumed, and how to measure progress versus the goals. He comments on resolution of these problems.

Edelstein discusses the importance of component goals and how to allocate system-level goals to the system components. He traces the evolution of methods to establish goals for the solar thermal technologies. He also assesses the procedures used to establish the recent program goals and evaluates the appropriateness of the results obtained.

The emphasis in this volume is on economic methods and the application of these methods to the issues of solar economics. One measure of progress in solar technologies is trends in the cost of solar heat systems. The federal solar energy program is only one of many factors contributing to these trends, so no causal relationship can be established. Nonetheless, a record of these trends is of great interest. Thus, in the concluding chapter of this volume, "Historical Cost Review," Charles Hansen and Wesley Tennant present historical trends in the cost of solar heat systems.

Hansen and Tennant have critically reviewed the literature and collected cost data for 1974 through 1982. Their experience shows the difficulty of finding complete and acceptably defined cost values. The authors use the installed system cost per unit of collector area, dollars per square foot (or per square meter) as the cost parameter, and group the results by temperature range and, where appropriate, by collector type, or energy use.

Trends in the cost data were examined with mixed results. The system cost, in 1985 dollars, remained essentially constant in three of the eight categories, whereas in four it decreased and in one case it increased. Hansen and Tennant discuss reasons for the observed trends, noting that experience, technical improvements, increased production rates, improved quality, and the relative size of systems all contribute to them.

References

EIA. 1986. *Annual Energy Outlook 1985*. DOE/EIA-0383(85). Washington, DC: U.S. Department of Energy, Energy Information Administration.

ERDA. 1975. *National Solar Energy Research, Development and Demonstration Program*. ERDA-49. Washington, DC: Energy Research & Development Administration, Division of Solar Energy, p. 1.

Herwig, L. O. 1981. "Perspectives of solar energy and its applications." *CRC Critical Reviews in Environmental Control* 11:301–403.

Lind, R. C., K. L. Arrow, G. R. Corey, P. Dasgupta, A. K. Sen, T. Stauffer, J. E. Stiglitz, J. A. Stockfisch, and R. Wilson. 1982. *Discounting for Time and Risk in Energy Policy*. Washington, DC: Resources for the Future, Inc.

Ruegg, R. 1986. *Life-Cycle Cost Manual for the Federal Energy Management Program, NBS Handbook 135 (Rev.)*. Gaithersburg, MD: National Bureau of Standards.

U.S. Department of Commerce. 1986. *Survey of Current Business* 66(3), Washington, DC: U.S. Department of Commerce, Bureau of Economic Analysis.

U.S. Department of Energy. 1980. *Solar Energy Program Summary Document FY 1981*. DOE/CS-0050. Washington, DC: U.S. Department of Energy.

U.S. Department of Energy. 1982. *Secretary's Annual Report to Congress*. DOE/S-0010(82). Washington, DC: U.S. Department of Energy.

2 Economic Methods

Rosalie T. Ruegg and Walter Short

This chapter covers the principal methods of economic evaluation that have been applied to solar thermal technologies; outlines major post-1972 developments, adaptations, and applications of economic methods to evaluating solar thermal technologies; and illustrates applications of these methods to solar thermal technologies through selected case studies from the literature.

The chapter is organized into eight sections. Section 2.1 identifies the solar energy economic issues that have prompted the need for economic methods. Section 2.2 is a brief synopsis of key economic evaluation methods used to address solar-related issues. Sections 2.3 through 2.6 draw from representative works published after 1972 that illustrate the application of economic methods to the different kinds of problems identified in section 2.1. Section 2.3 gives examples of feasibility studies of solar energy systems. Section 2.4 cites examples of marketing studies, particularly those that highlight factors important to the market acceptance of solar energy systems. Section 2.5 discusses economic impact studies, and section 2.6 discusses economic studies in support of formulating government programs and policies. To illustrate in more detail how analysts have applied economic methods to evaluate specific kinds of solar energy systems, section 2.7 critiques six economic case studies from the literature. Section 2.8 summarizes the chapter.

2.1 Solar Energy Economic Issues

Various solar energy issues have prompted the development, adaptation, and application of economic methods. The term "economic methods" here refers to systematic approaches to understanding economic phenomena. These methods, which are either newly developed or adapted from other areas of economics and related fields, reflect the economic phenomena to be investigated. Therefore a useful starting point in reviewing these methods is to summarize the major solar energy issues that prompted their development.

The solar economics literature reveals at least four major issues, each with several subdivisions. These issues with their respective subdivisions are economic feasibility (includes purchase decisions, financing decisions, and design/sizing decisions), marketing (includes factors important to market acceptance and potential market size), economic impact (includes employment, environment, and energy supply impacts), and government program and policy formulation (includes incentives and penalties, laws and regulations, demonstration projects, and efficient allocation of public research and development funds).

In the economic feasibility category such questions as the following have given rise to economic analyses: Do solar energy systems have the potential for cost-effectiveness (that is, for saving more than they cost over the long run)? Will they be made available in the marketplace and at what cost? Will individual consumers demand them and, if so, under what conditions? What designs and sizes are likely to be most cost-effective? Will systems be financed by lenders?

In the marketing category key questions that have prompted analyses are, What are the major determinants of demand for solar energy systems and the market barriers to widespread acceptance? How do we predict the sizes of various markets by geographical region, type of investor, building type, and in the aggregate?

In the economic impact category local, state, and federal governments have asked questions about the impacts of solar energy on local, regional, and national economies, including employment, the environment, and overall energy availability and costs.

Finally, the formulation of government programs and policy has caused investigation into the issue of what constitutes appropriate public policy regarding solar energy. Designing effective public programs to support these policies requires an understanding of the underlying economic relationships concerning solar energy and a predictive capability to test alternative actions.

2.2   Synopsis of Economic Methods

A variety of economic methods have been applied to solar-related phenomena to address the issues identified here. The methods vary across applications because of differences in objectives, perspectives, levels of sophistication and comprehensiveness, computation, and notation. Despite this diversity, certain economic methods are prominent in the solar economics literature. This section identifies the principal economic methods that were used to analyze solar-related issues and provides brief descriptions of the most frequently used methods.

To address the issues identified in section 2.1, researchers have drawn heavily from the existing literature for methods of evaluation. For example, they have drawn from the literature on benefit-cost analysis, such as Mishan (1976) and Dasgupta and Pearce (1972); from the literature on capital budgeting and financial analysis, such as Clark, Hindelang, and Pritchard (1984) and Weston and Brigham (1981); from the engineering economics literature, such as Grant, Ireson, and Leavenworth (1976), Smith (1973), and Au and Au (1983); from the operations research literature, such as Ackoff and Sasieni (1968) and Hillier and Lieberman (1980); and from the econometrics literature, such as Klein (1962) and Dorfman (1958).

Methods of measuring the economic performance of solar energy systems include life-cycle costing, net benefits (savings), required revenue, internal rate of return, benefit-to-cost (savings-to-investment) ratio, payback, levelized cost of energy, and break-even methods. These are briefly described in the following paragraphs, but first we need to define the nomenclature used.

The study period is designated by N. The discount rate is d, and the interest rate is i. The reinvestment rate is r. A_i denotes the energy system being evaluated. A_2 denotes a mutally exclusive alternative energy system against which A_1 is evaluated. Mutually exclusive alternatives are those for which accepting one automatically means not accepting the others. For a given project one mutually exclusive alternative is not to undertake the project, and it is against this alternative that potential investments are compared to determine their cost-effectiveness. The principal alternative to undertaking a project is referred to as a "base case." Alternative designs and sizes of a project for a given application are also mutually exclusive. $C_{A1,j}$ is the cost in year j for the system (A_1) being evaluated; $C_{A1,0}$ denotes the cost at the beginning of the period, and $C(x)_{A1,j}$ is the cost as a function of an unknown x. $C_{A2,j}$ is the cost in year j for the mutually exclusive alternative energy system (A_2); $C_{A2,0}$ denotes the cost at the beginning of the period. $B_{A1,j}$ is the benefits (positive cash flows including salvage values ($S_{A1,j}$)) in year j for A_1, and $B_{A2,j}$ is the benefits in year j for A_2. $A_1 : A_2$ denotes the evaluation of energy system A_1 relative to alternative A_2. The composite marginal income tax rate is designated by t. All cash flows are adjusted for income tax effects. The total annual energy load in Btu/yr (or J/yr) is denoted by L.

The first method for evaluating the economic performance of a solar energy system is the *life-cycle cost* (LCC) *method*. The unit of measure is the dollar. In this method the relevant present and future costs (less any positive cash flows such as salvage values) associated with an energy system are summed in present or annual value dollars over the study period. These costs include but are not limited to the facility's energy costs, acquisition costs of the energy system, operation maintenance, and repair and replacement costs. The formula for LCC is

$$\text{LCC}_{A1} = \sum_{j=0}^{N} (C_{A1} - B_{A1})_j / (1 + d)^j.$$

The LCC method can be applied to determine if a project is cost-effective and to find which combination of projects of variable design and size will minimize long-run costs of a given facility while meeting performance requirements. In order to determine whether an energy system is cost-effective, one must compare the LCC of the energy system to the LCC of the base case. If the LCC of the energy system is lower than that for the base case and the performance of the energy system in other aspects is equal, then the system is cost-effective. To use the LCC method to size and/or design an energy system, one must compare the LCCs of alternative sizes and/or designs of an energy system, choosing the design and size that meets the energy needs of the facility at the lowest LCC, other things being equal. To find the combination of projects that minimizes the overall LCC for a facility, one must compute and compare the LCCs of alternative combinations.

There are several special considerations that should be taken into account when using the LCC method. The method is most suitable when the focus is on costs rather than on benefits. When comparing alternatives based on their LCCs, benefits (positive cash flows such as salvage values) must be held constant or subtracted from costs. Second, effects that are not measured in dollars must be uniform across alternatives or should be weighted subjectively by the decision maker. Third, to be used as a decision tool, LCCs must be computed for two or more mutually exclusive alternatives over the same study period, such as LCC_{A1} and LCC_{A2}. Fourth, if the budget is limited, it may be necessary to forgo projects or increments to projects, even though they lower the LCC of the building or facility, in order to realize a greater savings from investing in competing projects or project increments with higher rates of return.

The second method that can be used to evaluate the economic performance of a solar energy system is the *net savings* (NS), or *net benefits* (NB), *method*. This method finds the net difference in present- or annual-value dollars between two alternative energy systems. The formula for finding the net savings (or benefits) in dollars is

$$NS_{A1:A2} = LCC_{A2} - LCC_{A1}$$

or

$$NB_{A1:A2} = \sum_{j=0}^{N} \frac{(B_{A1} - B_{A2})_j - (C_{A1} - C_{A2})_j}{(1 + d)^j}.$$

The NS or NB method can be applied to determine if a project is cost-effective. A project is cost-effective if NS or NB is positive. The method can also be used to find the best project design and size. Once again, the net savings or net benefits are computed for each design and size; the project size and design with the highest NS or NB is the optimal choice. And third, the NS or NB method can be used to determine the best project mix. In this application one wants to find either the combination of cost-reducing projects for a given facility that maximizes net savings while achieving a given level of benefits or the combination of benefit-producing projects that maximize net benefits.

There are three special considerations that should be kept in mind when using the net benefit or net savings method. First, effects not measured in dollars must be uniform across alternatives or should be weighted subjectively by the decision maker in comparing alternatives. Second, if there is no budget constraint, net savings (net benefits) will be maximized by accepting all independent projects and all increments to those projects having a positive net savings (net benefits). And finally, if the budget is limited, it may be necessary to forgo projects or increments to projects with positive net savings (net benefits) in order to realize a higher return from investing in competing projects or project increments.

The third method that can be used to evaluate the economic performance of a solar energy system is the *required revenue* (RR) *method*. The unit of measure is the dollar. The RR method divides the present-value or annual-value costs (less salvage values) associated with an energy system, adjusted for income taxes, by one minus the composite marginal income tax rate. This provides a measure of the before-tax revenue in present-value or annual-value dollars required to cover the costs on an after-tax basis. The formula used is

$$RR_{A1} = \left[\sum_{j=0}^{N} (C_{A1} - S_{A1})_j/(1 + d)^j \right] \bigg/ (1 - t).$$

The required revenue method can be used to decide between mutually exclusive alternatives by computing the required revenue for each alternative, such as RR_{A1} and RR_{A2}, and choosing the one with the lowest required revenue, other things being equal. Special considerations are that the RR method is commonly used in utility evaluations and that it is applicable only to revenue-generating investments.

The *internal rate of return* (IRR) *method* (unadjusted and adjusted) is the fourth method by which the economic performance of a solar energy system can be evaluated. The method is used to find the compound rate of interest (in percent) that will yield zero net savings (net benefits) when used to discount the differences in cash flows between two alternative energy systems. The unadjusted internal rate of return (UIRR) method is based on the assumption that returns from the project are reinvested at a rate of return equal to the rate of return on unrecovered investment funds. The adjusted (AIRR) version is based on the assumption that returns from the project are reinvested at a specified rate, which may differ from the rate on unrecovered investment funds.

For the unadjusted version one solves the following equation for the interest rate i:

$$\sum_{j=1}^{N} \frac{(B_{A1} - B_{A2})_j - (C_{A1} - C_{A2})_j}{(1 + i)^j} - (C_{A1_0} - C_{A2_0}) = 0.$$

Because of the nature of this equation, there may be multiple solutions or no solution for the unadjusted internal rate of return. For the adjusted version the equation to be solved for i is

$$\sum_{j=1}^{N} \frac{[(B_{A1} - B_{A2})_j - (C_{A1} - C_{A2})_j](1 + r_j)^{N-j}}{(1 + i)^N} - (C_{A1_0} - C_{A2_0}) = 0.$$

The approach used to estimate the cost-effectiveness of a project using the IRR method involves the comparison of the computed IRR of an energy system with the investor's minimum acceptable rate of return (MARR). The project is cost-effective if the IRR is greater than the MARR.

IRRs can be used for design and size determinations. In this application the decision maker computes the IRRs based on incremental cash flows by comparing each design and size of a project with the design and size of the same project just below it in cost. The more costly design or size is accepted (resp. rejected) if its *incremental IRR* is greater than (resp. less than) the MARR.

One can also assign priorities to independent projects under a budget constraint according to the descending order of their IRRs. Using the UIRRs to do this ususally leads to a *reasonably-satisfactory* group of projects within a budget constraint but may not maximize net benefits for the budget even if the budget can be totally expended. Setting priority among independent projects according to the descending order of the AIRRs will maximize net benefits provided that the budget can be totally expended. If project costs are "lumpy," so that the budget cannot be used up exactly, it may be necessary to depart from AIRR ranking to maximize net benefits from the budgeted expenditure.

The *savings-to-investment ratio* (SIR) [*benefit/cost ratio* (BCR or B/C)] *method* finds the ratio of the difference in future net cash flows of one energy system relative to an alternative system to the difference in their initial investment costs. The formulas to compute the ratios are

$$BCR_{A1:A2} = \sum_{j=1}^{N} \frac{[(B_{A1} - B_{A2})_j - (C_{A1} - C_{A2})_j]/(1 + d)^j}{(C_{A1_0} - C_{A2_0})},$$

$$SIR_{A1:A2} = \sum_{j=1}^{N} \frac{(C_{A2} - C_{A1})_j/(1 + d)^j}{(C_{A1_0} - C_{A2_0})}.$$

The SIR or BCR method can be used to estimate the cost-effectiveness of a project. The project is considered cost-effective if the ratio is greater than one.

The decision maker can use SIRs or BCRs for design and size determinations. As with the internal rate of return method, the analysis must be based on incremental cash flows, where each design and size of a project is compared with the design and size of the same project just below it in cost. One would accept (resp. reject) each more costly design and size if its incremental ratio is greater than (resp. less than) 1.0 and there is no budget constraint or if the incremental ratio is greater than (resp. less than) the cutoff ratio and there is a budget constraint.

As with the internal rate of return method, a decision maker can assign priorities to independent projects in descending order of the projects'

SIRs or BCRs. Setting priority among independent projects according to the descending order of their SIRs will maximize net benefits, provided that the budget can be used up exactly. If project costs are "lumpy," so that the budget cannot be used up exactly be adhering strictly to SIR rankings, it may be necessary to depart from the SIR ranking to maximize net benefits from the budgeted expenditure.

The sixth method for evaluating the economic performance of a solar system is the *payback* (PB) *method*, in either its simple (SPB) version or its discounted (DPB) form. The method finds the time (usually in years) required for the cumulative difference in future cash flows of one energy system relative to an alternative system to just equal the difference in their initial investment costs. Simple payback ignores the time value of money and hence misstates the true time to payback. Discounted payback includes the time value of money and hence is the more accurate method for finding the time to payback.

The method involves finding the minimum solution value of PB for which

$$\sum_{j=1}^{PB} \frac{(B_{A1} - B_{A2})_j - (C_{A1} - C_{A2})_j}{(1 + d)^j} = C_{A1_0} - C_{A2_0}.$$

For simple payback, $d = 0$.

The payback method can be used as a rough guide to cost-effectiveness. One solves for the time to payback, and, if this time is significantly less than the expected project life, the project is likely to be cost-effective. When project life is uncertain, one can perform a break-even analysis of project life. Again, the time to payback is calculated, and then the likelihood of the project's actual life exceeding the solution value is evaluated. The solution value is the break-even life. In addition, the payback method can be applied to protect the initial investment in the face of uncertainty. If risk is considered to be an increasing function of time, the maximum acceptable payback period (MAPP) may be constrained to eliminate projects that require a relatively long time to recover their costs.

It should be noted that the payback period method does not provide a comprehensive economic assessment. The investment with the shortest payback may not be the best investment. In addition, even if the payback measure shows the number of years to payback to be less than the project's life or the investor's MAPP, the investment may not provide the return expected by the investor.

The *levelized cost of energy* (LCOE) *method* can also be used to evaluate economic performance. This method finds the annualized cost (in dollars) per unit of load for an energy system. The equation is

$$\text{LCOE}_{A1} = \sum_{j=0}^{N} [(C_{A1} - B_{A1})_j/(1 + d)^j]\{d(1 + d)^N/[(1 + d)^N - 1]\}/L.$$

The LCOE method can be applied to determine a project's cost-effectiveness and its optimal design and size. For a cost-effectiveness determination, the LCOE is computed for a solar/auxiliary system and for the base-case energy system. The solar/auxiliary system is cost-effective if its LCOE is lower than that for the base case. To make design and size evaluations, one must compare the LCOEs of meeting the load by alternative combinations of solar/auxiliary energy and choose the system design and size that meets the requirements at the lowest unit costs, that is, the lowest LCOE.

With the LCOE method, one should take a life-cycle cost approach to developing the costs for both conventional and solar energy systems. Present-value life-cycle costs should be annualized using the appropriate discount rate, so that costs will correspond to the annual load. If L is fixed, the LCOE can be interpreted and used exactly like the LCC measure.

The eighth and final method included in this overview is the *break-even* (B-E) *method*. This is a flexible method in which the value of a designated input parameter is computed rather than given. The parameter occurs in the cost or benefit functions of an energy system, and its solution value causes net benefits (or net savings) to equal 0. For example find the value of x for which

$$\sum_{j=0}^{N} \{(B_{A1} - B_{A2})_j - [C(x)_{A1} - C_{A2}]_j\}/(1 + d)^j = 0.$$

The B-E method can be used to target specific cost reductions or performance improvements or to test for conditions of minimal acceptability. It is also used to reflect uncertainty as to the value that an input parameter should be assigned. In order to do either, one first specifies an evaluation model, setting net benefits equal to 0. The designated unknown for which a break-even calculation is desired is introduced into the model, and the solution value is calculated algebraically.

It should be noted that the break-even outcome is not an economically efficient outcome. Also, if the benefit functions and/or the cost functions are nonlinear, there may be more than one break-even point. Mathematical programming methods have been used to evaluate the optimal design of solar energy systems. Production theory and input/output theory have been used to explore the relationships between the inputs required to utilize solar energy and the resulting output. Marketing survey methods have been used to investigate the determinants of demand for solar energy systems. Market penetration analysis has been applied to predict solar markets. Mathematical, statistical, and other analytical methods have been used in the quantitative economic analysis of solar-related issues.

Most contributions to solar economic methodology have been adaptive rather than inventive; that is, the existing knowledge bases in economics, finance, business management, operations research, and engineering economics have provided the basis of approach in each of the four issue categories listed in section 2.1. The adaptive nature of the work is not surprising, because economic issues related to solar thermal systems are analogous to those for many other technologies for which evaluation methods are well established. In many cases the major challenge has been to choose the appropriate method and apply it correctly. This has entailed constructing detailed models, compiling necessary databases, making reasonable assumptions, carrying through on computations, and interpreting the results. In most cases the significant contribution has been the development of a body of literature and experience in applying existing economic methods to solar energy issues. For this reason, sections 2.3 through 2.6 give representative examples of works for each of the categories of issues, with little further description of the economic methods used, as they are introduced in a general way in this section and are covered in subsequent chapters of this book.

Although the contributions to solar economic methodology have been adaptive rather than inventive, they are nevertheless extensive and important. In the process of adapting existing methods to address solar energy issues and problems, we gained new insights, refined existing solutions, developed specialized and innovative models, significantly expanded the knowledge base pertaining to the economics of new technologies, and developed conclusions about the economic viability and sensitivities of solar energy systems under specific conditions.

2.3   Economic Feasibility Studies

Many solar economic feasibility studies were done during the 1970s using the general methods of section 2.2. Some of these assessed potential economic feasibility of solar energy based on analyses of hypothetical but representative cases; others assessed the economic feasibility of proposed, actual systems. The former tended to support government policy decisions regarding the allocation of public research and development resources; the latter supported individual investment decisions. Most of the studies used some type of life-cycle costing approach, which compared the economic performance of solar energy with alternative energy systems in terms of their life-cycle costs, net savings, cost per unit of energy delivered, rate of return on additional investment cost, savings-to-investment ratios, or payback periods. For example, residential economic feasibility studies were done by Löf and Tybout (1973), Duffie, Beckman, and Dekker (1976), Shams and Fichtenbaum (1976), Reid, Lumsdaine, and Albrecht (1977), and Bezdek, Hirshberg, and Babcock (1979).

Löf and Tybout (1973) performed one of the early landmark studies of solar economic feasibility. Löf, an engineer, and Tybout, an economist, solved for the least-cost combination of solar and conventional heating supplies in eight U.S. cities. They considered predicted performance of flat-plate collectors and heat storage systems in light of climate data and eight design parameters—house size, collector size, storage size, collector tilt, number of transparent surfaces in the collector, hot water demand, insulation on the storage unit, and thermal capacity of the collector. They derived capital costs for noncollector components based on "personal experience" and expressed cost information for five solar-heated buildings as functions of collector area. Collector costs, separated into fixed and variable components, were based on the costs experienced by several manufacturers. A 6% discount rate and a 20-year expected life were used in the analysis. Operating and maintenance (O&M) costs were excluded. A computer was used to calculate iteratively the least-cost combination of solar and conventional heat. The optimal combination was defined as that combination for which the marginal costs of solar and conventional energy are equal. Seven design and demand parameters were varied to determine the parameters to which optimal design was most sensitive. From this parametric analysis Löf and Tybout drew conclusions about the significance and effective ranges of values for the parameters tested. Design professionals and other researchers subsequently translated their results into widely used rules of thumb.

Löf and Tybout used economic methods that included amortization of capital costs, marginal cost analysis, sensitivity analysis, and optimization procedures, that is, procedures for finding the least-cost combination of solar and conventional energy that would satisfy the thermal requirements of the building. Later papers by Löf and Tybout (1974) and Pogany, Ward, and Löf (1975) expanded the Löf-Tybout equations of 1973, updated their analysis, and extended it to include combined heating and cooling systems in two regions.

The growing interest in solar energy and concern about economics in the mid-1970s led to the publication of several resource documents that provided guidance on how to perform economic analyses of solar energy systems. One was a National Bureau of Standards (NBS) publication by Ruegg (1975), which explained and illustrated the use of benefit-cost and life-cycle cost methods to evaluate and compare the economic efficiency of solar and conventional energy systems. This publication also demonstrated mathematically and graphically the conditions needed to optimize economically a solar energy system in conjunction with the thermal resistance of the building envelope.

This general reference was followed by an analytical review of solar economics (McGarity, 1976) for the Office of Energy Research and Development Policy of the National Science Foundation. McGarity gave present value and annual value evaluation methods and used these methods to study the economic feasibility of residential solar heating in twenty U.S. cities. His analysis included the effects on economic feasibility of future increases in conventional fuel prices.

Other reports providing general guidance for performing solar economic evaluations were Perino (1979) for solar-equipped buildings, Dickinson and Brown (1979) for industrial systems, Doane et al. (1976) and Rudasill (1977) for utility systems, and Ruegg (1986) for federal solar building projects. Kreith and Kreider (1976) provided another introduction to the economic analysis of solar energy systems. Their approach estimated the cost of solar energy in dollars per amount of energy supplied and compared this cost with that of a conventional system. They pointed out the importance of combining the economic analysis and optimization study with a thermal analysis of the building.

To simplify solar life-cycle cost models, Gershon Meckler Associates

(1976) developed procedures for grouping and regionalizing parameters to facilitate "scenario generation," cash-flow analysis for comparing parameters, and assessment of uncertainty in parameter ranges.

Alternative approaches to the revenue requirement (RR) method are documented in Rudasill (1977), Doane et al. (1976), Bennington (1976), and Phung (1977). RR methods compute the revenue requirements of the project such that the present value of the stream of after-tax revenue is just equal to the present value of the life-cycle costs. The example in section 2.7.6 is taken from Doane's report. The comparison of the different RR models identified three areas of disagreement: (1) inflation assumptions for calculating revenue stream; (2) calculation of fixed-charge rates to account for return of investment, depreciation, income taxes, and property taxes; and (3) formulation of the cost of capital.

Flaim et al. (1981) subsequently concluded that comparing solar and conventional utility investments based on required revenue per unit of energy delivered is faulty because such a comparison does not consider each technology's ability to meet peak utility loads—a primary consideration in any electric generation system.

Cassidy and Schirra (1977) expressed concern about inconsistent treatment of real and nominal price changes among evaluation methods. Their paper presented a systematic approach to treating price changes in the revenue requirements approach.

Perino (1979) illustrated that both constant and current dollar analyses yield the same present value costs if performed correctly, a point made earlier in the general economics literature. The example in section 2.7.1 is taken from Perino's report.

Brandemuehl and Beckman (1978) consolidated a number of costs into two multiplying factors in the life-cycle cost method. The first factor included any costs that were proportional to the first year's fuel costs; the second factor included costs that were proportional to system equipment costs. The first factor was equivalent to a uniform present worth factor adjusted to include a constant escalation rate and to incorporate income tax effects. The second factor incorporated financing and property tax effects, as well as maintenance, parasitic power, insurance, and salvage value, depending on the complexity of the problem. They used their simplified expression of costs to perform break-even analysis and to develop a tabular (or graphic) method for estimating optimal collector area and evaluating the economic efficiency of a solar energy system for locations and collector types that are frequently assessed. Their sensitivity analysis identified a tendency toward a relatively "broad optimum," in that near the optimum value relatively large changes in collector area resulted in relatively small changes in life-cycle savings.

A specific procedure for evaluating the economics of solar energy systems in federal buildings was established in the U.S. Code of Federal Regulations in 1979 based on conventional methods of economic analysis. A handbook for implementing the method and procedure was published in 1980, with subsequent revisions (Ruegg, 1986). The method calls for designing, sizing, and selecting active solar energy systems to minimize total life-cycle building costs; it also calls for assigning project priority according to the savings-to-investment ratio method. A distinctive feature of the handbook's method is that it provides special multiplicative factors, "modified uniform present worth factors" (UPW*), which combine DOE-projected real energy price escalation and adjustments for the time value of money (discounting). These UPW* factors are updated annually. The example in section 2.7.2 is taken from Ruegg's report.

Sav (1979) linked a solar life-cycle cost model as detailed by Ruegg and Sav (1981) with the Los Alamos National Laboratory's (LANL) solar performance model to develop a simplified method of determining the economically optimal size of a solar hot water system for a commercial building. Reducing the search for the optimally sized system to a single deterministic equation, Sav introduced the concept of universal economic optimization path, a locus of points describing the solar collector that maximizes total life-cycle savings from solar energy for a range of commercial hot water loads. The method allows the development of "families of curves" for different climate regions and for different economic assumptions. Sav concluded that, for a given climatic region and fixed values for economic parameters, the economically optimal solar fraction is uniquely determined and independent of the size of the hot water load and that the optimal collector size is directly proportional to the load. Sav recommended this method for evaluating the sensitivity of system design to economic parameters, in support of formulating government policies. This approach is discussed more in chapter 3.

Sedmak and Zampelli (1979) used the net present value method and principles of dynamic investment planning to analyze the optimal time of investment in solar energy, defined as the investment time for which net present value savings would be maximized. They demonstrated that under stated assumptions and constraints the optimal time is not when the net present value over the life-cycle is positive but rather the time after which there are no negative annual net cash flows, taking into account amortized capital costs, energy savings, and other cash flows. From a practical standpoint the analysis does not fully link the investment timing decision regarding a solar energy system with that regarding the building. If a solar investment is deferred until after a building is built, for example, the cost functions will change. Such considerations could alter the optimal timing decision to include up-front negative cash flows, since deferring the solar investment could lead to increased negative cash flows. Despite its shortcomings, this work was noteworthy because it introduces the dimension of investment timing to solar economic analysis.

Several analysts, such as Reiger (1978), Conopask, Fonash, and Easterly (1981), and Bezdek, Hirshberg, and Babcock (1979) called attention to the possible disparity between predicted purchaser behavior based on a particular measure of economic performance and actual behavior. Conopask, Fonash, and Easterly (1981) recommended pairing the investment criterion of net present value with the "apparent" purchase criterion of payback. They based the payback criterion on a previous survey of potential users in specific markets, which indicated that homeowners held a maximum payback criterion of eight years for energy-using durables in new houses and of two to five years in existing houses. The payback criterion of commercial building owners was estimated to be from five to seven years or longer. They concluded that solar energy systems with paybacks longer than those indicated by the survey are in a precarious market position even if they are life-cycle cost-effective. (Reiger's conclusions are treated in section 2.4 and Bezdek's in section 2.6.)

Researchers at the University of New Mexico (UNM) and LANL (Roach, Noll, and Ben-David, 1979; Noll and Thayer, 1979; and Balcomb, 1980) performed the major work in passive solar economics of residential buildings. LANL/UNM researchers developed a model to evaluate the economic performance of several passive solar residential design features, including thermal wall storage, direct gain, and attached sun space; their model allowed for varying the number and type of glazings, volume and type of storage, area of glazing, night insulation, and other features.

Their approach computed the life-cycle dollar costs and savings for

various combinations of features that provide the same solar contribution, identified the minimum-cost combination for each solar heating
contribution, traced the locus of all minimum-cost points for all solar
heating contributions to form an optimized expansion path, and determined the combination of passive design features and nonsolar energy
that results in maximum life-cycle savings. The example in section 2.7.4 is
taken from Balcomb's report.

Powell (1980) extended the passive evaluation methodology to commercial building analysis, including possible nonenergy benefits of passive solar design in the model. She used a benefit-cost evaluation method,
without optimization, and discussed in detail economic assumptions important to applying the model. The example in section 2.7.3 is taken from
Powell's report.

Until the mid-1970s economic studies of active solar energy systems
focused on residential buildings, and little generic study was done of
commercial solar applications, with the exception of single-family rental
housing and small nonresidential buildings whose energy requirements
were dominated by the building envelope. This residential application
focus is not surprising given the difficulty of generalizing solar economic
performance for large commercial buildings.

Ruegg et al. (1982) broadened the understanding of the economics of
active solar energy systems for commercial buildings by developing a detailed economic evaluation model and applying it in a series of case
studies for which they documented in detail their data, assumptions, and
findings. Their limited optimization model, which provided a computer
code, determined the least-cost combination of solar and conventional
energy needed to meet the commercial building's energy requirements
based on preestablished envelope and equipment characteristics (that is,
interdependence among the energy system, the envelope, and various
mechanical equipment was not evaluated). Using break-even analysis,
they estimated the minimum values of key parameters necessary for cost-
effectiveness.

The feasibility of using mathematical programming as the principal
tool for finding the most cost-effective design of a solar energy system
was explored by McGarity and Revelle in conjunction with the study by
Ruegg et al. (1982, appendix A) and by McGarity, Revelle, and Cohon
(1981). The authors questioned the prevailing practice of using a single
variate optimizing procedure, with collector area as the design variable,

and proposed multivariate optimizing as an improved approach. They pointed out that mathematical programming is appropriate for finding numerical values for a combination of variables that optimize an objective function; however, mathematical programming is subject to constraints imposed by required resource relationships, particularly when inequality constraints prohibit or make it difficult to use traditional economic optimization methods employing the Lagrange multiplier. (An objective function is the rule that states the relationship between the objective and the decision variables; a constraint is a relationship that specifies feasible values for decision variables.)

To facilitate the evaluation of solar energy systems, analysts have developed several computer simulation programs that combine thermal and economic analysis. These programs, for the most part, incorporate life-cycle costing and net present value methods. Some were written to search for the most cost-effective system design/size, whereas others were limited to evaluating single systems. Important among these were F-CHART (University of Wisconsin, Madison, 1978, 1980), SOLCOST (SOLCOST Service Center, 1980), BLAST (CYBERNET Services, 1980), DOE-2 (Lawrence Berkeley Laboratory, 1980), FEDSOL (Powell and Rodgers, 1981), and SOLCOM (Petersen, 1983).

Powell and Barnes (1982) compared the economic methods of the first five of these computer programs and assessed their differences in data and assumptions. Their comparison emphasized factors affecting the use of the different models for analyzing projects for federal buildings, but also provided an overview of the programs' capabilities for evaluating private sector solar investments. Table 2.1 compares the economic analysis methods of these programs. Table 2.2 compares their input variables. Chapter 3 gives more comparisons of these programs.

Petersen (1983) developed SOLCOM, the last of the six computer programs listed above, which was significant to the state of the art of solar energy economic analysis for commercial buildings. Petersen developed a life-cycle cost optimization algorithm for commercial buildings that determined the optimal overall conservation investment strategy, encompassing collector size for an active solar energy space and water heating system, modifications to the nonsolar heating and cooling plant, and modifications to the building envelope to reduce seasonal and peak-load heating and cooling requirements. His computer program integrated data on the building and building systems performance in order to find simul-

Table 2.1
Comparison of economic evaluation models in selected solar energy analysis programs: methods of analysis

	Measures of economic performance									
	Total life-cycle costs	Net savings	Discounted payback	Simple payback	Internal rate of return	Savings-to-investment ratio	Annual Btu savings	Cash flow analysis	Optimization of design variables	Break-even analysis
Required measures[a]		R		R		R				
FEDSOL (NBS)	×	×		×		×		×	×	×
F-CHART										
3.0		×	×	×[b]	×			×	×	
4.0	×	×		×[b]	×			×	×	
SOLCOST		×		×[b]	×		×	×	×	
BLAST	×							×		
DOE-2	×	×	×			×	×	×		

Source: Powell and Barnes (1982).
a. R = measure required by federal LCC rule; × = measure provided by model.
b. Calculated differently from federal simple payback measure.

Table 2.2
Comparison of economic evaluation models in selected solar energy analysis programs: input variables

	System costs				Energy costs		Financial variables				
	Investment costs	Annual nonfuel O&M costs	Nonannual repair and replacement costs	Salvage values	Base year unit price	Multiple Esc. rates	Discount rate	Inflation	Income, property taxes	Investment (tax) credits	Mortgage terms
Required input variables[a]	R	R			R	R	R			R	
FEDSOL (NBS)	X	X	X		X	X	X			X	
F-CHART											
3.0	X	X		X	X	X[b]	X	X	X	X[c]	X
4.0	X	X		X	X	X[b]	X	X	X	X	X
SOLCOST	X	X			X	X	X	X	X	X[d]	X
BLAST	X	X	X		X		X	X	X	X[c]	
DOE-2	X	X	X		X		X	X	X	X[c]	

Source: Powell and Barnes (1982).
a. R = variable required by federal LCC rule; × = variable contained in model.
b. Can be accounted for only by supplying a sequence of values for the estimated fuel price in each year of the study period.
c. Can be accounted for by adjusting values supplied for investment costs (BLAST and DOE-2) *and* annual O&M costs and salvage values (F-CHART 3.0 and SOLCOST).
d. Calculated differently from the investment cost adjustment allowed for federal solar energy projects.

taneously the minimum life-cycle cost solutions for the three investment categories. But unlike the other computer programs, Petersen's program did not incorporate the energy analysis and therefore must be used with an energy analysis program such as BLAST or DOE-2.

Petersen's model was based on previous optimization studies. For example, Sav (1978) described and depicted graphically the economic conditions for optimizing solar and energy conservation features in buildings. Balcomb (1980) presented a methodology for determining the optimal combination of investments in solar energy and energy conservation features in the residential building envelope. Noll and Thayer (1979) described graphically the trade-off among passive solar energy, auxiliary energy, equipment size, and energy conservation features in houses. Barley (1979) developed an algorithm for jointly optimizing solar equipment size and insulation levels in houses.

Bendt (1983) reexamined the life-cycle method of optimizing solar collector size, taking into account the considerable uncertainty in assumptions and future projections. He observed that, given the uncertainty, the range of design conditions for which life-cycle costs are minimized are broad and further that the optimal solar fraction for practical systems will tend to fall within the limited range of 30%–90%. From this observation he concluded that it is possible to approximate the optimal system while narrowing the search among candidate collector sizes. He derived a rule of thumb for selecting the candidate sizes in the attempt to reduce the number of iterations required to identify the least-cost size.

2.4 Marketing Studies

Another category of solar energy studies in which economic and related methods have been important is marketing studies. These studies, which have employed market surveys and analysis, real estate appraisal techniques, and financial analysis, have entailed primarily the adaptation of existing methodology.

Reiger (1978) called attention to marketplace perspectives, particularly the shortcomings of life-cycle cost analysis for predicting the market acceptance of solar energy systems. He pointed out the importance of first costs in market segmentation, the uncertainty of the resale market as an inhibiting factor to sales, the possibility of negative short-term cash flows (even on a life-cycle cost-effective system) as an unacceptable burden to prospective buyers, and the likelihood of a lender's tendency to under-assess the value of solar energy systems in calculating loan amounts.

As part of the HUD Residential Solar Energy Program, the Real Estate Research Corporation (1979) interviewed key participants in the solar construction and marketing process to identify trends and issues critical to market acceptance of solar energy. They used a general market acceptance model to identify important areas and organize the information. The survey identified the following actors as key to the market-ability of houses with solar energy: builders, purchasers, construction lenders, permanent lenders, utilities, insurance companies, tax assessors, planning/zoning officials, and building code officials. The survey data were analyzed to identify barriers and incentives to solar acceptance affecting each type of actor. The study was limited because its database did not provide a statistically valid sample for each type of actor.

To support market analyses of passive solar designs, Kirschner et al. (1982) developed a pre-1980 housing stock database for 220 solar regions in the United States. This database was significant for implementing pas-sive economic evaluation methods. The database includes housing num-bers, location, age, type, and space-heating fuel use.

Because it recognized the financial community's role in solar market acceptance, DOE sponsored workshops for lenders, appraisers, insurers, and tax consultants (DOE, 1979). These workshops informed the finan-cial community of, among other things, the methods, barriers, incentives, and system costs to use in evaluating the economics of solar energy prop-erties. This effort was significant because it indicated a growing aware-ness of the important role of financing in market acceptance and the need for greater dissemination of economic methods and data.

The Conference on Financial Issues for International Renewable En-ergy Opportunities (DOE and Brookhaven National Laboratory, 1981) investigated financing the transfer of solar technology to developing countries in the context of developing an international market for U.S. conservation and solar technologies. This investigation was concerned with sources of funds, long-term market forecasts, and economic barriers and incentives.

Market penetration studies of solar energy systems were undertaken to forecast the future extent of solar energy sales in various markets. They generally have been based on one of two methods: market surveys or life-cycle costing combined with diffusion analysis, that is, models for predicting the spread of the solar technology.

Schiffel et al. (1978) described and compared solar market penetration models. These included System for Projecting the Utilization of Renewable Resources (SPURR), developed for DOE by the MITRE Corporation (MITRE, 1978; Bennington et al., 1978); Market Oriented Program Planning System (MOPPS), developed for DOE by Energy and Environmental Analysis, Inc. (1977); Solar Heating and Cooling of Building Commercialization Model (SHACOB), developed by Arthur D. Little, Inc. (1977); and SRI Solar Penetration Model, developed for DOE by Stanford Research Institute (1977). SPURR used a life-cycle costing method; SHACOB, a payback method; and MOPPS and SRI, a cost per unit energy method. SPURR is generally considered the most comprehensive of these models; it used computer simulation with a database comprising economic, market, and climate data to evaluate the impact of alternative data and assumptions on solar market shares. Chapter 4 discusses market penetration studies.

Similar economic methods were applied to the study of photovoltaic market potential, as summarized by Dernburg, Depaso, and Fenton (1981).

Warren and Wahlig (1982) combined investment analysis and market penetration methods to derive cost goals for solar energy systems that corresponded to targeted market shares. From market studies of heat pumps, they identified relationships among the percentage of the market captured, the payback period, and the rate of return on investment, which allowed them to express the market penetration goal in terms of incremental solar system cost goals as a function of the year of purchase. This approach is discussed in chapter 8.

2.5 Impact Studies

Another broad area of solar research that has used economic methods of evaluation is impact assessment. These studies have assessed such diverse effects as the impact of international financial institutions on markets for solar energy systems, the impact of solar energy on oil imports, the impact of utility rates and service policies on solar market penetration, and the impact of solar energy on employment. Likewise, the economic methods employed in these studies have been diverse, comprising sources-of-funds assessments, institutional analyses, labor economics, market analyses, pricing policies, and input/output analysis, in addition to the methods of investment analysis discussed in the preceding sections and summarized in section 2.2.

DeLeon et al. (1980) investigated the costs of our dependence on foreign petroleum; the social, economic, and political costs and risks attributable to purchasing foreign oil; and, given estimated levels of market penetration by solar energy, the impact of solar energy on petroleum consumption. The authors concluded that taking into account the social, political, and economic costs results in a significantly higher cost of imported oil than the market price. They applied the SPURR computer simulation model to estimate the amount of oil that might be displaced by solar energy by the year 2000, under conditions stipulated by the National Energy Plan.

Feuerstein (1979) evaluated the impact of utility rates and service policies on solar economic feasibility, focusing on the issues of utility refusal to provide auxiliary service to solar users, utility rates that discriminate against auxiliary service, and utility refusal to purchase excess power generated by small-scale producers of electricity from solar technologies. The study pointed out that the validity of special utility rates to solar users depended on comparative costs of providing service, and Feuerstein conducted an economic analysis of selected rate-making practices.

Concerned with the impact of solar energy on employment, Ferris and Mason (1979) reviewed ten regional economic models for their capabilities in assessing employment impacts of solar energy commercialization and documented the steps required to develop the appropriate methods. They discussed the following general analytic methods that were used by the regional models: economic base analysis for multiplier estimation, demographic-economic interaction models for forecasting social or economic data, shift-share analysis for estimating and projecting regional economic growth, input/output analysis for tracing cash flows from producing to consuming sectors, and industrial location analysis for estimating the location of industry. The advantages and disadvantages of the various methods are discussed.

Kort (1980) estimated the regional economic and demographic impacts of commercializing solar technology by linking regional demand and resource information, a national input/output model, and an interregional econometric model. Solar impacts on output, employment, income, unemployment, and population were estimated for the fifty states.

Early et al. (1979) analyzed the macroeconomic and sector output and employment implications of a scenario provided by the DOE Office of Policy, Planning, and Evaluation, based on a simulation using the SPURR model. The scenario called for installing 2.2 million solar energy units during the 1976–1985 period.

To provide a basis for evaluating the economic impact of solar energy on the environment, Krawiec (1980) reviewed the literature on measuring the economic costs of air and water pollution, focusing in part on the evaluation methods. Krawiec concluded that market studies that include property value and wage differential assessments would be useful in evaluating the environmental impact of solar energy in economic terms.

Petersen (1977) used an augmented input/output method to identify the economic sectors that would be most affected by increased solar utilization and to derive estimated changes in sales and employment for 131 sectors of the U.S. economy. He concluded that no single sector of the economy would likely be extremely affected by solar energy development.

DOE sponsored a "Technology Assessment of Solar Energy" project aimed at providing policymakers with an analysis of potential health, environmental, and socioeconomic impacts related to large-scale commercialization of solar and other renewable energy technologies. To provide the economic basis for this effort, Mann and Neenan (1982) developed a method that used the engineering specifications of prototype systems in an economic input/output analysis, which gave cost profiles for selected solar energy technologies. The issue they addressed was how to incorporate the potential solar industrial sector within the existing economic input/output matrix, with the objective of providing a tool of analysis for tracing the impact of emerging solar energy technologies on other sectors of the economy.

Pleatsikas et al. (1979) used a model of national economic structure and growth (the Hudson-Jorgenson Energy/Economic Model) to assess the implications for the U.S. economy over twenty years of a large-scale investment program in solar energy. They described the analytical framework, estimated the capital and operating requirements for three levels of market penetration, and exercised the model to evaluate the macroeconomic effects of the specified levels of requirements.

2.6 Formulation of Government Programs and Policies

The economic methods of benefit-cost analysis, life-cycle costing, and related methods have been used to estimate the economic feasibility of competing energy technologies under alternative governmental policies. The results have been used by policymakers to guide policy formulation. These methods have also been used to identify promising areas of research and to guide the allocation of government research and development resources among competing programs.

A number of studies have been conducted of governmental incentives for solar energy. An early study of the impact of tax incentives on the utilization rate of solar energy for space conditioning was performed by Petersen (1976). This study took a macroeconomic approach and forecasted impacts of sales and property tax exemptions, income tax deductions and credits, rapid amortization provisions, and interest rate subsidies on future utilization rates of solar energy nationwide. Petersen used a market penetration method in a series of steps to estimate solar energy use at different points in time under different policy assumptions. One step optimized the system and determined the cost-effectiveness of the system with and without incentives. A second step characterized consumer preferences and estimated their propensity to purchase solar energy systems. A third step calculated the number of structures available for solar energy use. The final step estimated market shares according to a market diffusion model.

As Schiffel et al. (1978) point out, this approach has two major shortcomings: lack of a solid theoretical foundation for moving from the cost-effectiveness evaluations to the diffusion analysis and inadequate attention to actual economic decision criteria used by buyers in different markets.

RUPI, Inc. (1977) used a market survey method, utilizing field survey results of 1,500 households in 8 cities to develop baseline market projections of solar energy sales and homebuyer response to alternative incentives. Emphasis was placed on the relative rather than the absolute estimates of costs and impacts of incentive options. Scott (1977) estimated demand curves for solar energy by calculating consumer utility of various options based on survey data.

Ruegg (1976) used a microeconomic analysis model employing life-cycle costing to evaluate the impact of property tax exemptions, tax credits, grants, sales tax exemptions, and income tax deductions on system cost-effectiveness. Technology & Economics, Inc. (1979) performed a parametric analysis of investment tax credit programs for solar industrial process heat equipment, using a present value model. A disadvantage of this approach as compared with the market penetration approach is that it focuses on estimating cost-effectiveness rather than on the extent of market acceptance.

Battelle Pacific Northwest Laboratories (1978) developed a typology of incentives that has been used to increase production of energy from coal, gas, oil, nuclear, and hydro power. This contributed to an understanding of existing energy supply curves and provided useful background to analyzing incentives for solar energy.

Bos and Weingart (1983) described in detail the legal and financial impacts of the Economic Recovery Tax Act of 1981 on the commercialization of solar thermal electric technologies. Their report analyzed the role of limited partnerships and R&D partnerships in the financing of solar investments.

Bezdek, Hirshberg, and Babcock (1979) analyzed the economic feasibility in 1977 and 1978 of solar water and combined water and space heating for single-family detached residences and multifamily apartment buildings in four U.S. cities, assessing the impact of the federal tax credit in the National Energy Conservation Policy Act (NECPA) of 1978. This was not an optimization study, because collector size was predetermined. They applied regression analysis to data from residential solar demonstration projects to estimate cost functions. Their study emphasized that market acceptance depends on the particular decision criteria of the consumer. They identified three: a 10-year payback, a 5-year down payment recovery period, and a 3-year net positive cash flow. For commercial building owners they used a criterion of a 15% minimum internal rate of return. They compared calculated measures with and without tax incentives against the assumed decision criteria. The residential solar energy system was defined as economically feasible if two of the three decision criteria for homeowners were met.

The likely impact of alternative subsidy programs has been studied more than the issue of what is the magnitude of governmental subsidies to solar energy that is justifiable on economic grounds. In theory, to determine the socially justified level of federal investment in a particular area, we estimate the magnitude of external benefits and costs that are not reflected in market prices, determine the socially efficient level of investment using estimates of total social costs and benefits, estimate the level of private investment, and estimate the extent of underinvestment by comparing the socially efficient level with the level of private investment. A public incentives program to bring total investment up to the socially efficient level may then be attempted. This requires knowledge of the likely impact of different incentives. In practice, however, the quantitative estimation of externalities is difficult.

Cost assessments and economic evaluations of government demonstration programs comprise another category of economic studies that support government programs. Primarily these studies use the same economic methods as those discussed in section 2.3; however, they are distinguished by their focus on system costs and possibilities for cost reductions, and their greater use of empirical data.

Ormasa (1979) analyzed detailed cost relationships and potential means of cost reductions, focusing on the cost competitiveness of ten solar energy systems in the National Solar Heating and Cooling Demonstration Program, as indicated by the costs per unit of energy and by the discounted payback.

King et al. (1979) also analyzed costs using empirical construction costs from the Demonstration Program. They regressed system costs against collector area to determine how system costs were affected as a result of increased system size and assessed costs of collectors, support structures, piping, ductwork, insulation, controls, and storage by type of storage container.

2.7   Case Examples

This section presents six case examples of economic evaluations. These case examples, drawn from work completed under DOE sponsorship during the last several years, assess two active solar systems in different market sectors, one commercial space heating system, one residential passive solar system, one solar industrial process heat system, and one utility-owned solar electric system. They are summarized and accompanied by a critique and suggestions for possible alternative approaches.

The measures of merit used by the six examples include life-cycle costs, net present value, internal rate of return, and busbar energy cost. In spite of these different measures, there is a common use of discounted cash flow methods to determine the present value of cash flows such as investment cost (PVSYS), interest payments (PVINT), and depreciation (PVDEP). Because these cash flows are common to most of the examples, they are presented in summary form in the pages that follow with explanations or comments given only when their calculation departs from the standard procedures.

The monetary values in these examples are meant for illustrative purposes; they are not accurate and are certainly not up-to-date costs for these solar systems. In these examples dollar values have been converted to a 1985 basis.

We have modified the notation in these case examples to ensure consistency among the examples. We define all notation in the presentation of the inputs for each example, except for the following two discounting operations: $(P/A_{d,N})$ and $(A/P_{d,N})$.

$(P/A_{d,N})$ is the uniform multiple-payment present worth factor. It converts N equal payments made at the end of each of N consecutive periods to the equivalent amount in the present using the discount or interest rate d:

$$(P/A_{d,N}) = \frac{1 - [1/(1 + d)]^N}{d}.$$

$(A/P_{d,N})$ is the uniform capital recovery factor. It uses the discount rate d to convert a present value to the equivalent value of each of N equal payments made at the end of each of N consecutive periods:

$$(A/P_{d,N}) = \frac{1}{(P/A_{d,N})} = \frac{d}{1 - [1/(1 + d)]^N}.$$

2.7.1 An Active Solar Energy System for Residential Space Heating

This example comes from a report by Perino (1979) that presents a methodology for determining the economic feasibility of residential and commercial solar energy systems. The example compares the life-cycle cost (LCC) of a solar energy system with that of a conventional energy system. Other measures of economic performance are not included since the purpose of Perino's report is to present a methodology for calculating life-cycle costs. The example makes no attempt to examine optimal sizing, sensitivity to uncertainties, etc.

Example

A homeowner desires to assess the value of a solar space heating system with an electric backup unit for his new home. The conventional alternative is a natural gas furnace. Either system will be financed through the 30-yr home mortgage. All costs are estimated in 1978 dollars and converted to 1979 dollars in the analysis using the general inflation rate.

Data

Solar system cost (including backup)	$IC_s = \$6{,}000$
Backup electricity cost	$f_b = \$100/\text{yr}$
Gas furnace cost	$IC_c = \$1{,}800$
Conventional alternative natural gas cost	$f_c = \$250/\text{yr}$
First year of operation	$y_0 = 1979$
Price year	$y_p = 1978$
Analysis period	$N = 20 \text{ yrs}$
Mortgage period	$N_f = 30 \text{ yrs}$
General inflation rate	$g = 5\%$
Discount rate	$d = 10\%$
Mortgage interest rate	$i = 8.5\%$
Electricity price annual escalation rate	$e_e = 7\%$
Natural gas price annual escalation rate	$e_g = 10\%$
Down payment (% of initial cost)	$D = 20\%$
Annual O&M cost (% of initial cost)	$\text{O\&M} = 1.5\%$
Homeowner's marginal income tax rate	$t = 30\%$
Tax credits (% of initial solar system cost)	$TC = 0\%$
Property tax rate	$t_p = 0\%$

Computations

Perino calculates the LCC of both the solar energy system and the conventional alternative. The LCC of the solar energy system is subtracted from that of the conventional alternative to yield "solar savings."

LCC of the Solar Energy System

The present value of each component of the LCC is calculated and summed with the others to yield the LCC.

Commentary

1979 dollars are used as this is the year that Perino's report was published. At the conclusion of the example we will express LCC in 1985 dollars to be consistent with the rest of this report.

Parasitic energy costs are not explicitly accounted for.

All discount, interest, and escalation rates include inflation.

Recurring replacement costs are not explicitly included.

Federal solar tax credits did not exist when Perino's report was published.

Perino gives no reason for assuming a zero property tax rate. She presumably assumes that solar energy systems are exempt.

Perino's "solar savings" is simply the net present value (NPV) of the solar system investment. All present value calculations use nominal costs together with a nominal discount rate. This calculation is often preferred over using real costs and rates, since financing costs are unchanging in current dollars in the standard level payment loan.

This is the traditional approach to LCC calculation.

Example

· PV of system cost (PVSYS):

$$\text{PVSYS} = IC_s (1 + g)^{y_0 - y_p}$$

$$\times \{ D + (1 - D) \times [(A/P_{i,N_f}) \times (P/A_{d,N}) + \text{PVLB}(i, N_f, d, N)] \}$$

where $\text{PVLB}(i, N_f, d, N)$ = the present value of the balance on a loan of one dollar
at the end of the analysis period

$$= \frac{(1 + i)^{N_f} - (1 + i)^N}{(1 + d)^N [(1 + i)^{N_f} - 1]}$$

$$\text{PVSYS} = \$6,000(1.05)^1 \{ 0.2 + (1 - 0.2)[0.093 \times (8.51 + 0.091)] \}$$
$$= \$5,706$$

· PV of interest payments (PVINT):

$$\text{PVINT} = IC_s (1 + g)^{y_0 - y_p}$$

$$\times (1 - D) \left\{ \frac{(P/A_{\bar{i},N})}{1 + i} \times [i - (A/P_{i,N_f})] \right.$$

$$\left. + [(A/P_{i,N_f}) \times (P/A_{d,N})] \right\}$$

where $\bar{i} = \dfrac{1 + d}{1 + i} - 1$

$$\text{PVINT} = \$6,000(1.05)^1 (1 - 0.2) \{ 16.01 \times [0.085 - 0.093]$$
$$+ [0.093 \times 8.51] \}$$
$$= 3,344$$

· PV of recurring costs (PVRC):

$$\text{PVRC} = [O\&M \times IC_s (P/A_{\bar{g},N})] + [f_b \times (P/A_{e_e,N})]$$

where $\bar{g} = \dfrac{1 + d}{1 + g} - 1$ and $\bar{e}_e = \dfrac{1 + d}{1 + e_e} - 1$

$$\text{PVRC} = (0.015 \times 6,000 \times 12.72) + (100 \times 15.15) = 2,660$$

· $\text{LCC}_{\text{solar}} = \text{PVSYS} + \text{PVRC} - (t \times \text{PVINT})$
$$= 5,706 + 2,660 - [0.3 \times 3,344] = 7,364$$

LCC of the Conventional System

The same equations used for the solar system costs are also used for the present value of the
conventional system costs. Thus, only the numerical values are presented.

Commentary

The term $(1 + g)^{y_0 - y_p}$ is required to inflate the initial system cost expressed in 1978 (y_p) dollars to the cost in the first year of operation, 1979, which is the base year for the present value calculation.

The choice of an analysis period shorter than the financing period requires that the remaining loan balance at the end of the analysis period be accounted for. An alternative approach might be to extend the analysis period to equal the finance period and to adjust the O&M cost to reflect additional component replacement costs.

Capital investment analysis usually separates financing considerations (sources of funds) from project evaluations (uses of funds). The entire acquisition cost is generally treated as an initial cash outlay, regardless of whether the acquisition is financed or paid for in cash when evaluating the economic merits of a project. In this example, setting the discount rate (10%) higher than the mortgage interest rate (8.5%) has the effect of reducing the present value acquisition costs and making the solar energy project appear more cost effective than it would be otherwise. This should be an argument for borrowing when the loan interest rate is lower than one's opportunity cost, rather than support for buying solar energy.

Interest payments beyond the analysis period are ignored.

O&M costs are simply represented as a function of the initial system costs. If the data are available, a more accurate representation would to be separate out the replacement costs from routine maintenance costs.

In this example, property taxes, tax credits, and salvage value are assumed equal to zero.

Example

- PVSYS $= 1{,}800(1.05)^1\{0.2 + (1 - 0.2)[(0.093 \times 8.51) + 0.091]\}$
  $= 1{,}712$
- PVINT $= 1{,}800(1.05)^1(1 - 0.2)\{16.01 \times [0.085 - 0.093] + [0.093 \times 8.51]\} = 1{,}003$
- PVRC $= (0.015 \times 1800 \times 12.72) + (250 \times 20.0) = 5{,}343$

- $\text{LCC}_{\text{conventional}} = 1{,}712 + 5{,}343 - (0.3 \times 1{,}003) = 6.754$

Solar Savings

Solar savings is the difference between the two LCC values:

Solar savings $= \text{LCC}_{\text{conventional}} - \text{LCC}_{\text{solar}} = -610$

Thus the solar system is not economic.

Commentary
All interest rates and loan periods are the same as in the solar LCC calculation since the loan is the house mortgage.

Since the discount rate and the natural gas escalation rate are the same, the present value of annual natural gas payments is simply 20 times the first year payment.

NPV = -610. This NPV value is expressed in 1979 dollars. To convert it to 1985 dollars we can apply the GNP implicit price deflators between 1979 and 1985 to yield

$$NPV_{1985} = NPV_{1979} \times 1.421$$

$$= \$-610 \times 1.421 = \$-867.$$

Of course, this conversion to 1985 dollars accounts only for inflation. Any cost changes or technological innovations are not accounted for by such a conversion.

2.7.2   An Active Solar Energy System for Space Heating in a Government Building

In 1980 (revised 1982 and 1987) the National Bureau of Standards (NBS) under contract to DOE developed a manual for evaluating the cost-effectiveness of renewable energy and energy conservation projects in government buildings (Ruegg, 1987). The following hypothetical example, provided by the NBS report to illustrate the recommended evaluation method, determines the net present value of an active solar energy space heating system retrofitted in a government office building in Phoenix, Arizona. The example compares the net present value of two different solar energy system designs at the optimal collector size for each design, assuming the existing distillate-fired boiler is used as a backup system to the solar energy system, or, alternatively, is used to supply 100% of the heating requirements. All dollar values are given in 1985 dollars.

Example

This example is considerably simpler than the preceding residential example. Because it is a federal analysis, there are no taxes, tax deductions, tax credits, or financing considerations. This example also differs from the previous ones in that all calculations are made in constant dollars.

Data

Solar energy system fixed cost	$IC_{sf} = \$33,620$
Solar energy system variable cost	$IC_{sv} = \$26/ft^2$ ($\$280/m^2$)
Collector area	$CA = 1000\ ft^2$ (92.9 m^2)
Conventional system distillate fuel cost	$f_b = \$5.80/10^6$ Btu
Analysis period	$N = 20$ yr
Real discount rate	$d_r = 7\%$

Distillate price indices (I)

1985	1986	1987	1988	1989	1990	1991
1.00	0.97	0.98	1.01	1.06	1.10	1.17

1992	1993	1994	1995	1996	1997	1998
1.24	1.32	1.39	1.46	1.52	1.58	1.64

1999	2000	2001	2002	2003	2004	2005
1.71	1.78	1.83	1.89	1.95	2.02	2.08

Annual O&M cost (% of initial cost)	$O\&M = 1\%$
Salvage value (% of initial cost)	$SV = 20\%$
Solar fraction	$SF = 0.51$
Annual space heating load	$L = 500 \times 10^6$ Btu (473 GJ)
Distillate-fired furnace efficiency	$\eta_d = 0.60$

Computations

LCC of the Solar Energy System

· The initial capital investment for the solar energy system is

$IC_s = IC_{sf} + (IC_{sv} \times CA) = 33,620 + (26 \times 1000) = \$59,620$

· The LCC of the solar with backup is

$$LCC_{solar} = 0.9IC_s + [IC_s \times O\&M \times (P/A_{d_r, N})] - \frac{[IC_s \times SV]}{(1 + d_r)^N}$$

$$+ \frac{L}{\eta_d}(1 - SF)f_b(P/A_{I, d_r, N})$$

Commentary

Using constant dollars is a good choice in this example since no finance payments or depreciation deductions are fixed in current dollars. It is also the approach recommended by the U.S. Office of Management and Budget (OMB) for analyzing federal investments.

Results for other sizes are shown at the conclusion of this example. This size (1,000 ft^2) was estimated to be least-cost size (although it was not found to be cost-effective in comparison with the conventional system).

A 20-year study period is suggested by the Federal Solar Buildings Program guidelines. A 7% real discount rate (i.e., not including inflation) was required for federal solar analyses by the Energy Security Act (P.L. 96-294, Sec. 405, 1980).

These price escalation rates are projected by the Energy Information Administration of the U.S. Dept. of Energy and are intended to promote consistency of assumptions among federal energy analysts.

The O&M cost appears low since it must include parasitic energy costs and nonrecurring replacement and repair costs. The federal guideline is to use between 1% and 4% of initial cost for federal solar energy projects. The guideline is to assume a zero salvage value unless more definitive information is available.

The initial system cost has been reduced by 10% to account for the nonmonetary benefits of solar energy. This adjustment is patterned after the 10% tax credit for business investments in solar energy that was in effect at the time the federal methodology was developed.

Using the real discount rate in the O&M portion of the LCC computation implies that O&M costs are assumed to escalate at the same rate as inflation.

Example

---

where

$$P/A_{I,d,N} = \sum_{n=1}^{N} \frac{I(1985 + n)}{(1 + d)^n} \quad (I = \text{projected fuel price index}$$
$$n = \text{counter to designate each year})$$

$$LCC_{solar} = (0.9 \times \$59,620) + (\$59,620 \times 0.01 \times 10.59)$$
$$- (\$59,620 \times 0.20 \times 0.26)$$
$$+ \frac{500}{0.6}(1 - 0.51)\$5.80(14.30) = \$90,739$$

LCC of the Conventional Alternative

$$LCC_{conventional} = \frac{L}{\eta_d} \times f_b(P/A_{I,d,N})$$

$$LCC_{conventional} = \frac{500}{0.6} \times \$5.80(14.30) = \$69,116$$

Net Present Value

The NPV is the difference between the two LCC values:

$$NPV = LCC_{conventional} - LCC_{solar}$$
$$= \$69,116 - \$90,739$$
$$= -\$21,623$$

The example presents results for other sizes of this design (design A) and a second design (design B):

Size (ft^2)[a]	NPV, 1985$	
	Design A	Design B
0	0	0
450	−24,570	−31,877
675	−22,547	−30,463
1,000	−21,623	−30,034
1,500	−25,729	−35,647
2,275	−38,730	−52,434

a. (× 0.0929 = m^2).

None of the systems in this illustration yields a positive net present value. However, if a system is to be installed, the best one is design A with 1,000 ft^2 (92.9 m^2) of collector (the example given above.)

$P/A_{I,d,N}$ factors are published annually by the National Bureau of Standards based on updated projections of fuel price indices.

The hypothetical solar energy system is not economic even when the initial cost is reduced by 10% to account for externalities.

Another way of comparing would be to compare the cost of only the solar energy supplied (excluding backup) with the cost of only the distillate fuel replaced.

The example also calculates the savings-to-investment ratio for the preferred solar energy system to compare with other projects on other buildings. All dollars are 1985 dollars and the energy price projections are as of 1985.

2.7.3 Solar System for Commercial Space Heating

In a report prepared by NBS for DOE, Powell (1980) analyzed a vertical solar wall collector with rockbed storage for space heating retrofit in a commercial building. The solar energy system is compared against both oil-fired and electric conventional systems using life-cycle costs, internal rate of return, and the savings-to-investment ratio. Sensitivities to many of the economic assumptions are also presented.

Example

This example differs from the residential examples primarily in the fact that O&M expenses are deductible, as is depreciation of the system. Since the example was constructed prior to the passage of the tax bills of 1981 and 1982, some of the depreciation computations are no longer applicable. Powell makes all mortgage calculations on a monthly, rather than annual, basis. All first-year costs are presented in 1979 dollars. Furthermore, 1979 is also the base year for which all present value calculations are made. To conform with the rest of this document we also express the LCC in 1985 dollars at the end of this example. We examine only the LCC calculations.

Data

Solar energy system cost (excludes sales tax)	$IC_s = \$4,711$
Sales tax	$S = \$162$
Oil cost in first year	$f_c = \$67.28$
Analysis period	$N = 15 \text{ yr}$
Finance period	$N_f = 20 \text{ yrs}$
General inflation rate	$g = 6\%$
Discount rate (nominal)	$d = 12\%$
Mortgage interest rate (nominal)	$i = 3\%$

Commentary

Powell describes the system in this example as a passive system even though it employs a fan, a controlled air distribution system with dampers, and rockbed storage.

Powell derives this cost as the sum of the individual component costs.

As in the preceding example, the analysis period has arbitrarily been chosen shorter than the finance period.

Financing is assumed to be available at a subsidized rate. If the subsidized rate is only available for financing the solar energy system, the use of a higher discount rate to discount the cash flows is questionable given that the borrower does not have the opportunity to use the funds for other purposes that have a return of 12%.

Example		
Oil price escalation rate	1979–1985	$e_{01} = 11.1\%$
(annual compound rate)	1986–1990	$e_{01} = 8\%$
	1991–1994	$e_{03} = 11.2\%$
Oil furnace efficiency		$\eta_0 = 60\%$
Down payment (% of initial cost)		$D = 25\%$
Annual O&M cost (% of initial cost)		$O\&M = 1\%$
Combined state and federal income tax rate		$t = 34.9\%$
Combined state and federal capital gains tax rate		$t_g = 33\%$
Combined state and federal tax credits		$TC = 20\%$
Value of system at end of investment period		$EC_s = \$15,830$
Depreciation period		$N_D = 20$ yrs

Computations

Powell's example does not include the actual computations. The calculations shown here are recreated from Powell's methodology.

LCC of the Solar Energy System

The present value of each component of the LCC is calculated first. All mortgage payments are accounted for on a monthly basis (as opposed to annually).[a]

· Present value of system cost (PVSYS):

$$\text{PVSYS} = (\text{IC}_s + S)\left[D + (1 - D)\{[12 \times (A/P'_{i,N'_f})(P/A_{d,N})]\right.$$
$$\left. + \text{PVLB}(i', N'_f, d, N, N')\} - \frac{\text{TC}}{1 + d}\right]$$

where

$$\text{PVLB}(i', N'_f, d, N, N') = \frac{(1 + i')^{N'_f} - (1 + i')^{N'}}{(1 + d)^N[(1 + i')^{N'_f} - 1]}$$

$$\text{PVSYS} = (4711 + 162)\left\{0.25 + (1 - 0.25) \times [(12 \times 0.0055 \times 6.81) + 0.0564] - \frac{0.2}{1.12}\right\}$$

$$= 2,197$$

Commentary
These are based on 1979 price projections of the Energy Information Administration of the U.S. Dept. of Energy.

This is a low value for O&M especially since it includes parasitic energy costs and major replacement costs.

This assumes the system value is 90% of the initial value and that the current dollar cost has escalated by 9% annually. This assumption is extremely optimistic and is an unusual way of estimating the salvage value.

Straight line depreciation is assumed.

Accounting for monthly mortgage payments rather than annual payments produces less than a 1% change in the result.

Seventy-five percent of the pretax credit cost is financed.

Tax credits are discounted for one year and are applied to the sales tax as well as the system cost.

Example

- Present value of interest payments (PVINT):

$$\text{PVINT} = (\text{IC}_s + S)(1 - D)[\{(P/A_{\bar{i},N})[i' - (A/P_{i',N_i'})](P/A_{i',12})\}$$
$$+ [12 \times (A/P_{i',N_i'})(P/A_{d,N})]]$$

where $\bar{i} = \dfrac{1+d}{(1+i')^{12}} - 1$

$$\text{PVINT} = (4{,}711 + 162)(1 - 0.25)\{[8.2084(0.0025 - 0.0055)11.8073]$$
$$+ [12 \times (0.0055)(6.81)]\}$$
$$= \$577$$

- Present value of depreciation deduction (PVDEP):

$$\text{PVDEP} = \text{IC}_s \times t \times \frac{1}{N_D}(P/A_{d,N})$$

$$= 4{,}711 \times 0.349 \times \frac{1}{20}(6.81) = \$560$$

- Present value of recurring costs (PVRC):

$$\text{PVRC} = \text{O\&M}\,(\text{IC}_s + S)(P/A_{\bar{g},N})(1 - t)$$

where $\bar{g} = \dfrac{1+d}{1+g} - 1$

$$\text{PVRC} = 0.01(4{,}711 + 162)(9.931)(1 - 0.349) = \$315$$

- Present value of system resale less capital gains and recapture taxes (PVR):

$$\text{PVR} = \text{EC}_s - t_g\left[\text{EC}_s - \left(\frac{N_D - N}{N_D}\text{IC}_s\right)\right]\Big/(1 + d)^N$$

$$= 15{,}830 - 0.33\left[15830 - \left(\frac{20 - 15}{20}4{,}711\right)\right]\Big/(1.12)^{15}$$

$$= \$2{,}009$$

- $\text{LCC}_{\text{solar}} = \text{PVSYS} + \text{PVRC} - (t \times \text{PVINT}) - \text{PVDEP} - \text{PVR}$
$$= 2{,}197 + 315 - (0.349 \times 577) - 560 - 2{,}009$$
$$= \$-259$$

Commentary

Powell assumes straight line depreciation over 20 years. Accelerated depreciation over a shorter period is allowed by the 1986 Tax Reform Act.

Conventional fuel costs are included in the LCC of the conventional alternative. If this were a new system (not a retrofit), backup fuel costs should be included in the solar LCC so the oil and electric backup systems could be compared.

Where gains above the depreciated value but less than the initial purchase price are treated as ordinary income, PVR is calculated as follows:

$$\text{PVR} = \left[\text{EC}_s - t_d(\text{EC}_s - \text{IC}_s) - \left(t \times \text{IC}_s \frac{N}{N_D} \right) \right] \bigg/ (1 + d)^N = 2,014.$$

This formula is appropriate only for straight line depreciation,

$$\text{EC}_s > \text{IC}_s, \quad \text{and} \quad N_D > N.$$

This result differs slightly from Powell's (-295) as Powell's calculation includes an arithmetic error.

The unusual negative LCC value is a result of very optimistic assumptions for the loan interest rate and the salvage value of the system.

Example

LCC of the Conventional Alternative

$$LCC_{con} = \frac{(1 - t)f_c \cdot PV}{\eta_0}$$

$$\text{where } PV = \sum_{j=1}^{N_1} \left(\frac{1 + e_{01}}{1 + d}\right)^j + \left(\frac{1 + e_{01}}{1 + d}\right)^{N_1} \sum_{j=1}^{N_2} \left(\frac{1 + e_{02}}{1 + d}\right)^j$$

$$+ \cdots + \left(\frac{1 + e_{01}}{1 + d}\right)^{N_1} \cdots \left(\frac{1 + e_{0K-1}}{1 + d}\right)^{N_{K-1}} \sum_{j=1}^{N_K} \left(\frac{1 + e_{0K}}{1 + d}\right)^j$$

and K is the number of different fuel escalation rate periods.

$LCC_{con} = (1 - 0.349)67.28(13.23)/0.6 = 966$

Net Life-Cycle Benefits (NLCB)

$NLCB = LCC_{con} - LCC_{solar}$

$= 966 - (-259)$

$= \$1,225$

---

a. Monthly parameters are denoted in the equations by a prime.

Commentary

Since the solar system is a retrofit, the only conventional costs are the fuel costs of the replaced fuel. We examine the oil system costs.

NLCB is essentially the same as net present value (NPV). It is expressed in this example in 1979 dollars. To convert it to 1985 dollars we apply the GNP implicit price deflator between 1979 and 1985 to yield

$$NLCB_{1985} = NLCB_{1979} \times 1.421$$

$$= \$1,225 \times 1.421 = \$1,741.$$

The energy price projections, however, are as of 1979.

2.7.4

Example

For the house in Santa Fe, New Mexico, Balcomb determines the optimal level of expenditures in conservation and passive solar in the following example:

Data

Conventional alternative electricity cost
Electricity price escalation rate
Analysis period
Mortgage period
General inflation rate
Discount rate
Mortgage interest rate
Down payment fraction (% of initial cost)
Annual O&M cost (% of initial cost)
Property tax rate (% of initial cost)
Homeowner's marginal income tax rate
Resale value annual escalation rate
Annual space heating energy consumption
(assumes no conservation or solar improvements)

Computations

Balcomb first determines the maximum level of investment in conservation and passive that could be economically justified assuming 100% savings of the energy by solving the following equation for the initial cost (IC).

$$(P/A_{d,N}) \times \text{FCR} \times \text{IC} = L \times f_c \times (P/A_{\bar{e}_e,N})$$

where $\bar{e}_e = \dfrac{1 + d}{1 + e_e} - 1$

and FCR = fixed charge rate

Balcomb determines the following intermediate values:

$(P/A_{d,N}) = 7.0236$

$\text{FCR} = 0.0831$

$(P/A_{\bar{e}_e,N}) = 10.5283$

Solving for IC, Balcomb determines that the combined conservation and passive systems can cost as much as $26,650 if all the conventional fuel is replaced.

Commentary

$f_c = \$20.51/10^6$ Btu ($\$19.44/$GJ)

$e_e = 8\%$

$N = 10$ yrs

$N_f = 30$ yrs

$g = 6\%$

$d = 7\%$

$i = 10\%$

$D = 15\%$

O&M $= 1\%$

$t_p = 2\%$

$t = 25\%$

$e_r = 4.5\%$

$L = 71.8 \ 10^6$ Btu (75.7 GJ)

This equation equates the present value cost of the conservation and solar systems to the present value cost of conventional fuels if no investment is made, i.e., it produces the break-even investment level IC assuming 100% of the conventional fuel is replaced.

We have not been able to replicate Balcomb's calculations exactly. We calculate FCR = 0.0747 using the following formula:

$$FCR = (A/P_{d,N})[PVSYS - t \times PVINT + PVRC - PVSALVAGE]$$

where

$$PVSYS = D + (1 - D)\left\{\frac{(P/A_{d,N})}{(P/A_{i,N_f})} + \frac{(1 + i)^{N_f} - (1 + i)^N}{(1 + d)^N[(1 + i)^{N_f} - 1]}\right\}$$

$$PVSYS = 1.1735$$

$$PVINT = (1 - D)\left\{(P/A_{\bar{i},N})/(1 + i)[i - (A/P_{i,N_f})] + \frac{(P/A_{d,N})}{(P/A_{i,N_f})}\right\}$$

where $\bar{i} = \dfrac{1 + d}{1 + i} - 1$

$$PVINT = 0.5784$$

Example

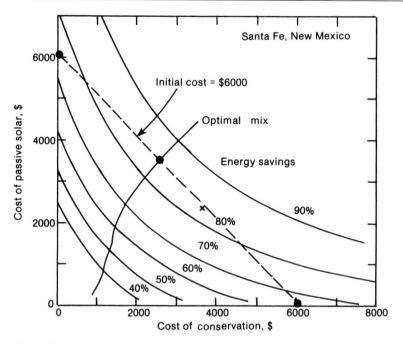

Figure 2.1
Trade-offs between conservation (added insulation, glazing, etc.) and passive solar (Trombe wall) investments. Cost in 1979 dollars ($\times 1.421 = 1985$ dollars). From Balcomb (1979).

To determine the optimal total investment amount, Balcomb examines the points along the optimal mix line of the graph of figure 2.1 and determines the point at which the maximum dollar savings occur.

For example, at an investment of $6,000 the optimal mix yields an energy savings fraction of $F(6,000) = 84.5\%$. The savings at this point is calculated as

$$\text{savings}_{6,000} = (26,650 \times 0.845) - 6,000$$
$$= \$16,519.$$

By examining several points Balcomb determines that the optimum total investment occurs at $7,160 ($3,110 for conservation and $4,050 for passive) with a total energy savings of $F(7,160) = 90.45\%$. At this point the total dollar savings is

$$\text{savings}_{7,160} = (26,650 \times 0.9054) - 7,160$$
$$= \$16,970.$$

Commentary

$$PVRC = (O\&M + t_p)(P/A_{\bar{g},N})$$

where $\bar{g} = \dfrac{1 + d}{1 + g} - 1$

$$PVRC = 0.2850$$

$$PVSALVAGE = \left(\frac{1 + e_r}{1 + d}\right)^N = 0.78945$$

Although Balcomb correctly calculates the optimal investment level, his total dollar savings is incorrect. The savings cannot exceed the present value of the conventional fuel costs if no investment was made; i.e. savings $\leqslant L \times f_c \times (P/A_{\bar{e},N}) = \$15,504$.

Savings should be calculated as follows:

$$\text{savings}_{7,160} = [15,504 \times F(IC)] - [IC \times FCR \times (P/A_{d,N})]$$
$$= (15,504 \times 0.9045) - (7,160 \times 0.0831 \times 7.0236)$$
$$= \$9,858$$

(in 1985\$ the savings are \$14,009).

2.7.5   A Solar Industrial Process Heat System

This example is drawn from a report by Dickinson and Brown (1979) prepared at Lawrence Livermore Laboratory for DOE. It is similar to the commercial sector example presented earlier in that the financing and tax considerations are comparable. It differs in that the measure of merit calculated is the internal rate of return and that all calculations are based on the revenue required to cover all costs (as opposed to the costs themselves).

Example	
Data	
Solar system cost per unit annual output	$IC_s/E = \$60/(10^6 \text{ Btu/yr})$ ($\$56.9/\text{GJ/yr}$)
Conventional alternative fuel cost in first year	$f_c = \$7.60/10^6 \text{ Btu}$ ($\$7.20/\text{GJ}$)
Analysis period	$N = 5 \text{ yr}$
Finance period	$N_f = 5 \text{ yr}$
Depreciation period	$N_D = 5 \text{ yr}$
General inflation rate	$g = 6\%$
Interest rate	$i = 9\%$
Conventional fuel price escalation rate	$e_f = 8\%$
Down payment (% of initial cost)	$D = 70\%$
Annual O&M cost (% of initial cost)	O&M $= 1\%$
Marginal income tax rate	$t = 50\%$
Tax credit (% of initial cost)	TC $= 20\%$
Conventional system efficiency)	$\eta_c = 0.714$
Replacement costs	0
Salvage value	0

Commentary

For illustration, Dickinson and Brown compute NPV on a year-by-year basis. To reduce the length of the example the analysis period was set to only 5 years.

Example

Computations

The internal rate of return is found by solving the following equation iteratively for d:

$$\left(FCR \times \frac{IC_s}{E} \right) - \frac{f_c(P/A_{\bar{e}_f,N})(A/P_{d,N})}{\eta_c} = 0$$

where $\bar{e}_f = \dfrac{1+d}{1+e_f} - 1$

To solve this equation the fixed charge rate (FCR) must be calculated:

$$FCR = (A/P_{d,N})\left\{ [O\&M(P/A_{\bar{g},N})] + \frac{1}{1-t}[PVSYS - (t \times PVINT) - PVDEP] \right\}$$

where $\bar{g} = \dfrac{1+d}{1+g} - 1$

and the present values for the system cost, interest and depreciation are calculated as shown below:

$$\cdot\ PVSYS = D + (1-D)\left\{ [(A/P_{i,N_f})(P/A_{d,N})] - \frac{TC}{1+d} \right\}$$

$$= 0.7 + (1-0.7)\left\{ [0.257(P/A_{d,N})] - \frac{0.2}{1+d} \right\}$$

$$\cdot\ PVINT = (1-D)\left[\left\{ \frac{(P/A_{\bar{i},N_f})}{1+i}[i - (A/P_{i,N_f})] + [(A/P_{i,N_f})(P/A_{d,N})] \right\} \right]$$

where $\bar{i} = \dfrac{1+d}{1+i} - 1$

$$= (1-0.7)\left\{ \left[\frac{(P/A_{\bar{i},N_f})}{1+0.09}(0.09 - 0.257) \right] + [0.257(P/A_{d,N})] \right\}$$

$$\cdot\ PVDEP = \frac{t}{N_D}(P/A_{d,N})$$

$$= \frac{0.5}{5}(P/A_{d,N})$$

Solving iteratively for d yields $d = 10\%$. Thus the internal rate of return is 10%.

Commentary

Dickinson and Brown chose to equate levelized annual costs rather than life-cycle costs since their methodology is centered around levelized revenue requirements. Using the intermediate value FCR (fixed charge rate) is common in utility economics but is not normally used in industrial evaluations.

In this example PVSYS, PVINT, and PVDEP are defined as the present value of the system, interest, and depreciation for a capital investment of one dollar. Their sum in the FCR formula is divided by $(1 - t)$. This accounts for the taxes paid on the revenue required to cover nondeductible capital costs.

Straight line depreciation is assumed. Accelerated depreciation as provided for under the 1986 Tax Reform Act would yield a slightly higher return on investment.

2.7.6   Utility-Owned Central Station Plant

This example is taken from a report by Doane et al. (1976) prepared for the Energy Research and Development Agency and the Electric Power Research Institute. The example illustrates Doane's methodology for computing annualized revenue requirements and levelized busbar energy costs. In addition to the utility-specific considerations, the example illustrates the treatment of an extended construction period and nonrecurring replacement costs. Although the example as presented by Doane is for a photovoltaic plant, Doane emphasizes that it is not specific to a particular system type.

Example

A utility intends to install a 200-MW solar power plant to be on line in 1990. All costs are expressed in 1975 dollars.

Data

Initial system costs

1985—Land, design engineering, etc.	$IC_L = \$50 \times 10^6$
1989—Construction costs	$IC_B = \$125 \times 10^6$
Nonrecurring replacement costs in 2005	$RC = \$60 \times 10^6$
First year of operation	$y_0 = 1,990$
Analysis period	$N = 30$ yr
Interest rate on debt	$i_d = 8\%$
Annual rate of return on common stock	$i_c = 12\%$
Annual rate of return on preferred stock	$i_p = 8\%$
Ratio of debt to total capitalization	$R_d = 0.5$
Ratio of common stock to total capitalization	$R_c = 0.4$
Ratio of preferred stock to total capitalization	$R_p = 0.1$
General inflation rate	$g = 5\%$
Escalation rate for capital costs	$g_c = 5\%$
Escalation rate for O&M costs	$g_c = 6\%$

Commentary

The capital costs are assumed to occur at the end of each of the years listed.

Example	
Annual property taxes (as a fraction of initial costs)	$t_p = 0.02$
Annual insurance premiums (as a fraction of initial costs)	$I = 0.0025$
Marginal income tax rate	$t = 40\%$
Annual O&M cost	$O\&M = \$3 \times 10^6$
Annual energy output	$E = 7 \times 10^5$ MWh

Computations

· The effective discount rate d and the fixed charge rate FCR are computed first.

$$d = [(1 - t)R_D i_d] + (R_c i_c) + (R_p i_p)$$

$$d = [(1 - 0.4)0.5 \times 0.08] + (0.4 \times 0.12) + (0.1 \times 0.08) = 0.08$$

$$FCR = \frac{1}{1 - t}[(A/P_{d,N}) - PVDEP] + t_p + I$$

PVDEP is the present value of depreciation

$$PVDEP = \frac{t}{N}$$

$$FCR = \frac{1}{1 - 0.4}\left(0.089 - \frac{0.4}{30}\right) + 0.02 + 0.0025$$

$$= 0.148$$

· The present (1990) value of the initial system cost (IC_s) expressed in dollars in the first year of operation (1990) is

$$IC_s = (1 + g_c)^{(1985-1975+1)}(1 + d)^{(1989-1985)}IC_L + (1 + g_c)^{(1989-1975+1)}IC_B$$
$$+ (1 + g_c)^{(2005-1975+1)}(1 + d)^{(1989-2005)}RC$$
$$= (1.05)^{11}(1.08)^4(50 \times 10^6) + (1.05)^{15}(125 \times 10^6) + (1.05)^{31}(1.08)^{-16}(60 \times 10^6)$$
$$= \$455.7 \times 10^6$$

· The present value of operating and maintenance (PVO&M) is

$$PVO\&M = (1 + g_{O\&M})^{(1990-1975)}O\&M(P/A_{\bar{g}_{O\&M},N})$$

where $\bar{g}_{O\&M} = \dfrac{1 + d}{1 + g_{O\&M}} - 1$

$$PVO\&M = (1.06)^{15}(3 \times 10^6)(22.75) = \$163.6 \times 10^6$$

Commentary

As noted in the text, the FCR as calculated here for a utility is different from an FCR calculated for an industry. The principal difference is in the use of an effective discount rate to account for tax deductibility of interest and the delayed repayment of principal. Note that the income tax deductibility of property taxes is not accounted for and that O&M costs are not included as part of the FCR.

Straight line depreciation is assumed with a depreciation period equal to the analysis period of 30 years. Although this depreciation schedule is legal, the 1986 Tax Reform Act permits double declining balance depreciation of utility investments.

Since the data give the costs in 1975, the costs must be escalated to the year in which they occur. For those costs that are not incurred at the beginning of the first year of commercial operation (the present in the present value calculations) a discount factor is applied. The " + 1" in the exponent accounts for the assumption that capital expenditures occur at the end of the year; the present value is assumed to be measured at the beginning of the year. Replacement costs are also included in the initial costs.

Example

· The annual revenues required (ARR) are

$$\text{ARR} = \frac{1}{(1 + g)^{1990 - 1975}} \{(\text{FCR} \times \text{IC}_s) + [(A/P_{d,N})\text{PVO\&M}]\}$$

$$= \frac{1}{(1.05)^{15}} [(0.148 \times 455.7 \times 10^6) + (0.089 \times 163.6 \times 10^6)]$$

$$= \$39.5 \times 10^6$$

· The levelized busbar energy cost in 1975 dollars based on a nominal discount rate (BBEC) is

$$\text{BBEC} = \frac{\text{ARR}}{E}$$

$$= \frac{\$39.5 \times 10^6}{7 \times 10^5 \text{ MWh}} = \$0.0564/\text{kWh}$$

· The levelized busbar energy cost in 1975 dollars based on a real discount rate (BBEC$_r$) is

$$\text{BBEC}_r = \text{BBEC}(P/A_{d,N})(A/P_{\bar{g},N})$$

$$\text{where } \bar{g} = \frac{1 + d}{1 + g} - 1$$

$$\text{BBEC}_r = 0.0564(11.26)(0.050)$$

$$= \$0.0318/\text{kWh}$$

Commentary

To reduce the annual required revenues to 1975 dollars, we discounted by the general inflation rate. (The general inflation rate is not a historical value, but a projected value since the example was constructed in 1976.)

If $0.0564 is paid for each kWh of electricity produced by the plant at the busbar, the full cost (including taxes and return on equity) will be exactly recovered during the analysis period.

Using the historical GNP implicit price deflator we can convert this levelized BBEC to constant 1985 dollars:

$$BBEC_{1985} = BBEC_{1975} \times 1.883$$
$$= \$0.0564 \times 1.883$$
$$= \$0.1062.$$

Alternatively, if in the year t

$$(1.05)^{t-1975} \times 0.0318$$

year t dollars are paid for each kWh, the full cost will be recovered during the analysis period.

Again, using the historical GNP implicit price deflator this levelized $BBEC_r$ can be converted to 1985 dollars:

$$BBEC_{r,1985} = BBEC_{r,1975} \times 1.883$$
$$= \$0.0318 \times 1.883$$
$$= \$0.0599.$$

2.8 Summary

This chapter began by identifying solar energy economic issues that prompted the need for economic methods. These issues concerned the economic feasibility, market potential, and economic impact of solar energy systems as well as the formulation of government policy and programs on solar-related issues.

The chapter then reviewed the principal methods of economic evaluation that have been applied to solar-related issues: life-cycle costing, net benefits, required revenue, internal rate of return, savings-to-investment ratio, payback, levelized cost of energy, and break-even analysis. Generalized formulas and brief instructions for using the methods were given. Other methods that have been used for solar evaluations—sensitivity analysis, expected value analysis, mathematical programming, production theory, input/output theory, market penetration analysis techniques, and marketing survey methods—were discussed in the context of the kinds of economic issues to which they have been applied.

Next, representative works were drawn from the literature to illustrate the developments, adaptations, and applications of economic methods for solar analysis since 1972. These selections were organized by type of issue addressed. Most of the advances have been in adapting existing economic methods to apply to solar-related issues, rather than in developing completely new methods.

Last, six case examples of economic evaluations of solar energy systems were presented in sufficient detail to allow the reader to follow the computations for several of the more important economic methods. The case examples included a life-cycle cost analysis of an active solar energy system for residential space heating; life-cycle cost and net savings analyses of an active solar energy system for a federal office building; life-cycle cost, internal rate of return, and savings-to-investment ratio analyses for a semipassive system for commercial space heating; an optimization analysis of combined conservation and passive solar investments for a residence; an internal rate of return analysis of a solar industrial process heat system, based on revenue required to cover all costs; and an analysis of the annualized revenue requirements and levelized bus bar energy costs for a utility-owned central station photovoltaic plant.

References

Ackoff, R. L., and M. W. Sasieni. 1968. *Fundamentals of Operation Research.* New York: Wiley.

Arthur D. Little, Inc., 1977. *Solar Heating and Cooling of Buildings (SHACOB) Commercialization Report,* vols. I–III. Cambridge, MA: Arthur D. Little, Inc., Engineering Sciences.

Au, T., and T. P. Au, 1983. *Engineering Economics for Capital Investment Analysis.* Boston: Allyn and Bacon.

Balcomb, J. D. 1979. *Conservation and Solar Working Together.* LA-UR-79-3195. Los Alamos, NM: Los Alamos Scientific Laboratory.

Balcomb, J. D., 1980. "Conservation and solar: Working together." In *Proceedings of the Fifth National Passive Solar Conference,* 44–50.

Barley, D. C. 1979. "Load optimization in solar space heating systems." *Solar Energy* 23: 149–156.

Battelle Pacific Northwest Laboratories. 1978. *An Analysis of Federal Incentives Used to Stimulate Energy Production.* PNL-2410-1. Richland, WA: Battelle Laboratories.

Bendt, P. 1983. "Appropriate sizing of solar heating systems." *Journal of Solar Energy Engineering* 105(1): 66–72.

Bennington, G. E. 1976. *Ten Easy Steps To Busbar Costs.* WP-11488. McLean, VA: MITRE Corporation, METREK Division.

Bennington, G., P. Curto, G. Miller, K. Rebibo, and P. Spewak. 1978. *Solar Energy: A Comparative Analysis to the Year 2020.* MTR-7579. McLean, VA: MITRE Corporation, METREK Division.

Bezdek, R., A. S. Hirshberg, and W. H. Babcock. 1979. "Economic feasibility of solar water and space heating." *Science* 203: 1214–1220.

Bos, P. B., and J. M. Weingart. 1983. *Impact of Tax Incentives on the Commercialization of Solar Thermal Electric Technologies.* SAND 83-8178. Menlo Park, CA: Polydyne, Inc.

Brandemuehl, M. J., and W. A. Beckman. 1978. "Economic evaluation and optimization of solar heating systems." *Journal of Solar Energy* 23: 1–10.

Cassidy, F., and G. W. Schirra. 1977. *Treatment of Inflation in Engineering Economic Analysis: Economic Analysis of Advanced Energy Technologies.* MTR-7611. McLean, VA: MITRE Corporation, METREK Division.

Clark, J., T. J. Hindelang, and R. E. Pritchard. 1984. *Capital Budgeting.* Englewood Cliffs, NJ: Prentice-Hall.

Conopask, J. V., P. Fonash, and J. Easterly. 1981. "Solar thermal residential and commercial economics: 1980 update." *ASHRAE Journal* 23: 39–44.

CYBERNET Services. 1980. *BLAST II. Vol. I, User Information Manual.* Minneapolis, MN: Control Data Corporation.

Dasgupta, A. K., and D. W. Pearce. 1972. *Cost-Benefit Analysis: Theory and Practice.* New York: Harper and Row.

deLeon, P., B. L. Jackson, R. F. McNown, and G. J. Mahrenholg. 1980. *The Potential Displacement of Petroleum Imports by Solar Energy Technologies.* SERI/TR-352-504. Golden, CO: Solar Energy Research Institute.

Dernburg, J., D. Depaso, and C. Fenton. 1981. *Market Development Potential for Photovoltaics: A Review of Recent Research.* NTIS DE83013877. Cambridge, MA: Technology and Economics, Inc.

Dickinson, W. C., and K. C. Brown. 1979. *Economic Analysis of Solar Industrial Process Heat Systems.* UCRL-52814. Livermore, CA: Lawrence Livermore Laboratory.

Doane, J. W., K. P. O'Toole, R. G. Chamberlain, P. B. Bos, and P. D. Maycock. 1976. *The Cost of Energy from Utility-Owned Solar Electric Systems.* JPL 5040-29. Pasadena, CA: Jet Propulsion Laboratory, California Institute of Technology.

Dorfman, R., P. A. Samuelson, and R. M. Solow. 1958. *Linear Programming and Economic Analysis.* New York: McGraw-Hill.

Duffie, J. A., W. A. Beckman, and J. G. Dekker. 1976. "Solar heating in the United States," in *Proceedings of the Winter Annual Meeting of the American Society of Mechanical Engineers.* New York: ASME Society.

Early, R. F., M. M. Mohtadi, E. L. Rossidivito, D. E. Serot, and H. Weisman. 1979. *Macroeconomic and Sector Implications of Installing 2.2 Million Residential Solar Units.* AM/IA/79-22. Washington, DC: U.S. Department of Energy.

Energy and Environmental Analysis, Inc. 1977. *The Market Oriented Program Planning Study (MOPPS).* DOE/ET-0010. Washington, DC: U.S. Department of Energy.

Ferris, G., and B. Mason. 1979. *A Review of Regional Economic Models with Special Reference to Labor Impact Assessment.* SERI/TR-53-100. Golden, CO: Solar Energy Research Institute.

Feuerstein, R. 1979. *Utility Rates and Service Policies as Potential Barriers to the Market Penetration of Decentralized Solar Technologies.* SERI/TR-62-274. Golden, CO: Solar Energy Research Institute.

Flaim, T., T. J. Considine, R. Witholder, and M. Edesees. 1981. *Economic Assessments of Intermittent Grid-Connected Solar Electric Technologies: A Review of Methods.* SERI/TR-353-474. Golden, CO: Solar Energy Research Institute.

Gershon Meckler Associates. 1976. *Parameter Grouping and Regionalizing to Simplify Solar Life Cycle Cost Models for Application to Solar Heating and Cooling of Buildings.* Washington, DC: Gershon Meckler Associates.

Grant, E. L., G. W. Ireson, and R. S. Leavenworth. 1976. *Principles of Engineering Economy,* sixth edition. New York: Ronald Press.

Hillier, F. S., and G. J. Lieberman. 1980. *Introduction to Operations Research.* San Francisco: Holden-Day.

King, T. A., P. E. Jefferson, G. Shingleton, P. A. Sabatuik, and J. B. Carlock, III. 1979. "Cost effectiveness: An assessment based on commercial demonstration projects.", in *Proceedings of the Second Annual Solar Heating and Cooling Systems Operational Results Conference.* Golden, CO: Solar Energy Research Institute.

Kirschner, C., D. Brunton, K. LaValle, and F. Roach. 1982. *Housing Stock Characteristics for Solar Market Analysis.* LA-9201-MS. Los Alamos, NM: Los Alamos National Laboratory.

Klein, L. R. 1962. *An Introduction to Econometrics.* Englewood Cliffs, NJ: Prentice-Hall.

Kort, J. R. 1980. *Estimating the Economic and Demographic Impacts of Solar Technology Commercialization on U.S. Regions.* DOE/CS/30504-T1. Washington, DC: U.S. Department of Energy.

Krawiec, F. 1980. *Economic Measurement of Environmental Damages.* SERI/RR-744-311. Golden, CO: Solar Energy Research Institute.

Kreith, F., and J. F. Kreider. 1976. "Preliminary design and economic analysis of solar-energy systems for heating and cooling of buildings." *Energy* 1:63–76.

Lawrence Berkeley Laboratory. 1980. *DOE-2 Users Guide*. LBL-8689. Berkeley, CA: Lawrence Berkeley Laboratory.

Löf, G. O. G., and R. A. Tybout. 1973. "Cost of house heating with solar energy." *Solar Energy* 14:253–278.

Löf, G. O. G., and R. A. Tybout. 1974. "The design and cost of optimized systems for residential heating and cooling by solar energy." *Solar Energy* 16:9–18.

Mann, G. S., and B. F. Neenan. 1982. *Economic Profiles of Selected Solar Energy Technologies for Use in Input-Output Analysis*. LA-9083-TASE. Los Alamos, NM: Los Alamos National Laboratory.

McGarity, A. E. 1976. *Solar Heating and Cooling: An Economic Assessment*. NSF 76-37. Washington, DC: National Science Foundation.

McGarity, A. E., C. S. Revelle, and J. L. Cohon. 1981. "Solar heating design with a performance requirement." in *Proceedings of the 1981 Annual Meeting of the American Section of the International Solar Energy Society*, Newark, DE: American Section of the International Solar Energy Society, 636–640.

Mishan, E. J. 1976. *Cost Benefit Analysis: An Introduction*. New York: Praeger.

MITRE Corporation. 1978. *SPURR Agricultural and Industrial Process Heat Component: Program Methodology and Documentation*. McLean, VA: MITRE Corporation, METREK Division.

Noll, S., and M. Thayer. 1979. "Passive solar, auxiliary heat and building conservation optimization: A graphical analysis," in *Fourth Passive Solar Conference Proceedings*. Newark, DE: American Section of the International Solar Energy Society, 128–131.

Ormasa, R. E. 1979. *An Economic Assessment of Solar Energy Systems in the Commercial Demonstration Program*. SOLAR/0823-79/01. PRC Energy Analysis Co.

Perino, A. M. 1979. *A Methodology for Determining the Economic Feasibility of Residential or Commercial Solar Energy Systems*. SAND 78-0931. Albuquerque, NM: Sandia National Laboratory.

Petersen, H. C. 1976. *Impact of Tax Incentives and Auxiliary Fuel Prices on the Utilization Rate of Solar Energy Space Conditioning*. Logan, UT: Utah State University.

Petersen, H. C. 1977. *Sector-Specific Output and Employment Impacts of a Solar Space and Water Heating Industry*. Logan, UT: Utah State University.

Petersen, S. R. 1983. *SOLCOM: A Computer Program to Integrate Solar and Conservation Economics for New Commercial Buildings*. NBSIR 83-2658. Washington, DC: National Bureau of Standards.

Phung, C. L. 1977. "IEA life-cycle costing methodology," in *Economic Analysis of Advanced Energy Technologies*. MTR-7611. McLean, VA: MITRE Corporation, METREK Division, 53–68.

Pleatsikas, C., E. A. Hudson, D. C. O'Connor, and D. H. Funkhouser. 1979. *A Study of Capital Requirements for Solar Energy. An Analysis of the Macroeconomic Effects of Increased Solar Energy Market Penetration*. DSE-4230-T1. Cambridge, MA: Urban Systems Research and Engineering, Inc.

Pogany, D., D. S. Ward, and G. O. G. Löf. 1975. "The economics of solar heating and cooling systems." in *Proceedings of the Annual Meeting of the International Solar Energy Society*.

Powell, J. W. 1980. *An Economic Model for Passive Solar Designs in Commercial Environments*. NBS BSS 125. Gaithersburg, MD: National Bureau of Standards.

Powell, J. W., and K. A. Barnes. 1982. *Comparative Analysis of Economic Models in Selected Solar Energy Computer Program*. NBSIR 81-2379. Gaithersburg, MD: National Bureau of Standards.

Powell, J. W., and R. C. Rodgers, Jr. 1981. *FEDSOL: Program User's Manual and Economic Optimization Guide for Solar Federal Building Projects*. NBSIR 81-2342. Gaithersburg, MD: National Bureau of Standards.

Real Estate Research Corporation. 1979. *Marketing and Market Acceptance Data from the Residential Solar Demonstration Program: 1979*, Vol. 1. Chicago, IL: Real Estate Research Corp.

Reid, R. L., E. Lumsdaine, and L. Albrecht. 1977. "Economics of solar heating with homeowner type financing." *Solar Energy* 19:513–517.

Reiger, A. J. 1978. "Marketplace realities and solar economics." in *SUN: Mankind's Future Source of Energy*. Washington, DC: U.S. Department of Housing and Urban Development, 248–252.

Roach, F., S. Noll, and S. Ben-David. 1979. "Passive and active residential solar heating: A comparative economic analysis of select designs." *Energy: The International Journal* 4(4): 623–644.

Rudasill, C. 1977. "Revenue requirements calculation for utility systems analysis," in *Economic Analysis of Advanced Energy Technologies*. MTR-7611. McLean, VA: MITRE Corporation, METREK Division.

Ruegg, R. T. 1975. *Solar Heating and Cooling in Buildings: Methods of Economic Evaluation*. NBSIR 75-712. Gaithersburg, MD: National Bureau of Standards.

Ruegg, R. T. 1976. *Evaluating Incentives for Solar Heating*. NBSIR 76-1127. Gaithersburg, MD: National Bureau of Standards.

Ruegg, R. T. 1987. *Life-Cycle Cost Manual for the Federal Energy Management Program, NBS Handbook 135 (Rev.)*. Gaithersburg, MD: National Bureau of Standards.

Ruegg, R. T., and G. T. Sav. 1981. "Microeconomics of solar energy," in *Solar Energy Handbook*. J. F. Kreider and F. Kreith, eds. New York: McGraw-Hill, Ch. 28, 28-1–28-42.

Ruegg, R. T., G. T. Sav, J. W. Powell, and E. T. Pierce. 1982. *Economic Evaluation of Solar Energy Systems in Commercial Buildings; Methodology and Case Studies*. NBSIR 82-2540. Gaithersburg, MD: National Bureau of Standards.

RUPI, Inc. 1977. *Federal Incentives for Solar Houses: An Assessment of Program Options*. Cambridge, MA: Regional and Urban Planning Implementation, Inc.

Sav, G. T. 1978. "Economic optimization of solar energy and energy conservation in commercial buildings," in *Proceedings of the Conference on Systems Simulations and Economic Analysis for Solar Heating and Cooling*. Washington, DC: U.S. Department of Energy, 88–90.

Sav, G. T. 1979. "Universal economic optimization paths for solar hot water systems in commercial buildings." *Energy: The International Journal* 4:415–427.

Schiffel, D., D. Costello, D. Posner, and R. Witholder. 1978. *The Market Penetration of Solar Energy: A Model Review Workshop Summary*. SERI/16. Golden, CO: Solar Energy Research Institute.

Schiffel, D., D. Posner, K. Hillhouse, J. Doane, and P. Weis. 1978. *Solar Incentives Planning and Development: A State-of-the-Art Review and Research Agenda*. SERI/TR-51-059. Golden, CO: Solar Energy Research Institute.

Scott, J. 1977. *Solar Water Heating, Economic Feasibility, Capture Potential, and Incentives.* Newark, NJ: Delaware University, College of Business and Economics.

Sedmak, M. R., and E. Zampelli. 1979. *The Determination of the Optimum Solar Investment Decision Criteria.* McLean, VA: PRC Energy Analysis Co.

Shams, A., and R. Fichtenbaum. 1976. "The feasibility of solar house heating: A study in applied economics," in *Sharing the Sun: Solar Technology in the Seventies.* Cape Canaveral, FL: American Section of the International Solar Energy Society, 32–50.

Smith, G. W. 1973. *Engineering Economy: Analysis of Capital Expenditures.* Ames, IA: Iowa State University Press.

SOLCOST Service Center. 1980. *SOLCOST—Solar Energy Design Program for Non-Thermal Specialists.* SERI/SP-751-686. Golden, CO: Solar Energy Research Institute.

Stanford Research Institute. 1977. *Solar Energy in America's Future.* Washington, DC: Energy Research and Development Administration.

Technology and Economics, Inc. 1979. *A Parametric Analysis of Investment Tax Credit Programs for Solar Industrial Process Heat Equipment.* Cambridge, MA: Technology and Economics, Inc.

University of Wisconsin, Madison. 1978. *F-Chart Version 3.0 Users Manual: A Design Program for Solar Heating Systems.* SERI/SP-35-124. Golden, CO: Solar Energy Research Institute.

University of Wisconsin, Madison. 1980. *F-Chart Version 4.0 Users Manual: A Design Program for Solar Heating Systems.* SERI/SP-35-124 R1. Golden, CO: Solar Energy Research Institute.

U.S. Code of Federal Regulations. 1986. *Federal Energy Management and Planning Programs, Subpart A—Methodology and Procedures for Life Cycle Cost Analyses.* 10 CFR, part 436.

U.S. Department of Energy, 1979. *Workshop Manual on Technical, Economic and Legal Considerations for Evaluating Solar Heated Buildings for Lenders, Appraisers, Insurers, and Tax Consultants.* DOE/CS-0083. Washington, DC: U.S. Department of Energy.

U.S. Department of Energy and Brookhaven National Laboratory. 1981. *Proceedings of the Conference on Financial Issues for International Renewable Energy Opportunities.* BNL 51490. Brookhaven, NY: Brookhaven National Laboratory.

Warren, M., and M. Wahlig. 1982. *Cost and Performance Goal Methodology for Active Solar Cooling Systems.* LBL-12753. Berkeley, CA: Lawrence Berkeley Laboratory.

Weston, J. F., and E. F. Brigham. 1981. *Managerial Finance,* seventh edition. Hinsdale, IL: Dryden Press.

3 Economic Models

G. Thomas Sav

Since the 1973 oil embargo, there has been a renewed and increasing interest in substituting solar energy for conventional nonrenewable energy resources. The interest in solar energy was stimulated by increasing costs of conventional energy sources and the desire to use alternative energy resources that were environmentally acceptable. Solar energy is accepted as a safe environmental alternative and is attractive because it is, for all practical purposes, an inexhaustible source of energy that is technologically feasible in many applications, including space heating and cooling of buildings, domestic hot water, and a variety of industrial process applications. The widespread application of solar energy, however, depends ultimately on its economic feasibility and cost relative to alternative energy sources.

Economic models were developed to evaluate the cost of solar energy, to design economically optimal solar energy systems, and to compare the cost of such systems to alternative means of producing energy. Moreover, these models were expanded to incorporate and analyze the effects of a variety of financial incentives for solar energy, such as tax incentives. Financial incentives directly affect the relative cost of solar energy and the economically optimal design of systems. Economic models that have successfully incorporated such effects have played an important role in shaping public policy as it relates to solar energy at the federal, state, and local levels of government.

In this chapter I discuss the economic models developed to analyze and design solar energy systems as alternatives to conventional nonrenewable energy sources in producing building energy services. In section 3.1 I review the foundations of these models. To be useful, economic models must be empirically linked to solar thermal performance models. Empirical approaches to modeling solar energy have taken several forms, ranging from sophisticated computer simulation models to simple, single-equation deterministic models. In section 3.2 I review the alternative models and the advantages and disadvantages of each. Section 3.3 is a summary and conclusion.

3.1 Foundations of Economic Models

3.1.1 Modeling Production of Energy Services

Residential, commercial, and industrial energy services, such as domestic hot water, space heating and cooling, and process steam, are produced on site rather than purchased directly. Production is accomplished by combining durable capital equipment, such as hot water heaters, furnaces, and air conditioners, with nonrenewable fuels, for example, electricity, natural gas, and oil. Innovative energy sources, such as solar energy processes, can reduce the consumption of nonrenewable conventional fuels in the production of these energy services. However, solar-produced energy generally cannot completely substitute for conventionally produced energy given the current state of technology. A backup conventional energy process is usually required because storing solar-produced energy over extended periods (for example, during inclement weather) is prohibitively expensive. Thus, in modeling the technical substitution possibilities between solar and conventionally produced energy, economists assume a conventional backup energy system capable of meeting the full energy load in the absence of a solar process.

In modeling the production of energy services, many economists begin with fundamental microeconomic theory and assume that an energy service subfunction E is embedded in and separable from a master production function [see Sav (1984a–c) for a review]:

$$y = F(E, X),   \tag{3.1}$$

where y is final output, E is energy input, and X is a vector of inputs not entering E. Using an energy balance equation, we can express the energy service subfunction:

$$E = E_c + E_s,   \tag{3.2}$$

where E_c is the quantity of energy produced from a conventional energy process and E_s is the quantity of energy produced from a solar energy process. Given the conventional backup energy process constraint, E_c can be viewed as being produced with a conventional fuel (the capital equipment already in place):

$$E_c = g(F),   \tag{3.3}$$

where F is the quantity of fuel. Similarly, E_s can be viewed as being produced with the two primary solar design variables:

$$E_s = h(A, V),\tag{3.4}$$

where A is solar collector area and V is solar storage volume. Substituting equations (3.3) and (3.4) into equation (3.2) results in a stock-flow production function:

$$E = g(F) + h(A, V),\tag{3.5}$$

where durable capital equipment A and V can be substituted for a conventional fuel input F in the production of an energy service E.

Many of the solar energy process models, such as the well-known F-CHART model developed by Beckman, Klein, and Duffie (1977) used A (solar collector area) as the single design variable with the understanding that storage volume V varied in fixed proportion to A; that is, $V = bA$, where b is usually set between 0.050 and 0.100 m³/m² of A, depending on the solar process application. Given the fixed proportion relationship, equation (3.5) is reduced to

$$E = g(F) + h(A),\tag{3.6}$$

or, more generally,

$$E = F(F, A).\tag{3.7}$$

However, even assuming fixed proportions does not preclude one from analyzing the effect of different size storage on system performance. Beckman, Klein, and Duffie (1977) demonstrate a method for proceeding along these lines but also conclude that the optimal relationship is such that b lies in the range of 0.050 to 0.100 m³/m² of A.

3.1.2   Optimal Design Models

Many factors drive individuals to consider and eventually adopt solar as an alternative energy source. Yet one of the most attractive features of solar energy from an economical standpoint is its ability to reduce the total costs of producing energy services, such as space heating and/or cooling and domestic hot water services. As such, research has focused on designing optimal solar energy systems that minimize total energy costs.

The basic framework for much of the work in optimal design models was set forth in the seminal work of Löf and Tybout (1973). Since their study more sophisticated models have been developed. Ruegg's work (1975, 1976) has probably been the largest contributor to this sophistication.

Ruegg's optimal design model (1975, 1976) relies on the so-called Life-cycle method. (Chapter 2 gives more background on the life-cycle method.) In this method the objective is to minimize the total life-cycle costs (LCC_T) of producing a given level energy service:

$$LCC_T = LCC_c + LCC_s, \tag{3.8}$$

where LCC represents the life-cycle costs of owning and operating a system discounted to the present and the subscripts c and s represent the conventional energy system and the solar energy system, respectively. Normally it is assumed that a 100% conventional energy backup system is required with a solar energy system; thus LCC_c is comprised solely of conventional fuel costs.

As the solar energy system increases in size, its life-cycle costs LCC_s increase. Simultaneously, as LCC_s increases, conventional fuel requirements and the associated life-cycle costs of conventional fuel LCC_c decrease. Thus, the objective is to choose a system size that minimizes LCC_T.

Many studies were conducted on the economic feasibility and design of solar energy systems using this basic framework and extending it to account for solar tax incentives, residential and commercial comparisons, and regional factors. Beckman, Klein, and Duffie (1977) incorporated the basic model in their F-CHART method for designing solar facilities. Ben-David et al. (1977) examined the economic feasibility of solar energy on a state-by-state basis using a basic variation of the life-cycle cost model. MITRE Corporation (1976), Nicholls (1977), O'Neal, Carney, and Hirst (1978), Bezdek (1978), and Lameiro and Bendt (1978) all expanded in some way on the life-cycle cost model in analyzing and designing solar energy systems for residential and commercial applications, as well as industrial and government facility applications.

Although the exact specification of the terms in the life-cycle cost model varies among the many studies, a general approach was provided by Ruegg and Sav (1981) and Perino (1979). Considering the efficiency of the conventional energy system and the energy content of conventional fuel, the LCCc term is generally specified as

$$\mathrm{LCC}_c = P_c(1 + t_c)(1 - t_y)L(1 - f)\delta^{-1}B^{-1}P/A_{d,e,l}, \tag{3.9}$$

where P_c is the present year cost per unit of conventional fuel, t_c is the sales tax rate on conventional fuel, t_y is the federal and state composite income tax rate, L is the annual energy load, f is the annual fraction of L supplied by solar energy, δ is the efficiency of conventional energy equipment, B is the energy content of conventional energy source, and $P/A_{d,e,l}$ is a uniform present worth factor for a nominal discount rate d, conventional energy escalation rate e, and a solar energy system life l. Here

$$P/A_{d,e,l} = \frac{1 + e}{d - e} \frac{(1 + d)^l - (1 + e)^l}{(1 + d)^l} \quad \text{for} \quad d \neq e \tag{3.10}$$

and

$$P/A_{d,e,l} = l \quad \text{for} \quad d = e. \tag{3.11}$$

Both nominal and real values of d and e are used in computing $P/A_{d,e,l}$. The choice was merely one of the researcher's preference, perhaps constrained by the available data. For residential building owners conventional fuel expenditures are not deducted from taxable income; thus $t_y = 0$ in the LCC equation. Bezdek (1978) investigated the effect of the absence of the deduction of fuel expenses.

Similarly, determining life-cycle costs of the solar energy system LCC_s varied among the many studies but generally included

$$\mathrm{LCC}_s = K = + R + P - D - M, \tag{3.12}$$

where K is the present discounted value (PDV) of capital costs associated with acquiring and installing a solar energy system, R is the PDV of annual recurring costs associated with maintenance and repair, P is the PDV of property tax payments arising from the solar energy system, D is the PDV of tax savings resulting from depreciation deductions from taxable income, and M is the PDV of tax savings resulting from loan interest deductions from taxable income.

Terms comprising the solar energy life-cycle costs accounted for site, system, and region-specific factors, that is, on the availability of governmental grants and tax incentives K, periodicity of recurring costs R, property tax assessments P, and income tax laws governing capital depreciation methods D. Early analyses, such as that of Löf and Tybout (1973), were less inclusive than the later studies of MITRE (1976), Ruegg (1976), and Ben-David et al. (1977) in this respect. Part of the movement from simpler modeling to detailed incorporation of these factors can be attributed to the need to account for solar energy tax incentives explicitly in economic feasibility and design studies. For example, following Ruegg's (1976) analysis it became a fairly well-accepted practice to include the availability of cash grants and tax credits in determining solar energy capital costs K as follows:

$$K = (F_x + vA)[q + (1 - q)A/P_{i,m}P/A_{d,m}] - G - T(1 + d)^{-1}, \qquad (3.13)$$

where F_x is the fixed costs of collector and noncollector components, v is the variable costs of collector and noncollector components per unit of collector area A, A is the collector area, $F_x + vA$ is the contract cost, q is the fraction of the total system contract cost $(F_x + vA)$ placed as a downpayment, G is the governmental cash grant assumed here to be available after contractual payment, T is the governmental tax credit assumed here to be available one year after installation and having a value of $t_R(F_x + vA)$ (where t_R is the fraction of the contract cost available as a tax credit), and $A/P_{i,n}$ is a capital recovery factor for a nominal loan rate i and loan life m:

$$A/P_{i,n} = \frac{i(1 + i)^m}{(1 + i)^m - 1}. \qquad (3.14)$$

$A/P_{d,n}$ is a present worth factor for a nominal discount rate d and loan life m:

$$P/A_{d,n} = \frac{(1 + d)^m - 1}{d(1 + d)^m}. \qquad (3.15)$$

Usually, as presented here, the governmental cash grant and tax credit become available after contractual payment. Perino (1979) considered different variations on this theme, as well as the case when $i = d$ and the product $A/P_{i,m}$ and $A/P_{d,m}$ equals unity. In this latter situation the present value cost is the same whether the system is financed through a loan or is paid for at the outset from equity funds.

Both the early modeling efforts and the more recent sophisticated studies faced the persistent problem of realistically capturing that part of solar energy systems costs associated with long-term maintenance and repair. In large part this problem arises from the absence of long-term empirical data on the maintenance and repair of solar energy systems in various applications. As a result, such optimization models as F-CHART (Beckman, Klein, and Duffie, 1977) and GFL (Lameiro and Bendt, 1978) and such general approaches as Ruegg (1976) and Ruegg and Sav (1981) assumed that costs associated with maintenance and repair occur on an annual basis as some fixed percentage of the initial contract cost $(F_x + vA)$. These costs included, for example, the yearly cost of cleaning collectors, replacing or replenishing antifreeze, and the yearly cost of insurance premiums plus uninsured damage. Many studies assumed that repair costs could be described as a recurring cost in proportion to the total system cost, so that the PDV of the recurring cost factor R in equation (3.12) is simply

$$R = r_R(1 - t_y)(F_x + vA)P/A_{d,g,l}, \tag{3.16}$$

where r_R is the annual recurring cost as a fraction of $F_x + vA$, t_y is a federal and state composite income tax rate, and g is the general rate of inflation. Then

$$P/A_{d,g,l} = \frac{1 + g}{d - g} \frac{(1 + d) - (1 + g)^l}{(1 + d)} \quad \text{for} \quad d \neq g \tag{3.17}$$

and

$$P/A_{d,g,l} = l \quad \text{for} \quad d = g. \tag{3.18}$$

The t_y term in equation (3.16) was included in those studies concentrating on the commercial building sector (Ruegg et al., 1982) and accounted for the deduction of annual recurring costs $r_R(F_x + rA)$ that commercial building owners can take. For residential building owners $t_y = 0$ in equation (3.16).

In several taxing jurisdictions in the United States, solar energy systems were exempt from property tax, and many studies explicitly accounted for this by comparing the economic feasibility and optimal design of solar energy systems with and without property taxes. The value of P in equation (3.12) is zero when the system is exempt. Where property taxes are applicable, the value of P depends on local assessment practices. Generally, P is approximated (Ruegg et al., 1982) by basing it on the initial contract cost $(F_x + vA)$. Then

$$P = t_p(1 - t_y)(F_x + vA)P/A_{d,g,l}, \tag{3.19}$$

where t_p is the effective property tax rate [that is, the assessment rate multiplied by the nominal property tax rate; see Ruegg (1976)].

Capital-depreciation deductions are specific to commercial building owners and depend on tax laws regarding the portion of solar energy system costs allowable as a depreciable expense. Allowable practices differ with the type of commercial building. Most studies assumed that the total contract cost $(F_x + vA)$ was allowable as a depreciable expense and that the straight-line depreciation method was used. For example, Ruegg and Sav (1981) and Perino (1979) show that D can be computed as follows:

$$D = t_y n^{-1}(F_x + vA)P/A_{d,n}, \tag{3.20}$$

where n is the number of allowable depreciation years, $P/A_{d,n}$ is a present worth factor for a discount rate d and n depreciation years, and $t_y, f_x, v,$ and A are as previously defined. Sav (1979) incorporates a declining balance depreciation method into an optimization model and computes D as follows:

$$D = t_y r(F_x + vA)(r + dn)^{-1} \frac{n^n(1 + d)^n - (n - r)^n}{n^n(1 + d)^n}, \tag{3.21}$$

where r is the declining balance rate (for example, $r = 2$ for 200% declining balance, $r = 1.5$ for 150% declining balance).

A few models and corresponding studies include the PDV of income tax deductions accruing from loan interest in the life-cycle costs of the solar energy system. Perino (1979) and Sav (1979) do so and in general show that

$$M = \frac{t_y(1 - q)(F_x + vA)}{d - i} \left\{ i - \frac{i(1 + i)^m}{(1 + d)^m} \right.$$

$$\left. - A/P_{i,m}\left[1 - \left(\frac{1 + i}{i + d}\right)^m - \frac{d - i}{d} + \frac{d - i}{d(1 + d)^m} \right]\right\}$$

$$\text{for} \quad d \neq i \quad (3.22)$$

and

$$M = t_y(1 - q)(F_x + vA)\left[\frac{mi}{1 + i} - A/P_{i,m}\left(\frac{m}{1 + i} - \frac{1}{i} + \frac{1}{i(1 + i)^m}\right)\right]$$

$$\text{for} \quad d = i. \quad (3.23)$$

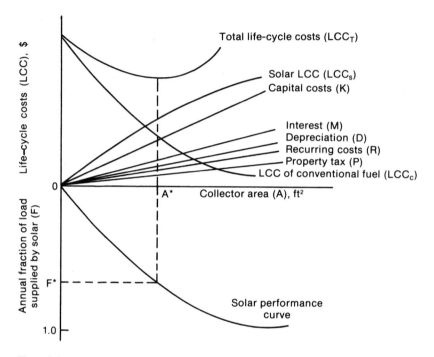

Figure 3.1
Optimization under LCC_T minimization. From Sav (1979).

The LCC_T optimization model is completed by substituting equations
(3.13) through (3.23) into equation (3.12) and substituting equations (3.9)
and (3.12) into equation (3.8). The result expresses LCC_T as a function of
the annual fraction f of the energy load L supplied by a solar energy
system, with the size of the solar energy system as measured by the area of
solar collector surface in square meters.

The upper portion of figure 3.1 shows the generic shape of a LCC_T
curve, which results when individual life-cycle costs that comprise LCC_T
are plotted against collector area. The illustration applies to a commercial building. For residential buildings some terms are excluded as noted
previously, whereas others have different shapes. The lower portion of
figure 3.1 shows the generic shape of the solar performance curve. The
most economical and optimally sized collector is indicated by A^*, where
LCC_T reaches a minimum point. From the value of A^*, the economically
optimal load fraction (f^*) supplied by solar is determined in the lower

part of figure 3.1. If, however, the LCC_T cost curve rises continuously from its origin on the vertical axis, then solar energy is not economically feasible under the LCC_T minimization criterion; that is, $A^* = 0$, and the solar investment should not be undertaken based on the criterion of cost minimization.

As an alternative to the LCC_T minimization approach many studies of solar energy feasibility and design inverted the objective function and used the maximization of total life-cycle savings LCS_T as an alternative optimization. LCS_T are the difference between what it would cost in life-cycle terms to provide all the energy demands using 100% conventional energy $LCC_{100\%}$ and what it would cost to provide the same energy demands using a combination of conventional energy LCC_c and solar energy LCC_s. LCS_T can be expressed as

$$LCS_T = LCC_{100\%} - (LCC_c + LCC_s), \tag{3.24}$$

where $LCC_{100\%}$ is obtained by setting f equal to zero in equation (3.9). Maximizing equation (3.24) is identical with minimizing equation (3.8), because LCS_T is at a positive maximum point when LCC_T is at a minimum point. Figure 3.2 illustrates graphically the equivalence of the two concepts. As with figure 3.1, however, if LCS_T are everywhere negative in figure 3.2, then the optimal collector area is zero (that is, the solar investment should not be undertaken). Ruegg (1976) and Ruegg and Sav (1981) detailed the conditions under which the two approaches lead to the same result and applied the two methods to both commercial and residential building applications of solar energy.

The approaches of minimizing LCC_T and maximizing LCS_T are rooted in the basic principles of microeconomic theory. To some extent these principles were masked and incorporated into the general life-cycle cost model. More general approaches to analyzing solar energy feasibility and optimal design directly incorporated the microeconomic model of input choice and cost minimization. For example, Ruegg and Sav (1981), Sav (1984a), Nicholls (1977), and Peterson (1979) begin by minimizing the total costs of producing energy services as follows:

$$\min C = P_F F + P_A A + P_v V, \tag{3.25}$$

subject to

$$E = g(F) + h(A, V), \tag{3.26}$$

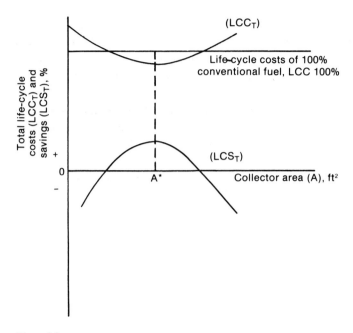

Figure 3.2
Minimization of LCC_T equals maximization of LCS_T. From Sav (1979).

where P_i is the price of the ith input and E is a given level of energy service from equation (3.5).

In this approach the economically efficient combination of solar (A and V) and conventional (F) energy inputs depends both on the marginal productivity of those inputs and their relative prices. A necessary condition for arriving at the minimum cost of providing a given level of energy service is that the marginal productivity per dollar spent is equal for all inputs. If we assume continuous and smooth functions, we can express this necessary condition as

$$\frac{P_F}{MP_F} = \frac{P_A}{MP_A} = \frac{P_V}{MP_V},\tag{3.27}$$

where MP_i is the marginal product of the ith input in producing the energy service. Equation (3.27) states that the minimum cost combination of conventional fuel F, solar collector area A, and solar storage volume V is achieved when the marginal costs of all inputs are equal. Similarly, using equation (3.6), the objective is to minimize total cost C:

$$C = P_f F + P_A A, \tag{3.28}$$

subject to

$$E = g(F) + h(A), \tag{3.29}$$

where it is now understood that the optimally sized (cost-minimizing) solar storage volume V is determined in fixed proportion to the optimally sized solar collector area (that is, $V^* = bA^*$). Again, the necessary condition for an optimal design is

$$\frac{P_F}{MP_F} = \frac{P_A}{MP_A}. \tag{3.30}$$

This latter condition is illustrated graphically in figure 3.3. The lines C_1, C_2, and C_3 illustrate three of a family of total cost (isocost) equations. Each line indicates a specific total cost ($C_1 < C_2 < C_3$) and the combinations of fuel F and collector area A that may be purchased for this specific total cost outlay. The slope of each total cost line is the relative price of inputs (that is, P_F/P_A). The curve $\overline{E}$ (the isoquant) gives the different combinations of inputs, F and A, that can be used to produce this given level of energy service, $\overline{E}_1$. The slope of the isoquant represents the technical trade-offs between inputs in producing a given level of energy service and is equal to the ratio of the marginal products (MP_F/MP_A).

Given the relative costs of F and A, as depicted by the slopes of the isocost lines, and the technical rate at which A may be substituted for F, as depicted by the slope of the isoquant, the minimum cost combination of F and A for producing E occurs at the point of tangency between C_2 and $\overline{E}_1$; that is, the minimum cost combination F^* and A^* occurs where

$$\frac{P_F}{P_A} = \frac{MP_F}{MP_A}. \tag{3.31}$$

However, if the tangent between the isocost cost curve and the isoquant occurs at the intercept of the isoquant with the vertical axis, then a corner solution is attained and the energy service $\overline{E}_1$ should be produced with solely conventional fuel; that is, the solar investment should not be undertaken based on the economic criterion of cost minimization.

To implement this approach empirically for an optimal solar energy system design requires an estimate of the isoquant $\overline{E}$; that is, empirical es-

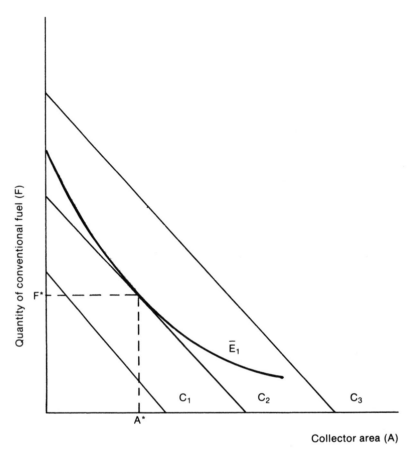

Figure 3.3
Optimization using microeconomic theory.

timates of the technical substitution possibilities between conventional fuel and solar collector area must be obtained. Few studies have taken such an approach because of the difficulty and expense of estimating these relationships. The exceptions include a rigorous evaluation by Peterson (1979) and several by Sav (1984a–c). In each of these studies an estimate of the technological substitution possibilities relied on simulation models of the type developed by Beckman, Klein, and Duffie (1977).

3.1.3 Optimal Timing of Investments

In section 3.1.2 I presented economic models that were used to investigate the optimal design of solar energy systems and to answer the question of whether to invest in solar energy systems. These models, however, did not take into account the dynamic aspect of investments in solar energy. Specifically they did not address the problem of timing solar investments. That is, although the criteria for LCC_T minimization or LCS_T maximization may indicate that a solar investment is economically feasible in the current analysis period, a lower cost or greater savings may result from deferring the investment to some future date. The investment deferment or timing decision is truly a dynamic one.

Many formal economic models were developed to handle the timing decisions of investments in such capital equipment as solar energy systems. For example, Ben-David et al. (1977) and Sav (1984a) provide an in-depth theoretical approach and empirical application of the method. A general model and solution to the timing problem can be formulated as an optimal control model using the calculus of variations. This approach provides the foundations for studies that have been conducted in market penetration relating to solar energy. Moreover, this approach has the added advantage of being capable of handling simultaneous investment decisions, for example, in sizing a conventional energy system in conjunction with a solar energy system.

To simplify, it is generally assumed that the planning horizon is a finite T years and that the energy requirements $E(t)$ in each time period t over $0 < t < T$ are known with certainty. The planning horizon is divided into two parts, t_1 and t_2, of undetermined length. The firm or household invests in conventional energy equipment at $t = 0$ and must decide on the scale of this investment. The firm may also undertake a second investment in solar energy equipment at $t = t_1$, where t_1 is determined along with the scale of the investment. The energy requirements in t are described by the following stock-flow production functions:

$$E(t) = f_1[F_1(t), X_1] \quad \text{for} \quad 0 \le t \le t_1, \tag{3.32}$$

$$E(t) = f_1[F_1(t), X_1] + f_2[F_2(t), A_2; X_1] \quad \text{for} \quad t_1 \le t \le T, \tag{3.33}$$

where $F_1(t)$ is the quantity of conventional fuel used in period 1, $X_1 > 0$ is the quantity of the conventional energy equipment used in period 1, and $F_2(t)$ is the quantity of fuel used in period 2 given that a solar energy system of size $A_2 \ge 0$ is installed and used in period 2. The firm or household must decide (1) the quantity of X_1 to initially install, (2) when to install A_2, if at all, (3) the quantity of A_2 to install if it is optimal to undertake the investment, and (4) the quantity of fuel to use in each period.

We assume that the objective is to minimize the following present discounted cost of production over the planning horizon:

$$C = \int_0^t P_f F_1(t_1) e^{-rt} \, dt + \phi P_x X_1 + \phi P_A A_2 e^{-dt_1}$$

$$+ \int_{t_1}^T [P_f F_1(t) + P_f F_2(t)] e^{-rt} \, dt, \tag{3.34}$$

subject to the production function constraints in equations (3.32) and (3.33). The price P_f of fuel is assumed to remain constant over time. The unit prices of the two capital inputs are P_x and P_A, and ϕ is a user cost-of-capital term, which converts all future costs (from t_1 to T) per dollar of solar investment to a present discounted value at time t_1. Future costs are discounted at a real rate of r, assumed constant over time.

An equivalent problem is to minimize this function with respect to $F_1(t)$, $F_2(t)$, X_1, A_2, and t_1 (the timing of the solar investment) using Lagrangian multipliers. Optimizing will lead to conditions (Sav, 1984a) that determine the optimal scale of investment with respect to X_1 and A_2 and the optimal timing of the investment in solar energy. Because A_2 is nonessential, its optimal quantity may be zero; that is, the solar investment is never undertaken. The investment in solar is deferred so long as the present discounted savings in production costs resulting from the installation of solar are less than the present discounted cost of solar (including interest, maintenance, and other operating costs). It is also possible to incorporate time rates of acceleration on the input prices in the problem. For example, if the fuel price is expected to accelerate at a rate of k per unit of time, then e^{-rt} within the integrals becomes $e^{(k-r)t}$. Then the timing of the solar energy process depends on the rate k at which fuel prices accelerate over time.

The general problem of timing is illustrated in figure 3.4. In figure 3.4a (top) the present discounted cost C of producing energy services is shown as depending on the time t at which the solar investment is undertaken. In figure 3.4b (bottom) optimal collector area or system size and optimal conventional fuel are shown as corresponding to the timing of the solar investment. As illustrated, the optimal time t^* to undertake the investment occurs when C is at a minimum. The corresponding optimally sized solar energy system at t^* is $(A*)^{t*}$.

3.1.4   Integrated Optimization of Solar Energy and Energy Conservation

In addition to installing solar energy systems to reduce the consumption of conventional fuels and lower the costs of producing energy services, energy conservation (that is, improving the thermal integrity of the building envelope or efficiency of the energy delivery system) options also exist. Generally the primary economic objective in designing or retrofitting the heating and cooling components of a building is to provide, at the lowest possible life-cycle cost, a desired level of thermal comfort, comprising temperature, humidity, and other related attributes (and considering related factors such as lighting). The minimum cost search considers the technical substitution of inputs and the relative prices of inputs. Holding all inputs constant, except those concerning the building envelope and the energy system(s), trade-offs exist between (1) energy conservation alternatives that improve the thermal integrity of the building envelope, and therefore reduce the thermal load, and (2) energy system alternatives, for example, nonsolar and solar energy systems that satisfy given thermal loads. Ruegg et al. (1982) and Sav (1978) constructed a general model showing that the economically optimal system configuration is attained when the marginal dollar expenditure for each input per marginal unit of thermal comfort obtained from that input is equal for all inputs. This optimal condition can be stated algebraically:

$$\frac{MC_c}{MP_c} = \frac{MC_s}{MP_s} = \frac{MC_{LR}}{MP_{LR}}, \tag{3.35}$$

where MC represents marginal cost, MP marginal product, and the sub-

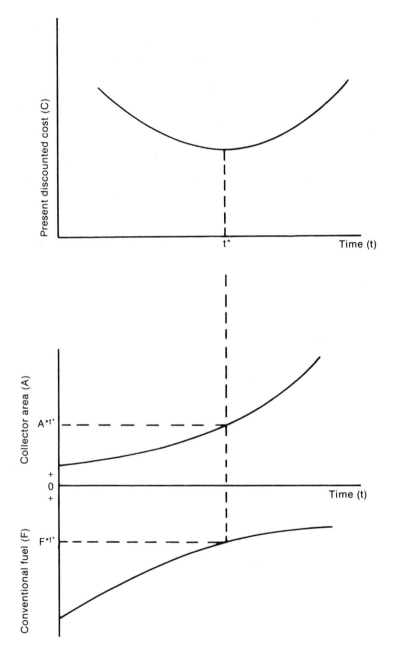

Figure 3.4
Optimal timing of solar investments.

scripts c, s, and LR the nonsolar (conventional) energy inputs, the solar energy inputs, and the load reduction (energy conservation) inputs, respectively.

In other words, the optimal search is for a combination of energy conservation inputs (load reduction options) and energy system inputs, including solar thermal processes that will minimize the LCC_T of achieving a desired level of thermal comfort (or maximize the LCS_T from the total investment).

Holding all factors constant, except thermal load L as determined by the thermal integrity of the building envelope, and noting that alternative energy systems can satisfy L, thermal comfort k can be expressed as

$$k = k(E_c, E_s, E_{LR}), \tag{3.36}$$

where E_c represents a quantity of nonsolar (conventional) energy input, E_s a quantity of solar energy input, and E_{LR} a quantity of energy reduction obtained by upgrading the thermal integrity of the building envelope. All inputs are expressed in a common unit of measure, such as the joule.

The cost C of achieving various levels of k can be described by a family of isocost curves described by

$$C = P_c(E_c)E_c + P_s(E_s)E_s + P_{LR}(E_{LR})E_{LR}, \tag{3.37}$$

where $P_i(E_i)$ expresses the price of the ith input as a function of the level of the ith input used.

Once a target level of thermal comfort (say, $\bar{k}$) is determined, the economic objective is to minimize C. The minimum cost combination of inputs E_c, E_s, and E_{LR} can be obtained by using the Lagrange multiplier. Thus

$$C = P_c(E_c)E_c + P_s(E_s)E_s + P_{LR}(E_{LR})E_{LR} - \lambda[\bar{k} - h(E_c, E_s, E_{LR})], \tag{3.38}$$

where λ is the Lagrange multiplier and is interpreted as the marginal cost of producing thermal comfort k.

From equation (3.38), the optimal condition is found to be

$$\frac{P_c + P_c'E_c}{k_c'} = \frac{P_s + P_s'E_s}{k_s'} = \frac{P_{LR} + P_{LR}'E_{LR}}{k_{LR}'}, \tag{3.39}$$

where P_i' is the first derivative of $P_i(E_i)$ with respect to E_i and k_i' is the first derivative of $k(E_c, E_s, E_{LR})$ with respect to E_i.

The numerator of each ratio is the marginal cost MC of the respective input. The denominator is the marginal product MP of the input in producing thermal comfort k. Hence, this is consistent with the optimal condition stated at the outset in equation (3.35).

Because initially all inputs except thermal load were held constant, it follows that marginal products are all expressed in a common unit, for example, the joule. The ratios of the marginal products are therefore unity, and the optimal condition can be restated in the familiar form

$$MC_c = MC_s = MC_{LR}. \tag{3.40}$$

An optimal economic condition is attained when the marginal costs of all inputs are equal. When this optimal condition is achieved, LCC_T of maintaining a given level of thermal comfort are minimized. That is, the following LCC_T equation is a minimum:

$$LCC_T^{Lj} = LCC_c^{Lj} + LCC_s^{Lj} + LCC_{LR}^{Lj}, \tag{3.41}$$

where the superscript Lj represents alternative loads.

Equation (3.41) describes a family of LCC_T curves, each corresponding to a different load Lj. The empirical form of equation (3.41) differs depending on the nature of economic and technical trade-offs among the size (capacity) of the solar energy system, the conventional energy system, and energy conservation options. If, for simplicity, no capacity reductions in the nonsolar conventional energy system are assumed as the size of the solar energy system is expanded, LCC_c consists only of conventional fuel costs. LCC_{LR} is building design specific and depends on the initial design load against which the costs of load reductions are evaluated.

In this integrated framework, the alternative to the LCC_T minimization criterion (that is, the LCS_T maximization criterion), is achieved by maximizing

$$LCS_T^{Lj} = LCC^{L0} - (LCC_c^{Lj} - LCC_s^{Lj} - LCC_{LR}^{Lj}), \tag{3.42}$$

where LCC^{L0} represents the sum of the initial space heating L_{H0} and hot water L_W loads from which load reduction options are evaluated. (Note that $Lj = L_{Hj} + L_W$, where L_W represents a fixed hot water load, which is unaffected by energy conservation design in the building envelope.)

From the Ruegg et al. (1982) study, figures 3.5 and 3.6 graphically depict the integrated optimization procedure. Figure 3.5 shows a ge-

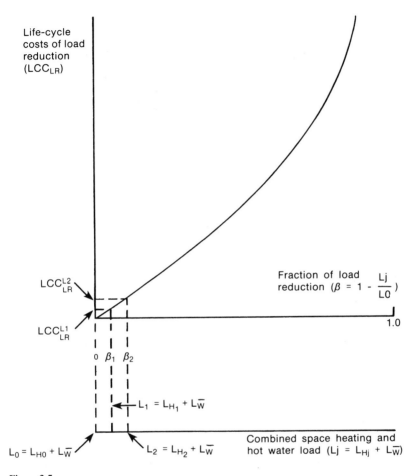

Figure 3.5
Integrated optimization procedure: load reductions. From Sav (1978) and Ruegg et al. (1982).

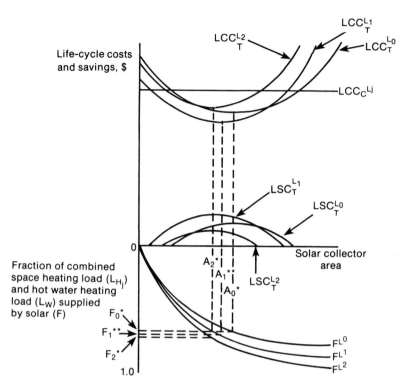

Figure 3.6
Integrated optimization procedure: solving for the combination of conservation, solar, and nonsolar inputs to thermal comfort. From Sav (1978) and Ruegg et al. (1982).

neric LCC_{LR} function. The horizontal axis measures the fraction β by which the initial space heating load L_{H0} is reduced. LCC_{LR} for various load reductions are measured along the vertical axis. The lower scale in figure 3.5 shows the combined space heating and hot water load ($Lj = L_{Hj} + L_W$) corresponding to each β.

Figure 3.6 shows in the upper portion a family of LCC_T curves; in the middle portion are LCS_T curves; and in the lower portion, solar performance curves for alternative loads (Lj). The figure also shows a life-cycle cost curve (LCC^{L0}) for a nonsolar energy system providing all the energy to meet the initial load of L_0, used to derive the LCS_T curves. For any given load, the optimal collector area A_j^*, which is indicated on the horizontal axis, and the fraction of load supplied by solar F_j^*, which is indicated on the lower segment of the vertical axis, occur at the minimum point on the $LCC_T^{L,j}$ curve or the corresponding maximum point on the $LCS_T^{L,j}$ curve. The optimal amount of nonsolar energy input is simply $(1 - F_j^*)$. The optimal combination of all inputs, including load reduction options, occurs at the minimum point of the lowest $LCC_T^{L,j}$ curve among the family of $LCC_T^{L,j}$ curves or the corresponding maximum point on the highest $LCS_T^{L,j}$ curve. For example, the optimal combination of all inputs in figure 3.6 occurs at the minimum point on the $LCC_T^{L,j}$ curve, which is the same as the maximum point on the $LCS_T^{L,j}$ curve. The optimum is found to be A_1^{**} of solar collector area, F_1^{**} fraction of the load by solar, $1 - F_1^{**}$ fraction of the load met by nonsolar energy input, and β_1^{**} of load reduction, corresponding to a combined thermal load of $L_1 = L_{H1} + L_W$.

3.2 Alternative Empirical Approaches to Economic Modeling

The previous section reviewed the foundations of economic models used in solar energy design and economic feasibility studies. However, to be useful, these economic models must be linked to solar thermal performance models to determine empirically the optimal solar energy system design and its economic feasibility. A variety of empirical approaches were developed for this purpose, ranging from simple and inexpensive mathematical models to complex and expensive computer simulation models. This section reviews the alternative empirical approaches to modeling that were developed to evaluate the economics of solar-produced energy.

3.2.1 Computer Simulation Models

A number of computer simulation models were developed to evaluate the thermal and economic performance of solar energy systems. Table 3.1 summarizes the most frequently used computer packages available for combined economic and solar thermal analyses. Most of these programs contain the same basic component models but vary in their complexity. General subroutines within the programs represent the physical pieces of equipment that comprise a combined conventional solar energy system. The user is responsible for instructing the programs as to how the components are connected and what the design parameters are for each component. The programs generally perform the necessary simulations by

Table 3.1
Summary of existing solar energy computer programs with economic analysis capability

Program name	Latest version	Purchase ($)	Time share	Special arrange-ments	Comments	User manual	Service hot water	Space heating	Space cooling	Process heat
BLAST[a]	1980	Nom.	yes	no	Training available	yes	yes	yes	yes	no
DOE-2[a]	1980	400	yes	no		yes	yes	yes	yes	no
EMPSS	1978	500	no	yes	Consulting with Arthur D. Little	yes	yes	yes	yes	no
F-CHART	1978	100	yes	no	Training available	yes	yes	yes	no	no
FREHEAT	1979	150	no	no	Limited documentation	no	no	yes	no	no
HUD-RSVP/2	1979	175	yes	no	Based on F-CHART	yes	yes	yes	no	no
PACE	1980	100	yes	no	Based on F-CHART, SLR	yes	yes	yes	no	no
SHASP	1978	Available on request	no	no		yes	yes	yes	yes	no
SOLCOST	1979	300	yes	no		yes	yes	yes	no	no
SOLFIN 2	1980	Nom.	no	yes	Documentation costs $6	yes	yes	yes	yes	yes
SOLOPT	1978	20	no	no		yes	yes	yes	no	no
SUNCAT	1979	Nom.	no	no	Limited documentation	yes	no	yes	no	no
SUNSYM[R]	1979		yes	yes	Offered as service only	yes	yes	yes	yes	no
SYRSOL	1978	Nom.	no	no	Available but not actively marketed	no	yes	yes	yes	no
TRACE SOLAR[a]	1980		yes	yes	Offered as service only	yes	yes	yes	yes	yes
TRNSYS	1979	200	yes	no	Training required	yes	yes	yes	yes	yes
TWO ZONE	1977	No charge	no	yes		yes	no	yes	yes	no
WATSUN II, III	1980	Contact author	no	no		yes	yes	yes	no	no

Source: SERI (1980) and Powell and Barnes (1982).
a. Programs are primarily developed for large-scale, multizone applications.
b. ANSI 1966 Standard Fortran.

| | Intended users | | | | Computation interval | | | | |
Active system	Passive system	Research engineers	Architect/ engineers	Builders	Hour	Month	Computer versions available	Economic analysis	Sponsor
yes	yes	yes	yes	no	yes	no	CDC	yes	USAF, USA, GSA
yes	yes	yes	yes	no	yes	no	CDC	yes	LASL, DOE
yes	Being added	yes	yes	no	yes	no	IBM	yes	EPRI
yes	no	no	yes	yes	no	yes	CDC, IBM, UNIVAC[b]	yes	DOE
no	yes	yes	no	no	yes	no	CDC	yes	DOE
yes	no	no	yes	yes	no	yes	CDC, UNIVAC	yes	HUD
yes	yes	no	yes	yes	no	yes	CDC, UNIVAC	yes	DOE, SERI
yes	no	yes	no	no	yes	no	UNIVAC	yes	DOE
yes	yes	yes	yes	yes	no	yes	CDC, IBM, UNIVAC	yes	DOE
yes	yes	yes	yes	yes	Annual		IBM, CDC	yes	California Energy Commission
yes	no	yes	no	no	no	yes	AMDAHL	yes	Texas A&M Univ.
no	yes	yes	yes	no	yes	no	Data General Eclipse	yes	NCAT
yes	no	no	yes	no	yes	no	IBM	yes	Sunworks Computer Systems
yes	no	yes	yes	yes	yes	no	IBM	yes	ERDA, NSF, DOE
yes	no	yes	yes	no	yes	yes	IBM	yes	The Trane Co.
yes	yes	yes	no	no	yes	no	CDC, IBM, UNIVAC[b]	Being added	DOE
yes	yes	yes	no	no	yes	no	CDC	yes	LBL
yes	no	yes	no	no	yes	no	IBM	yes	National Research Center of Canada

solving a system of algebraic and differential equations. All the programs are in the public domain and were developed with federal funds.

Because of the general complexity of these programs, a detailed review is not possible for the present discussion. However, a recent study conducted by the National Bureau of Standards (Powell and Barnes, 1982) compared six of the most widely used programs in terms of their similarities and differences and advantages and disadvantages. An earlier study by Freeman, Maybaum, and Chandra (1978) compared the noneconomic aspects of four solar simulation programs. The following programs were included in the Powell and Barnes study: SOLCOST, F-CHART 3.0, F-CHART 4.0, BLAST, DOE-2, and FEDSOL. Table 3.2 summarizes their findings on the advantages and disadvantages of each program. Although the analysis focused on using these programs for evaluating federal solar energy projects, it provided useful insight into the general strengths and weaknesses of the programs for a wide variety of applications.

The six computer programs were also analyzed using test cases to compare predictions of economic performance and optimal system size. The analysis was performed for a representative single-family detached residence and a low-rise office building. Figures 3.7 and 3.8 summarize the residential building test case differences found among four of the programs. Figures 3.9 through 3.11 summarize the commercial building test case differences found among six of the programs. Note that in each case, the designs show a negative net savings; that is, solar was not cost-effective for the economic data used in the study. As illustrated in figures 3.7 and 3.8, the optimal collector area (maximum net savings, even though net savings are negative) for a residential building located in Washington, D.C., ranged from a low of 152 ft^2 (14.1 m^2) using F-CHART 3.0 to a high of 209 ft^2 (19.4 m^2) using F-CHART 4.0. For Bismarck, North Dakota, the range was 244 ft^2 (22.7 m^2) using F-CHART 4.0 to 365 ft^2 (33.9 m^2) using FEDSOL. With respect to thermal performance the programs predicted annual solar fraction for a given size collector area to within 2% of each other in Bismarck, North Dakota, and within 5% in Washington, D.C. Predictions of optimal solar fraction varied slightly more, differing by 8% for both cities. Similar results were obtained in comparing an office building in Washington, D.C., as illustrated in figures 3.9, 3.10, and 3.11. Differences in the economically optimal collector area and solar fraction can be attributed to differences in the solar thermal performance predictions, in the economic variables included in the programs, and in the optimization routines employed in the programs.

Table 3.2
Advantages and disadvantages of selected simulation programs

Program	Advantages	Disadvantages
SOLCOST	Moderate cost Can save data files Moderately easy to use Includes tax and mortgage analysis Prints table of results for nine system sizes Includes limited optimization of collector area and building load variables Performs simplified energy analysis for space and water heating (ASHRAE method) if desired	Thermal analysis not validated for multizone buildings Does not calculate SIR (savings-to-investment ratio) (federal requirement) Does not allow for salvage or resale value
F-CHART 3.0	Low cost Easy to use Includes tax and mortgage analysis Performs simplified energy analysis for space and water heating (ASHRAE method) if desired Optimizes collector area	Cannot save data files User must input sequence of values for energy prices in each year of study period to assume different rates of fuel price escalation for different time periods Does not calculate SIR Thermal analysis not validated for multizone buildings No investment credit variables No analysis of nonenergy costs of reference and auxiliary systems
F-CHART 4.0	Moderately low cost Moderately easy to use Includes tax and mortgage analysis Performs simplified energy analysis for space and water heating (ASHRAE methods) if desired Optimizes a number of design variables	User must input sequence of values for energy prices in each year of study period to assume different rates of fuel price escalation for different time periods Does not calculate SIR Thermal analysis not validated for multizone buildings
BLAST	Performs comprehensive analysis of costs for energy, capital equipment, and overhaul for all heating and cooling plant components (auxiliary, solar, and reference)	High cost Difficult to use Does not calculate net savings or SIR

Table 3.2 (continued)

Program	Advantages	Disadvantages
	Can save data files Thermal analysis applicable to multizone buildings	No provision for different rates of fuel price escalation for different periods (federal requirement) No mortgage or tax analysis Does not allow for salvage or resale value
DOE-2	Performs comprehensive analysis of costs for energy, capital equipment, and overhaul for all heating and cooling plant components (auxiliary, solar, and reference) Can save data files Thermal analysis applicable to multizone buildings	High cost Difficult to use No provision for different rates of fuel price escalation for different periods No mortgage or tax analysis No investment credit variables No provision for salvage or resale value Must perform baseline (reference) building analysis in separate computer run from solar energy analysis
FEDSOL	Low cost Easy to use Meets all federal LCC requirements Contains data required under federal LCC rule as default values Can perform an economic analysis independently of thermal analysis (that is, with solar performance data from another source) Optimizes collector area Prints table of results for ten system sizes Performs break-even analysis (under conditions of negative net savings) Can save data files Solar energy performance analysis applicable to multizone commercial buildings and single-zone residential buildings	Thermal analysis applicable only to standard active (flat-plate) systems for space heating or combined space and service water heating No provision for mortgage or tax analysis

Source: Powell and Barnes (1982).

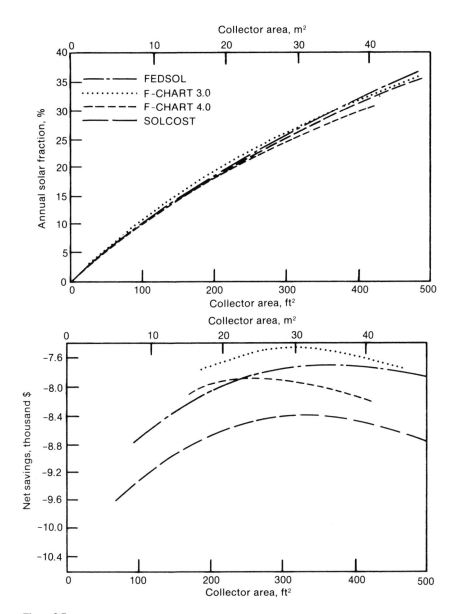

Figure 3.7
Economic and thermal performance curves derived from FEDSOL, F-CHART 3.0, F-CHART 4.0, and SOLCOST for a residential space heating system in Bismarck, North Dakota (savings reported in 1980 dollars; 1985 dollars = 1.30 × 1980 dollars). From Powell and Barnes (1982).

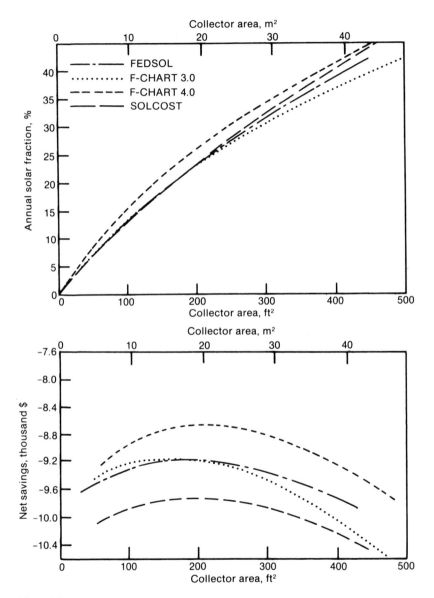

Figure 3.8
Economic and thermal performance curves derived from FEDSOL, F-CHART 3.0, F-CHART 4.0, and SOLCOST for a residential space heating system in Washington, D.C. (savings reported in 1980 dollars; 1985 dollars = 1.30 × 1980 dollars). From Powell and Barnes (1982).

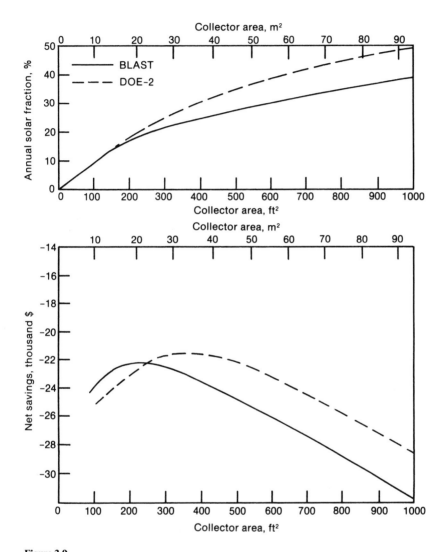

Figure 3.9
Economic and thermal performance curves derived from BLAST and DOE-2 for a space and
service water heating system for a federal office building in Washington, D.C. (savings
reported in 1980 dollars; 1985 dollars = 1.30 × 1980 dollars). From Powell and Barnes
(1982).

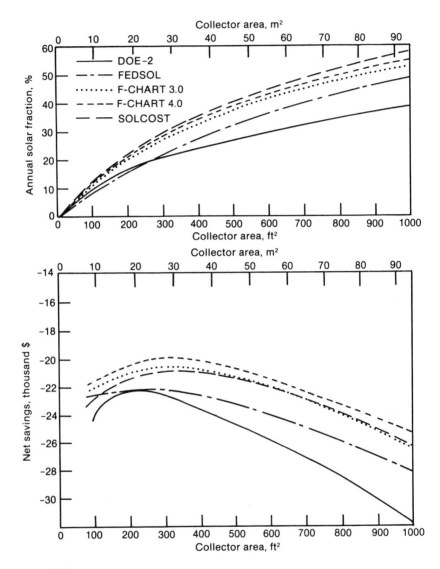

Figure 3.10
Economic and thermal performance curves derived from BLAST, FEDSOL, F-CHART 3.0,
F-CHART 4.0, and SOLCOST for a space and service water heating system for a federal office
building in Washington, D.C. (savings reported in 1980 dollars; 1985 dollars = 1.30 × 1980
dollars). From Powell and Barnes (1982).

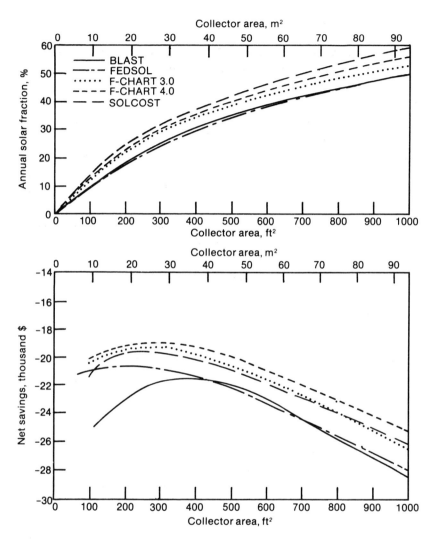

Figure 3.11
Economic and thermal performance curves derived from DOE-2, FEDSOL, F-CHART 3.0, F-CHART 4.0, and SOLCOST for a space and service water heating system for a federal office building in Washington, D.C. (savings reported in 1980 dollars; 1985 dollars = 1.30 × 1980 dollars). From Powell and Barnes (1982).

3.2.2 Deterministic Models

Several simplified models were developed and used to design solar energy space heating and domestic hot water systems. These models evaluate several simple algebraic equations. The calculations can be performed by hand or by using a pocket calculator; a computer is not necessary.

For example, generalized solar performance curves for solar domestic hot water systems in commercial buildings were developed by the Los Alamos National Laboratory (LANL). The curves were based on simulations of solar performances in eight U.S. cities using a domestic hot water profile based on American Society of Heating, Refrigeration and Air-Conditioning Engineers (ASHRAE) data. The simulations indicate that the monthly fraction f_j of hot water load met by a solar energy system can be expressed as a function of the ratio AI_j/L_j, where A is the collector area, I_j is the solar radiation incident on a tilted collector surface in month j, and L_j is hot water load in month j. Curve-fitting equations, which explicitly describe the relationship between f_j and AI_j/L_j, are of the general form

$$f_j = \begin{cases} aAI_j/L_j & \text{for} \quad 0 \le AI_j/L_j < 0.8 \\ 1 - b \cdot \exp(-cAI_j/L_j) & \text{for} \quad 0.8 \le AI_j/L_j \le 5.0, \end{cases} \tag{3.43}$$

where a, b, and c are parameters whose values depend on hot water design temperature. Table 3.3 provides the values of a, b, and c for four hot water design temperatures. From the appropriate form of equation (3.43), the monthly fraction of hot water load met by solar heating is determined for given values of collector area, monthly insolation, and monthly hot water load. An annual fraction f of hot water load met by solar energy can be determined from monthly performance calculations using the relation

$$f = \sum_{j=1}^{12} L_j f_j / L, \tag{3.44}$$

or, substituting equation (3.43) for f_j,

$$f = \begin{cases} aA/L \sum_{j=1}^{12} I_j & \text{for} \quad 0 \le AI_j/L_j < 0.8 \\[2mm] 1 - b/L \sum_{j=1}^{12} L_j \exp(-cAI_j/L_j) & \text{for} \quad 0.8 \le AI_j/L_j \le 5.0, \end{cases} \tag{3.45}$$

where L is the annual hot water load ($L = \sum L_j$).

Table 3.3
Parameters corresponding to universal curve fit equation

Curve fit identification number	Hot water design temperature		a	b	c
	°F	°C			
I	110	43.3	0.568	1.153	0.933
II	130	54.4	0.499	1.080	0.729
III	150	65.6	0.440	0.978	0.514
IV	170	76.7	0.348	0.966	0.365

The universal performance curves may be simplified for hot water loads that are uniformly distributed throughout the year. For uniform loads an annual value of solar radiation can be employed to predict the annual fraction of load met by solar energy. Thus, equations (3.44) and (3.45) simplify to equations (3.46) and (3.47), respectively:

$$f = aIA/L \qquad \text{for} \quad 0 \leq AJ/L < 0.8, \tag{3.46}$$

$$f = 1 - b \cdot \exp(-cAI/L) \qquad \text{for} \quad 0.8 \leq AI/L \leq 5.0, \tag{3.47}$$

where $I = \sum I_j$. A typical universal solar performance curve corresponding to equations (3.46) and (3.47) is illustrated in figure 3.12.

The LANL universal solar performance curves can be easily linked to the LCC_T or LCS_T model presented in equation (3.8) or (3.24), respectively. Ruegg et al. (1982) and Sav (1978) use the LANL solar performance model with an economic model and showed that, under the LCC_T minimization criterion or the LCS_T maximization criterion, the minimum or maximum point always occurs in the nonlinear range of the LANL universal performance curves, where $A \geq 0.8L/I$ in figure 3.12. Using this result, they demonstrated that the search for the most economical and optimally sized solar energy system can be reduced to a single deterministic equation:

$$A^* = \frac{L}{cI} \ln \frac{bcI(1 - t_y)\overline{X}}{v\overline{Y}}, \tag{3.48}$$

where from equations (3.8) through (3.24)

$$\overline{X} = P_c(1 + t_c)(1 - t_y)\delta^{-1}B^{-1}P/A_{d,e,l}, \tag{3.49}$$

and $\overline{Y}$ can assume one of four possible forms (Sav, 1978) depending on

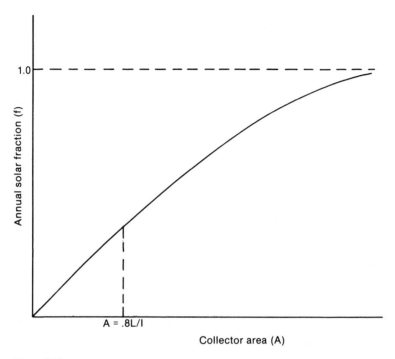

Figure 3.12
General form of a universal performance curve.

the method of depreciation (straight line or declining balance) and whether $d \neq i$ or $d = i$. For a declining balance depreciation method and $d \neq i$,

$$\bar{Y} = q + (1 - q)(A/P_{i,m})(P/A_{d,m}) - t_g(1 + d)^{-1}$$

$$+ (r_r + t_p)(1 - t_y)(P/A_{d,l}) - rt_y(r + dn)^{-1}\frac{n^n - (n - r)^n}{n^n(1 + d)^n}$$

$$- \frac{t_y(1 - q)}{di}\left[i - i\left(\frac{1 + i}{1 + d}\right)^n\right]$$

$$- (A/P_{i,n})\left(1 - \left(\frac{1 + i}{1 + d}\right)^n - \frac{d - i}{d} + \frac{d - i}{d(1 + d)^n}\right], \qquad (3.50)$$

where all terms are as defined in equations (3.8) through (3.24). (Here it is assumed that the loan life is equal to the number of allowable depreciation years.)

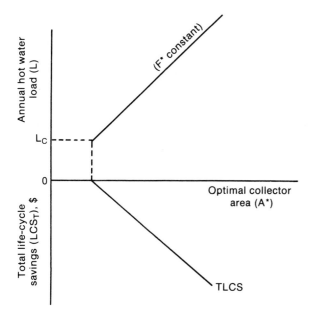

Figure 3.13
General form of a universal economic optimization path.

The general form of equation (3.48) using the LCS_T maximization criterion is illustrated in figure 3.13, where L_c is the critical hot water load, below which solar is not cost-effective.

Similar methods for sizing and determining the economic feasibility of solar energy systems were developed along the lines of the LANL approach. Hittle et al. (1977), Manning and Rees (1982), and Lameiro and Bendt (1978) developed particularly useful and simple approaches using this general method. The Lameiro and Bendt (1978) method, developed at the Solar Energy Research Institute (SERI), was dubbed the GFL method and is probably the closest in methodology to the LANL universal curve approach. The GFL method essentially correlates a large number of F-CHART runs. To calculate the annual fraction f of load met by a solar energy system, the GFL method relies on the following equation:

$$f = 1 - \exp(-RY - SY^2), \tag{3.51}$$

where

$$R = A + BX + CX^2 \tag{3.52}$$

and

$$S = D + EX + FX^2. \tag{3.53}$$

X and Y are calculated according to the collector design parameters, and A through F are tabulated in Lameiro and Bendt (1978) by climatic location and air or liquid systems. The performance curve corresponding to equation (3.53) is similar to that illustrated in figure 3.12, except that there is no linear section.

As with the LANL universal performance curves or equations, the GFL method can be easily linked to one of the economic performance models reviewed in section 3.1.2. The result is a single deterministic equation from which an economically optimal design can be derived quite easily and without the use of an expensive computer.

3.2.3 Econometric Models

Econometric models apply statistics to economics and have long played an important role in economic modeling. Although econometrics is a broad area of study, the primary focus here is on using multiple regression analysis to estimate solar energy system performance models. This analysis has not been widely used but offers sufficiently important advantages to warrant continued and increased attention.

Unlike engineering approaches to modeling production relationships, the econometric approach requires prior specification of an explicit functional form. Statistical data are then used to estimate the parameters of the function. One of the most popular functional forms for empirical estimation is the Cobb-Douglas production function of the form

$$y = A \prod_{i=1}^{n} X_i^{\alpha_i}, \tag{3.54}$$

where A is an efficiency parameter, α_i is the elasticity of output y with respect to solar input X_i, and the returns-to-scale parameter is the sum of the α_i. Returns to scale may be either decreasing, constant, or increasing, depending on whether this sum is less than unity, equal to unity, or greater than unity. Returns to scale have important economic implications for the underlying solar production process. For decreasing returns to scale an equal proportionate increase in all solar inputs will result in less than a proportionate increase in output. Likewise, an equal proportionate increase in all inputs will result in the same proportionate increase in output when the production process exhibits constant returns to scale and a greater than proportionate increase when an increasing returns to scale exists.

Returns to scale also carry dual implications with respect to the costs of producing different levels of output. Given fixed input prices, decreasing returns to scale imply that a doubling, for example, of output from a solar process will more than double the total cost of producing solar energy. Similarly, a doubling of output will just double total costs when the underlying production process is of the constant returns-to-scale variety and less than double total costs in the increasing returns-to-scale case.

Econometric estimation of a specific form such as equation (3.54) requires adequate statistical data on observed solar process inputs and resulting energy outputs. In this respect the lack of econometric estimates of solar energy production processes can be attributed to the absence of statistical data on historically observed solar inputs and outputs. This is, of course, because solar energy applications are relatively new or, at least, they have not been adopted on a wide scale. On the other hand, it is possible to supplant statistical data with data generated from a solar energy simulation model. That is, we may choose one of the models discussed in section 3.2.1 to determine the solar energy outputs that arise from different size systems and use these data to supplant statistically observed data in the econometric estimation of equation (3.54).

Peterson (1979) used this latter technique in estimating the following Cobb-Douglas production function for solar heating, cooling, and domestic hot water:

$$Q_s = aL^b A^c V^d, \tag{3.55}$$

where Q_s is useful energy supplied by the solar energy system to meet the total load, L is the total load, A is the collector area, and V is storage volume. The parameters a through d were estimated using data obtained from repeated runs on TRNSYS. The result was

$$Q_s = 1.3478 L^{0.4809} A^{0.4801} V^{0.0550}. \tag{3.56}$$

Equation (3.56) suggests that, for a given energy load L, equal proportionate increases in collector area and storage volume increase solar energy output by less than those proportionate increases; that is, the sum of the exponents on A and V sum to less than unity, thereby indicating decreasing returns to scale.

Such econometric representations of solar production processes can be easily incorporated into the cost models presented in section 3.1.2 and therefore can be used to evaluate optimal design strategies and alternative tax incentive packages on the economic feasibility of solar energy.

3.3 Summary

In this chapter I reviewed the development and use of economic models that evaluate solar energy systems in residential and commercial buildings. After reviewing the theoretical development and the foundations on which economic models of energy analyses are based, several optimal design models were presented. These models were generally of the so-called life-cycle cost type, although traditional models based on microeconomic theory were also presented. Applying these models proved both feasible and fruitful in the cost analyses and optimal design of solar energy systems. The ability of these models to incorporate financial incentives for solar energy systems also proved to be a valuable tool for building designers, engineers, and public policymakers at all levels of government.

In this chapter I also reviewed extensions of the basic models in the context of more sophisticated analysis related to the optimal timing of solar investments and the more general problem of integrating the optimal design of solar energy systems with energy conservation investments. Although the basic framework for integrating simultaneous investment decisions was established, more work is needed in this area to establish empirically workable models.

Numerous empirical approaches to economic models were developed to analyze the cost-effectiveness of solar energy systems. I reviewed the more widely used approaches and specific models. These ranged from large-scale computer-simulated models to simple deterministic models and econometric models. Each has its own particular strengths and weaknesses depending on the user's needs. Certainly all have played a critical role in establishing the technological and economic feasibility of solar energy under a variety of conditions.

Most important, economic models have been and will continue to be an essential part of the increasing interest in determining the market penetration of solar energy.

References

Beckman, W. A., S. A. Klein, and J. A. Duffie. 1977. *Solar Heating Design by the F-Chart Method.* New York: Wiley.

Ben-David, S., W. D. Schulze, J. D. Balcomb, R. Katson, S. Noll, F. Roach, and M. Thayer. 1977. "Near term prospects for solar energy: An economic analysis." *Natural Resources Journal* 17:169–207.

Bezdek, R. H. 1978. *An Analysis of the Current Feasibility of Solar Water and Space Heating.* DOE/CS-0023. Washington, DC: U.S. Department of Energy.

Freeman, T. L., M. W. Maybaum, and S. Chandra. 1978. "A comparison of four solar system simulation programs in solving a solar heating problem," in *Proceedings of the Conference on Systems Simulation and Economic Analysis for Solar Heating and Cooling*, SAND 78-1927. Washington, DC: U.S. Department of Energy, 4–7.

Hittle, D. C., G. N. Walton, D. F. Holshouser, and D. J. Leverenz. 1977. *Predicting the Performance of Solar Energy Systems.* Interim Report E-98. Champaign, IL: Construction Engineering Research Laboratory.

Lameiro, G., and P. Bendt. 1978. *The GFL Method for Sizing Solar Energy Space and Water Heating Systems.* SERI-30. Golden, CO: Solar Energy Research Institute.

Löf, G. O. G., and R. A. Tybout. 1973. "Cost of house heating with solar energy." *Solar Energy* 19:253–278.

Los Alamos National Laboratory. 1977. *ERDA Facilities Solar Design Handbook.* ERDA 77-65. Washington, DC: Energy Research and Development Administration.

Manning, N., and R. Rees. 1982. "Synthetic demand functions for solar energy." *Energy Economics* 4(4):225–231.

MITRE Corporation. 1976. *An Economic Analysis of Solar Water and Space Heating.* M 76-79. McLean, VA: MITRE Corporation METREK Division.

Nicholls, R. L. 1977. "Optimal sizing of solar heating components by equating marginal costs of suboptimal investment paths." *Solar Energy* 19:747–750.

O'Neal, D., J. Carney, and E. Hirst. 1978. *Regional Analysis of Residential Water Heating Options: Energy Use and Economics.* ORNL/CON-31. Oak Ridge, TN: Oak Ridge National Laboratory.

Perino, A. M. 1979. *A Methodology for Determining the Economic Feasibility of Residential or Commercial Solar Energy Systems.* SAND 78-0931. Albuquerque, NM: Sandia Laboratories.

Peterson, H. C.. 1979. "Simulation of the input of financial incentives on solar energy utilization for space conditioning and water heating: 1985." *Energy and Buildings* 2(1):77–84.

Powell, J. W., and K. A. Barnes. 1982. *Comparative Analysis of Economic Models in Selected Solar Energy Computer Programs.* NBSIR 81-2379. Washington, DC: National Bureau of Standards.

Ruegg, R. T. 1975. *Solar Heating and Cooling in Buildings: Methods of Economic Evaluation.* NBSIR 75-712. Washington, DC: National Bureau of Standards.

Ruegg, R. T. 1976. *Evaluating Incentives for Solar Heating.* NBSIR 76-1127. Washington, DC: National Bureau of Standards.

Ruegg, R. T., and G. T. Sav. 1981. "The microeconomics of solar energy," in *Solar Energy Handbook*, J. F. Kreider and F. Kreith, eds. New York: McGraw-Hill, 28-1–28-42.

Ruegg, R. T., G. T. Sav, J. W. Powell, and E. T. Pierce. 1982. *Economic Evaluation of Solar Energy Systems in Commercial Buildings: Methodology and Case Studies.* NBSIR 82-2540. Washington, DC: National Bureau of Standards.

Sav, G. T. 1978. "Economic optimization of solar energy and energy conservation in commercial building," in *Proceedings of the Conference on Systems Simulation and Economic Analysis for Solar Heating and Cooling,* SAND 78-1927. Washington, DC: U.S. Department of Energy, 88–90.

Sav, G. T. 1979. "Universal economic optimization paths for solar hot water systems in commercial buildings." *Energy: The International Journal* 4:415–427.

Sav, G. T. 1984a. *The Dynamic Demand for Energy Stocks: An Analysis of Tax Policy Options for Solar Processes.* Greenwich, CT: JAI Press.

Sav, G. T. 1984b. "The engineering approach to economic production functions revisited: An application to solar processes." *Journal of Industrial Economics* 33:21–35.

Sav, G. T., 1984c. "Micro engineering foundations of energy-capital complementary: Solar domestic water heaters." *Review of Economics and Statistics* 66:334–338.

Solar Energy Research Institute. 1980. *Analysis Methods for Solar Heating and Cooling Applications.* SERI/SP-35-232R. Golden, CO: Solar Energy Research Institute.

4 Assessing Market Potential

Gerald E. Bennington

In this chapter I describe studies and modeling efforts undertaken between 1973 and 1983 to support the federal government's solar energy program. These studies assessed market potential and impacts of the use of solar technologies. Two primary uses for the solar energy market projections were planning and promotion.

Planners involved in selecting and managing research and development (R&D) of solar technologies, such as the intensive management of high-growth trees and plants for energy use, quickly recognized that it would take years and even decades before the products of their programs would have significant impacts on the nation's energy budget. To compare the trade-offs of different technical approaches, they needed to consider the timing, cost, and performance of each element of the programs. Planners also realized that, if solar energy was to have a major impact in this century, the federal government would have to take an active role in promoting the rapid use of a wide range of options that were loosely grouped under the government's solar program. Planners of these incentives programs needed to analyze energy user's responses to solar technologies available in the marketplace. To do this analysis, they needed to consider the full range of options available to users, including existing and new uses of gas, oil, and electricity competing with the commercial products resulting from solar research and development activities. To further complicate the analysis, investments in capital-intensive solar technologies would be strongly affected by interest rates, inflation, tax treatment, and the risks perceived by users and their financiers. Thus project managers for solar programs had to deal not only with the risks inherent in research and development of complex equipment but also with the uncertainties of trying to predict investment alternatives twenty years in the future.

Market assessments also promoted a particular policy or program. This advocacy is a healthy and necessary part of any program, whether it is in industry or government. Without champions, many of today's technological achievements, such as telephones and airplanes, would not be available. Solar market assessments were particularly interesting because the widespread use of solar energy had the potential to touch every segment of the economy from heating water in the home to beaming energy to earth from orbiting solar generators. Moreover, a significant portion of the populace identified with solar energy as a way to a simpler and more self-sufficient lifestyle. A conflict between long-term R&D planning and the public policy issues of self-reliance and ecology pervaded the development and use of the models and market assessments throughout the period considered here.

In the following sections I discuss a multilevel structure for classifying the different types of models and assessments, provide a chronological overview of the major market assessments, and assess the various formal models used in these federally funded studies. The primary emphasis is on the models and their assumptions rather than on the political motivations.

4.1   A Multilevel Structure for Using Market Potential Estimates in Federal Planning

The stated objectives of much of the federally funded solar market penetration (potential) studies were policy analysis and goal setting; research, development, and demonstration (RD&D) planning; incentives analysis; commercialization analysis; and impacts assessment. Each of these objectives addressed different aspects of the federal government's role in supporting the development and use of energy technologies. The goals of policy analysis were to answer the questions of what was achievable and what was best for the country. These goals were then transformed into specific programs, which initially concentrated on developing specific technologies and demonstrating their technical feasibility. Incentives analysis investigated the federal and state government inducements that could encourage the development and use of solar technologies, such as the development of standards, tax benefits to producers and users, low-interest loans, and similar incentives. The incentives concept was later broadened to analyze the appropriate federal role in accelerating the commercialization of solar technologies. Finally, impacts assessments investigated the effects that widespread use of solar energy would have on labor, environment, economic growth, and other nonenergy sectors.

As the nation's energy program grew and matured, a multilevel structure for estimating the market potential and estimated energy contributions of solar energy developed. This structure considered at least three dimensions:

1. The solar technology being used (important because the funding and programs were initially organized by technology)

2. The application of the energy being used (important because it determined the nature of the energy demand and the price of the competing energy sources)

3. The type of fuel displaced by the technology (initial assessments were one-for-one fuel displacement trade-offs)

Estimating the contributions of solar energy can be viewed as a tree-like process that consolidates detailed results from smaller analyses, combining fuel types and applications. In this section this hierarchical process is organized around physical energy flows in a bottom-up approach from individual uses to the aggregated totals for the country. The levels are national and international supply-demand-price models, industry development models, major use categories (for example, residential), market segment application (for example, single-family hot water), and technology-specific price/performance comparisons.

The models in this section are discussed by major category. Many of the models contain submodels or estimating parameters and thus can operate without the support of other estimates and models.

At the bottom of this hierarchy, technology-specific price/performance comparisons, are trade-offs between similar design choices, for example, the size of the solar collector on a hot water system in Denver. These submodels were surrogates for the design process that would take place as individual applications were evaluated.

Individual market segment models typically included the user's choice to buy new equipment or retrofit equipment to reduce energy bills. These individual segments were then aggregated by the next level of major use categories, for which standard reports were available.

The growth of a solar energy industry was an important input to several models as well as an important benefit. No organized solar industry existed in 1973, and it was expected that the widespread development of solar energy would create jobs as sales volume would reduce the early high product cost.

In this chapter I concentrate on national energy supply and demand models, national solar models, and market segment and technology models. Other levels of the hierarchy are represented with different degrees of detail in these models.

1. National energy supply and demand models

   a. Project Independence Evaluation System/Mid-Term Energy Forecasting System (PIES/MEFS)

   b. Brookhaven National Labs Model

   c. FOSSIL1 and FOSSIL2

2. National solar models

   a. System for Projecting the Utilization of Renewable Resources (SPURR)

   b. Solar Working Group

3. Market segment and technology models

   a. Arthur D. Little (Heating and Cooling)

   b. Energy and Environmental Analysis, Inc. Industrial Process Heat Model (EEA)

   c. Oak Ridge Conservation Model

This list of models is not complete because it is limited to formal government-funded models. The models included had to be relatively well documented and able to create consistent and reproducible estimates. Therefore literally thousands of qualitative or less formal estimates of the solar energy contributions were excluded. Many of these estimates were at least as good as the ones produced by the more formal models. In fact, the formal models cited here were always calibrated to what was "reasonable." This is common in estimating future outcomes where the essential parameters—inflation, R&D progress, energy demand and prices, consumer response, and industry growth—were all subject to rapid and uncharted changes.

4.2   Overview of Market Potential Estimates

The scope and emphasis of the models continually changed and evolved with the solar energy program between 1973 and 1983. The federal program had its beginnings in the National Science Foundation (NSF) and the National Aeronautics and Space Administration (NASA). The initial emphasis tended to be on heating and cooling technology and on electricity production.

Thus the original market potential estimates concentrated on the impacts of residential heating and cooling and the production of electricity by solar technologies that were patterned after conventional forms of electric power generation. The primary emphasis was in trying to understand the technical trade-offs that would form the basis for the large R&D programs that were being formulated in the wake of the Arab oil embargo. Typical questions of interest were, Is the cost performance of a trickle-down collector better than that of an air collector? What is the value of power from a solar electric power system with no storage and only statistically reliable sunlight? Will the eventual installed cost of flat-plate collectors reach $4.00/ft^2 ($43.44/m^2) if they are mass produced? If so, what will that do to the large-scale use of solar heating?

Such questions were important as the nation grappled with the terminology, technology, economics, and politics of energy. Issues were raised by the questions that identified problems that the general public and most of the government had abandoned to the oil, gas, and coal companies and the utilities. In 1973 the tools available to the solar energy planner were conceptual engineering models of the components of solar systems, for example, double-walled heat exchangers and fossil and electric planning models. The field was so newly rediscovered that no infrastructure was developed to understand the cost/performance of "real systems" or to predict the actual impact these would have on other sources of energy. Moreover, no concept was established that the demand for energy was elastic and subject to choice by the consumer.

Out of thousands of projections, guesses, and surveys on the eventual use of solar energy, the following events and studies are of special importance because they marked the use of a particular methodology or had a major impact on federal policy or programs. The results and references are covered in more detail in the next sections.

Date	Event or study
Fall 1973	OPEC oil embargo
November 1973	Nixon energy independence speech
December 1973	National Science Foundation Plan
November 1974	Project Independence Report
June 1975	Energy Research & Development Administration Heating & Cooling Program (ERDA-4a)

Date	Event or study
January 1976	Centralized solar projections
February 1976	National Energy Outlook
April 1976	Creating choices for the future (ERDA 76-1)
November 1976	Analysis of solar water & space heating
January 1977	Solar Energy in America's Future
April 1977	The National Energy Plan
Spring 1977	*Soft Energy Paths* (Lovins, 1977)
March 1978	Comparative analysis to 2020
April 1978	Solar Energy: Progress and Promise
June 1978	Office of Technology Assessment Study (1978)
October 1978	Domestic Policy Review (DPR)
May 1979	National Energy Plan II
December 1979	National Academy of Sciences Study (Committee on Nuclear and Alternative Energy Systems, 1979)
January 1980	National Plan for the Commercialization of Solar Energy
Spring 1980	Harvard Business School Energy Project
April 1981	Building a Sustainable Future
Fall 1982	America's Solar Potential: A National Study of the Residential Solar Consumer

The following sections in this chapter look at the chronological progression of the policy, technical, and economic issues and responses as they developed. The chronology is important because the data and tools became more sophisticated as time went on and the focus of the issues shifted from technical to social and economic.

4.2.1 Embargo through 1976

Shortly after the Arab oil embargo the National Science Foundation released a report of a study (MITRE Corporation, 1973) that identified seven solar technologies: heating and cooling of buildings, process heat for agriculture and industry, solar thermal electric generation, ocean thermal gradient systems, photovoltaic systems, wind energy systems, and utilization of organic materials. Table 4.1 gives the solar potential identified in the NSF study for each of these technologies for 1980, 1985,

Table 4.1
Solar potential from the NSF study

Solar potential (% of U.S. requirement)	1980	1985	2000	2020
Total U.S. requirement, quads/yr[a]	90	120	180	180
Heating and cooling of buildings	0.01	0.2	2	5
Process heat	0	0.1	1	10
Solar thermal	0	0.01	2–5	10
Ocean thermal	0	0.01	1–5	10
Photovoltaic	0	0.01	2–5	10
Wind energy	0.01	1–3	5–10	15
Organic materials	0.01	0.7	7	10
Total	0.03	2–4	20–35	70

Source: MITRE Corporation (1973).
a. 1 quad $= 10^{15}$ Btu $= 1.055 \times 10^8$ TJ.

2000, and 2020. The study defined a general program structure, reviewed the status of current research and applications, assessed the relative merits of potential applications, and recommended research and proof-of-concept experiments. In addition, the study developed two implementation scenarios (minimum production and accelerated production) for each technology.

The NSF program described in this study provided the structure for the solar program that eventually moved to the Energy Research and Development Administration (ERDA) and finally to the U.S. Department of Energy (DOE). Although the energy estimates in the study were based on the ability to develop and produce the equipment, they provided an early claim that solar energy could compete with other conventional methods of energy production. The analysis contained many of the elements that were to be formalized in the models developed later. The scenarios considered energy demand by market segment, competing fuel prices, and a measure of economic merit—the economic viability ratio. Many NSF staff members played important roles in subsequent years, for example, D. Beattie, D. Blieden, R. Cohen, L. Divone, R. Fields, and L. Herwig. These programs and many of the people involved became the nucleus of the solar research programs and initiated the demonstration programs as well.

While NSF was developing its research and proof-of-concept experiments, the Federal Energy Administration (FEA) was creating the Project Independence Blueprint (FEA, 1974) using the Project Independence

Table 4.2
Solar goals from the 1975 ERDA study

Goal	1985	2000	2020
Total U.S. demand, quads/yr	100	150	180
Conversion technology			
Direct thermal	0.2	3	20
Solar electric	0.07	5	15
Fuels from biomass	0.5	3	10
Total solar contribution	0.77	11	45
Estimated % of demand	0.08	7	25

Source: ERDA (1975a).

Evaluation System (PIES), which consisted of an immense collection of interrelated models of energy production, refining, processing, conversion, distribution, transportation, and consumption. It utilized price-supply curves for each energy source and region of the country. The solar program provided inputs for heating and cooling of buildings and electricity production. Estimates for a set of years—1980, 1985, 1990—were obtained by solving a large linear program. The primary emphasis was on reducing oil imports through fuel switching and on increasing domestic oil production. Because of the short-term outlook of the study, it is not surprising that solar and conservation received little attention and that the primary contributors to import reductions were increased oil production, coal and nuclear power generation, and transportation initiatives.

In 1974 ERDA was formed and took over many of the energy programs that had existed in the Atomic Energy Commission (AEC), Department of Defense (DOD), FEA, and NSF. The solar program was moved to ERDA with a mandate to pursue vigorously the "goal of providing the option of utilizing solar energy as a viable source for the nation's future energy needs" (ERDA, 1975a). Table 4.2 identifies the goals of the ERDA study, which were ad hoc and subjective in nature. Although formal models were not used, the total demand and many other parameters were made consistent with the Project Independence blueprint. For the most part, the programs were carried over from NSF in structure and content.

Although ERDA had the primary responsibility for technology and understanding the potential impacts of their programs, the primary responsibility for energy forecasting and policy assessment rested with FEA. FEA continued the work that began with the *Project Independence*

Report (FEA, 1974) and produced the *National Energy Outlook* (FEA, 1976). The primary emphasis was on the supply and demand for energy (oil, gas, and electricity) through 1985. The critical factors considered were dependent on the price of oil and the effects of differing assumptions about reserves. The PIES model was being used throughout the study to balance and integrate the estimates for each region and market sector. Because the emphasis was on short-term energy contributions, new technology had only a minor role in the formal analysis. The executive summary noted, "Solar, geothermal, and synthetic fuels will make only a small contribution to domestic energy supplies by 1985 (about 1%)." The summary also emphasized that "the major contribution from solar, geothermal and synthetic fuels will not be felt until after 1990." About 2 quads from new technology were included in their "reference scenario" for 1990. Of this amount only a small portion (approximately 0.2 quad) was attributed to solar with the vast majority of the potential being assigned to synthetics. A second estimate, "total under most optimistic assumptions," estimated 6 quads in 1990 with no allocation among the technologies.

A wide variety of conservation alternatives were considered. Two of the major conclusions were, "An active conservation program could further reduce energy demand by the equivalent of 3 million barrels per day, reducing the annual energy growth rate to 2.2 percent through 1985," and "Actions which improve automobile efficiency and the efficiency of homes and office buildings would have the greatest impact in the next ten years" (FEA, 1986). Although this study did little to foster the understanding of the potential for solar technologies, the price and energy demand data formed the basis for many of the subsequent technology studies.

In support of the FEA studies, the MITRE Corporation developed projections for the centralized solar energy conversion systems (Bennington, Rebibo, and Vitray, 1976). Seven scenarios were presented based on two reference scenarios: business as usual (BAU) and high solar development. This represented the initial published study using the utility model of what would become a major building block for the SPURR model (Bennington et al., 1977). Using the PIES BAU demands, the solar technologies were compared with coal, nuclear, and oil power plants competing for new generation capacity. The projected ranges included a delayed R&D schedule with no major incentives, up to high estimates represent-

Table 4.3
Central solar potential from the 1976 MITRE study

Solar potential	1980	1985	1990
Total electric demand, quads/yr[a]	1.1	8.3	21
Conversion technology			
Wind	0–0.007	0.07–0.8	0.3–4.2
Photovoltaics	—	—	0–0.006
Solar thermal electric	—	0–0.01	0.01–0.06
Ocean thermal	—	0–0.004	0.008–0.234
Biomass	0.1–0.3	0.2–0.5	0.3–0.7
Total solar	0.1–0.3	0.3–1.3	0.6–5.2
Estimated % of demand	9–27	4–16	3–25

Source: Bennington, Rebibo, and Vitray (1976).
a. Equivalent energy input.

ing the maximum rate of implementation with incentives sufficient to make wind and solar thermal cost-effective. Solar potentials for five conversion technologies discussed in the study are given in table 4.3. Although the projections were replete with technical problems, this study brought together the initial technology database that was expanded and used in several of the later studies by MITRE and others. The inclusion of on-site energy using biomass (wood burning) highlighted the difficulties of accounting for energy production that was not reported as purchased from a utility or its equivalent.

ERDA released its plan later in 1976 (ERDA, 1976). It was similar to the 1976 FEA study in that it stressed the role of the private sector in commercializing new energy technologies, gave a high priority to energy conservation, and recommended increased funding for RD&D by 30%. The major winners of funding were nuclear, with a 73% increase, and conservation, with a 64% increase. A 35% increase was recommended for the solar energy budget.

The ERDA plan was notable in avoiding outright energy projections. The plan correctly stressed specific development tasks, milestones, and possible applications. The underlying theme for the plan as stated in the executive summary was, "Decisions . . . must be made today in the face of uncertainty, without foreclosing future options. Indeed, the basis for undertaking a program of energy RD&D is to broaden the Nation's range of available energy options—to create energy choices for the future" (ERDA, 1976).

Table 4.4
Solar projections from the 1976 ERDA plan[a]

Conversion technology	1985	2000	Beyond 2000
Heating and cooling of buildings	0.2	2	
Agricultural and industrial Process heat	0.2	3	
Solar thermal electric	[b]	[b]	
Photovoltaics	nil	0.004	
Wind	nil	0.002	
Ocean thermal	0	0.002	
Fuels from biomass	0.2	2–5	10–30

Source: ERDA (1976).
a. No demand estimates were given; all data in quads/yr.
b. No estimates.

In addition, the dialogue was broadened to include analyses of several new technical issues that were later assimilated in more formal studies. These issues were understanding the relationship among energy, the economy, and the environment; calculating the net energy contributions of energy processes; and studying market penetration by specific markets (for example, electric utilities).

The detailed discussions of the solar technologies contained the "goals" or projections of potential solar energy contributions shown in table 4.4. These projections emphasize the disparity of the individual program estimates.

In line with the desire to emphasize specific applications and market sectors, ERDA released *An Analysis of Solar Water and Space Heating* in late 1976 (Bennington et al., 1976). Although this study included no market penetration estimates, it used the F-CHART computer program developed by the University of Wisconsin (Klein, Beckman, and Duffie, 1975) to estimate the economic viability of solar space and water heating in thirteen selected cities using actual fuel costs in those cities. The assertion that "solar water heating and space heating installed at an equivalent system cost of $20 per square foot of collector is competitive today against electric resistance systems throughout most of the U.S. (Bennington et al., 1976)" was the subject of considerable public discussion. More important, the assertion illustrated that solar technologies are likely to have widely different acceptance based on location, fuel availability and price, and installed price. Furthermore, this level of analysis

Table 4.5
Solar projections from the Solar Working Group

Projection	1985	2000	2020
Total U.S. energy demand, quads/yr	93	153	183
Conversion technology			
Heating and cooling of buildings	1	3	10
Agricultural and industrial process heat	nil	nil	nil
Solar thermal electric	nil	nil	nil
Photovoltaics		a	
Wind		a	
Fuels from biomass	1	1	1
Total solar contribution	2	6	11
Estimated % of demand	2	4	6

Source: SRI (1977).
a. No estimates.

became one of the building blocks for the detailed market potential studies that would follow.

As a means of assessing the impacts of the ERDA program, a Solar Working Group was established with participation by government, industry, and academic leaders. Stanford Research Institute (SRI) was awarded a contract to provide technical assistance to the group. A market assessment model (SRI, 1977) was developed and used to provide estimates for solar energy contribution in 5-year increments through 2020. Table 4.5 gives some of the results from the base-case scenario. This study provided two valuable insights into the potential for solar energy. First, the use of the SRI National Model provided a creditable and easily used method to relate changes in fuel price with the size and competitiveness of a large number of disaggregated market sectors. In fact, the low demand case of 75 quads in 1975 was one of the early discussions of what solar's role would be in a future with high-priced fuel and low-energy demand. Second, the market penetration methodology sparked discussions among the builders and users of solar market penetration models (Schiffel et al., 1978).

4.2.2   National Energy Plan, 1977

On April 29, 1977, President Carter presented the National Energy Plan (NEP) that had been under development for some months (Executive Office of the President, 1977). The plan stressed conservation and fuel

Table 4.6
Solar projections from the 1978 MITRE study

Projection	1985	2000	2020
Total U.S. energy demand, quads/yr	86	115	189
Conversion technology			
Heating and cooling of buildings	0.15	1.6	3–5
Agricultural and industrial process heat	0.02	2.0	13
Solar thermal electric	[a]	0.3	2.9
Photovoltaics	[a]	[a]	0.2
Wind	[a]	1.7	6.6
Ocean thermal	[a]	[a]	2.4
Fuels from biomass	0.03	0.4	4.4
Total solar contribution	0.2	6.0	34.0
Estimated % of demand	0.2	5	18

Source: Bennington et al. (1978).
a. An estimate of less than 0.01 quad.

efficiency, rational energy pricing and production, stability in government policies, use of abundant energy sources (coal), and development of nonconventional technologies for the future.

The plan addressed solar energy in two ways: a federal income tax credit was proposed for the use of most solar technologies, and a goal was established to use solar in more than 2.5 million homes by 1985. The proposed changes in energy pricing, tax credits, and fuel-use regulations created such potentially wide-ranging changes in the energy sector as to invalidate most of the scenarios previously prepared. Thus the turmoil in the energy sector that began with the oil embargo continued. This turmoil made it difficult to obtain a consensus on energy demand and price by sector, investment criteria in new technology, and buying preferences of the energy user. These are a few of the essentials needed for a believable estimate of the likely acceptance of a capital-intensive technology like solar energy.

The MITRE estimates (Bennington et al., 1978) provided two scenarios: the National Energy Plan and continuing the trends since 1973. The study used the SPURR model extensively and developed a technology database in coordination with the ERDA programs. Table 4.6 presents the NEP scenario projections. The second scenario is not presented here, but was used to create a basis for calculating the incremental benefits derived from the NEP initiatives.

Table 4.7
Solar goals from 1978 Council on Environmental Quality study

Goal	1977	2000	2020
Total U.S. energy demand, quads/yr	76	80–120	70–140
Conversion technology			
Heating and cooling of buildings	Small	2–4	5–10
Agricultural and industrial process heat	None	2–5	5–15
Solar thermal electric	None	0–2	5–10
Photovoltaics	Small	2–8	10–30
Wind	Small	4–8	8–12
Ocean thermal	None	1–3	5–10
Fuels from biomass	1.3	3–5	5–10
Hydropower[a]	3	4–6	4–6
Total solar contribution[b]	4	20–30	35–70
Estimated % of demand	5	25	50

Source: CEQ (1978).
a. Hydropower had not previously been generally considered to be a solar technology.
b. The total contribution is calculated from the goals as a percent of total demand. The individual contributions do not add because the technologies were estimated separately and will compete with each other.

The Council on Environmental Quality report *Solar Energy, Progress and Promise* (1978) was optimistic about the rate and dimension of solar energy's contribution to the U.S. energy problem. Its estimates were stated as being attainable goals: "Our conclusion is that with a strong national commitment to accelerated solar development and use, it should be possible to derive a quarter of the U.S. energy from solar by the year 2000, with major growth occurring after 1985."

The report reviewed the progress of solar development to date and proposed a federal program that would presumably support the goals established earlier in the report. Although the report references other studies, no attempt was made to integrate the individual estimates or to estimate specific paths leading to the goals cited. Table 4.7 presents their estimates of the maximum solar contribution under accelerated development.

4.2.3 Domestic Policy Review of Solar Energy and the 20% Solar Goal

On June 25, 1978, President Carter called for a Domestic Policy Review (DPR) of solar energy as part of a message to Congress on solar energy. The DPR acted as a melting pot for many of the issues and studies underway at the time. The DPR Impacts Panel published a report (DOE, 1978)

Table 4.8
Solar projections from the domestic policy review of solar energy

Projection	1977	2000	2020
Total U.S. energy demand, quads/yr	76	95–114	Technical limit
Conversion technology			
Heating and cooling of buildings	Small	0.9–2.0	3.8
Passive heating[a]	Small	0.2–1.0	1.7
Industrial and agricultural[b]	d	1.0–2.6	3.5
Solar thermal electric	d	0.1–0.4	1.5
Photovoltaics	d	0.1–1.0	2.5
Wind	d	0.6–1.7	3.0
Ocean thermal	d	0.0–0.1	1.0
Biomass	1.8	3.1–5.4	7.0
Hydro[c]	2.4	3.9–4.3	4.5
Total solar contribution	4.2	9.9–18.5	28.5
Estimated % of demand	5	9–20	25–30

Source: DOE (1978).
a. Passive heating would typically be in heating and cooling of buildings but had significant estimates of potential.
b. Industrial and agriculture processes include process heat and on-site electricity.
c. Hydro includes conventional high-head systems as well as smaller low-head systems.
d. Indicates an estimate of less than 0.01 quad.

that contained three scenarios: a base case with oil at $25/bbl, a maximum practical case with oil at $32/bbl, and a technical limit case. Table 4.8 gives the results for the year 2000. The first two scenarios were used to develop the range shown in the third column, whereas the third scenario is provided in the fourth column.

The second National Energy Plan (NEP-II), released in May 1979, contained an extensive analysis of energy futures and their relationship to the economy and environment (DOE, 1979). The study used the descendant of the PIES model renamed the Mid-Term Energy Forecasting System (MEFS), introduced a new model (FOSSIL2), and used several other models previously available. The study also emphasized the basic difference between end-use energy and the equivalent of primary energy that was needed to produce the same end-use energy. Although projections were run for intervening years, only the year 2000 estimates were presented in detail; they are shown in table 4.9. Note that the primary energy varied greatly depending on the technology and fuel displaced.

On December 25, 1979, NAS released their long-awaited *Energy in Transition 1985–2010* as the final report of the Committee on Nuclear

Table 4.9
Solar projections from the National Energy Plan II

	Year 2000	
Projection	End-use energy	Primary energy
Total for U.S., quads/yr	85	119
Decentralized		
Biomass	2.78	2.92
Solar heating and cooling	0.49	0.70
Passive heating and cooling	0.44	0.63
Agriculture and industrial	1.00	1.43
Wind	0.10	0.30
Photovoltaics	0.03	0.08
Total decentralized	4.84	6.06
Centralized		
Hydro/geothermal[a]	1.40	4.30
Photovoltaics	0	0
Wind	0.03	0.10
OTEC	0	0
Solar thermal	0.03	0.10
Biomass	0.10	0.30
Total centralized	1.56	4.80
Total renewables	6.40	10.86
Estimated % of U.S. total	8	9

Source: DOE (1979).
a. Geothermal had not been included in previous solar energy estimates. It had been regrouped under the renewable energy.

and Alternative Energy Systems (CONAES, 1979). This study was commissioned in April 1975 as a "comprehensive and objective study of the role of nuclear power in the context of alternative energy systems." Understandably, the report concentrated on national supply-demand issues and the major alternatives—coal, synthetics, and oil shale. The findings stressed conservation to reduce demand, the immediate problem of liquid fuel supplies, and the need to balance the support for coal and nuclear options. Solar energy was the subject of considerable analysis but was not strongly emphasized: "Because of their higher economic costs, solar energy technologies other than hydroelectric power will probably not contribute much more than 5 percent to energy supply in this century unless there is a massive government intervention in the market to penalize the use of nonrenewable fuels and subsidize the use of renewal energy sources" (CONAES, 1979).

Table 4.10
Solar projections from the 1979 National Academy of Sciences study

Projection	1975	1990	2000	2010
Total U.S. energy supply, quads/yr	60.5	82.5	113.2	142.5
Energy form				
Crude oil	19.6	20.0	18.0	16.0
Natural gas	19.4	15.8	15.0	14.0
Oil shale	0	0.7	1.0	1.5
Coal	16.4	26.6	37.2	49.5
Geothermal	0	0.6	1.6	4.1
Solar	0	1.7	5.9	10.7
Nuclear	2.7	13.0	29.5	41.7
Hydroelectric	2.4	4.1	5.0	5.0
Solar as % of supply	—	2	5	8

Source: CONAES (1979).

Draft reports were widely circulated and discussed for some time before their official release. Table 4.10 shows the supply estimates for the enhanced-supply assumptions intending to represent policies and regulatory practices to encourage energy production. Note that the format for comparing national energy use is used without particular emphasis on specific technologies.

In January 1980 the MITRE Corporation released a report on accelerated commercialization of solar energy (Bennington et al., 1980). The report examined national goals for solar energy by providing detailed technology goals by region, assessing physical, institutional, and other requirements related to achieving each goal, and providing "growth trajectories" as checkpoints for measuring progress. Table 4.11 gives the solar projections from this report.

The study used a reference case and three accelerated levels corresponding to 19, 22, and 26 quads by the year 2000. In contrast to the earlier MITRE study in 1978, the primary emphasis of the 1980 study was on analyzing the impacts of varying levels of solar use. In particular, the study investigated the physical growth rates, investment requirements, and implied government subsidies. Table 4.12 shows the energy savings by market and technology for the reference case. The reference case is shown because the energy displacement is in the same range as the 1978 estimates during this period. However, the total energy demand was assumed to be higher.

Table 4.11
Solar projections from the 1980 MITRE study

Projection	1978	1990	2000
Total U.S. demand, quads/yr	77	98	115
Residential			
Thermal	a	0.31	1.06
Passive	a	0.04	0.02
Wind	a	0.05	0.31
Photovoltaics	a	0.02	0.18
Wood stoves	0.30	0.47	0.60
Commercial			
Thermal	a	0.27	0.66
Passive	a	0.01	0.02
Wind	a	a	0.02
Photovoltaics	a	a	0.03
Industrial			
Solar thermal	a	0.50	2.18
Biomass	1.6	2.00	2.20
Wind	a	a	0.05
Photovoltaics	a	a	0.03
Thermal electric	a	a	0.02
Total energy	a	a	0.01
Small-scale hydro	a	a	0.15
Electric utility			
Wind	a	0.15	1.32
Solar thermal	a	0.05	0.96
Photovoltaics	a	a	0.01
Ocean thermal	a	a	0.10
Biomass electric	a	0.01	0.03
Hydro	3.0	3.40	3.80
Fuels and chemicals			
Wood	a	a	0.45
Animal waste	a	a	0.20
Total solar contribution	4.90	7.28	14.59
Estimated % of demand	6	7	13

Source: Bennington et al. (1980).
a. Denotes a value less than 0.01 quad.

Table 4.12
Solar projections from the Harvard Business School Energy Project[a]

Projection	2000
Total U.S. energy demand, quads/yr	105
Energy source	
Active and passive heating	4.2
Other on-site technologies	4.2
Wood and waste	8.4
Hydropower (large scale)	4.2
Total solar contribution	21
Estimate as % of demand	20

Source: Stobaugh and Yergin (1980).
a. Original estimates converted to quads from millions of barrels of oil per day equivalent.

Stobaugh and Yergin's book *Energy Future: The Report of the Harvard Business School Energy Project* (1980) provides an excellent counterpoint to the excessive detail of the preceding studies. After an insightful review of much of the work and debate that took place, Stobaugh and Yergin reemphasize what most of the advocates had forgotten:

The truth is that no one really knows what the contribution from solar energy will be in the year 2000. Solar's contribution depends on at least six other complex and uncertain variables, each difficult to forecast.

1. Prices of competing energy sources

2. Overall levels of domestic energy consumption

3. Level of federal involvement in solar energy

4. Rate of advancement of solar technologies

5. Rate at which institutional barriers to solar will be overcome

6. Reliability of energy supply

Nonetheless, Stobaugh and Yergin could not resist presenting an estimate of how an aggressive program might obtain a 20% solar contribution by the year 2000 (table 4.12).

4.2.4   1981 and Later

The Solar Energy Research Institute (SERI) *Report on Building a Sustainable Future* (1981) comprehensively assessed a strategy for combining energy conservation and solar energy as a major energy contributor. To the surprise of many who considered SERI a solar advocate, the primary em-

Table 4.13
Solar projections from the 1981 SERI study

Projection	2000
Total U.S. energy demand, quads/yr	62–66
Residential	
Thermal	1.6–1.9
Biomass	1.0
Wind	0.8–1.1
Photovoltaics	0.3–0.45
Commercial	
Thermal	0.3–0.4
Photovoltaics	0.1–0.25
Industrial	
Thermal	0.5–2.0
Biomass	3.5–5.5
Agriculture	
Biomass	0.1–0.7
Transporation	
Biomass	0.4–5.5
Utilities	
Wind	0.5–3.4
Hydro	3.4–3.7
Total solar contribution	12.3–22.5
Estimated % of demand	20–34

Source: SERI (1981).

phasis of the assessment was the potential of energy conservation as the keystone to creating an energy program that has long-term stability.

On a closer look the study and its projections emphasize the major impacts that might arise from the efficient use of energy and the matching of type and form of energy to the end-use. Each Btu of energy saved from better insulation of an electrically heated home or the redesign of a refrigerator saves roughly 3 Btu of fossil fuel that would be needed to generate the electricity. The basic idea driving this analysis is that the real leverage is to reduce the energy demand while maintaining lifestyles by efficient use of the energy. Solar energy's role was for additional heating and generating capacity for this reduced demand. The solar technologies included in the study were solar thermal, biomass, wind, photovoltaics, and hydroelectric. Only those technologies projected to have a measurable impact are shown in table 4.13.

A major assertion, based on the demand reductions projected, is that the electric generation capacity already installed or under construction will satisfy the country's needs in 2000. From this and other considerations, no contributions are anticipated from the large-scale solar electric technologies, that is, solar thermal electric, wind, ocean thermal energy conversion, and photovoltaics.

4.3   Solar Market Assessment Models

In section 4.2 I reviewed the sequence of estimates; now I shall discuss the methods used to produce these estimates. As seen from the preceding sections, market potential assessments were a game that it seemed anyone could play. Although the aggregate energy projections quickly polarized the various constituencies in the ongoing policy debates, the actual work in these studies consisted of the unglamorous task of estimating the effects of budgets on R&D programs and then on the schedule and cost/ performance of advanced technologies. Try as they might, the engineers could not turn this into a "science." Although the models could not predict the future, they could do the next best thing to assist in the planning of a national program for renewable resources. Based on rational interpretations of the R&D trade-offs, these models could

1. Rank the cost/performance of existing and proposed technologies in representative applications

2. Tabulate the resources used and energy effects of different rates of technology substitution

3. Measure interactions among the numerous factors in the development, financing, and use of energy technologies

4. Update projections as conditions changed

5. Produce reproducible, internally consistent, and plausible energy scenarios that could be used to assess policy and program alternatives

Although the primary emphasis of these models was to assess solar thermal energy technologies, the entire gambit of models—from national supply and demand models to individual technology models—had to be considered. The national models set the context for alternative prices and market segment demands, whereas the individual models established the basic cost/performance information.

4.3.1 National Supply and Demand Models

The Project Independence Evaluation System (PIES) was developed as an integrating model to support the initial Project Independence Blueprint Report initiated by President Nixon immediately following the 1973 oil embargo. The PIES model is static, representing equilibrium conditions at the end of a particular planning horizon, for example, 5, 10, and 15 years. The system consists of a set of models: a demand model, specialized supply models, and an integrating model. The integrating model is the most relevant for analyzing solar potential.

The integrating model uses a set of linear equations to represent the supply, demand, and other restrictions while minimizing (by means of linear programming) the national cost of balancing supply and demand. This is accomplished by determining market clearing prices. The clearing prices describe a static market equilibrium for which the clearing prices will balance supply and demand. Under these circumstances the suppliers, who attempt to maximize profit, will be willing to supply exactly the same quantities as those demanded by the consumers, who are attempting to minimize their costs. The framework for the integrating model is depicted in figure 4.1.

The PIES model uses a variety of regions to match the established sources of data for each producing sector. The model used the nine census regions for estimating demand and electric utility supply but had twelve coal, seven refinery, three shale, thirteen oil, and fourteen gas regions.

The integrating model generates a complete energy scenario using an iterative process. Given a scenario, which included a Data Resources Incorporated (DRI) economic forecast, the model estimates a set of demand prices. Using these prices, a set of demands are then determined for each use and region, for example, transportation gasoline for the Mid-Atlantic census region consisting of New York, New Jersey, and Pennsylvania. Next the linear programming solution of the supply model, which provides consumers with the prespecified quantities of the end fuels at minimum cost, is obtained. The supply model generates a set of marginal production costs for each energy demand as a by-product. These supply side costs are compared with the starting demand prices. If the supply and demand prices are the same, the solution is in equilibrium and the process is complete. If the supply and demand prices are different, a new set of prices is selected, and the process is repeated. In practice, only a couple of iterations are required to find a reasonable solution.

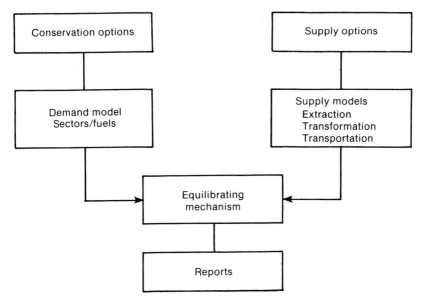

Figure 4.1
Integrating model framework (FEA, 1974, appendix A).

PIES attempted to approximate the supply-demand trade-offs for the nation's entire production, transportation, and use of energy. Given the impact and importance of the previous embargo and the subsequent increase in the price of imported oil, a tremendous amount of detail relating to the near-term impacts of oil pricing was included. A key parameter in the individual scenarios was the price of imported oil. "From this perspective, the entire system can be viewed as a procedure for calculating adjustments in all sectors and the resulting demand for imports" (FEA, 1974, appendix A).

The supply model attempted to analyze the near-term impact of export pricing, natural gas deregulation, and similar issues that have billion dollar impacts on existing industries and consumers; hence it is not surprising that it was not ideal for analyzing solar technologies that were still under development, would not make a significant impact in the near term, and did not resemble existing technologies. The supply model illustrates one of the simplest examples of the basic differences between evaluating resource-based fuels and solar technologies. "The model assumes a competitive economic structure with upward sloping supply curves and downward sloping demand curves" (FEA, 1974). An upward sloping supply curve means the more that is produced, the more expensive it gets, which is true for natural resources such as oil. Although it is certainly true that the burgeoning solar industry would produce less in the *short run* according to an upward sloping demand curve, one of the oft-repeated promises of solar technology proponents is that, if the solar industry can get into mass production, then the consumer's cost will go down. In fact, the underlying cost goals for several of the solar programs were based on *long-term* supply curves that were downward sloping and included cost reductions of a factor of 100 or more.

Despite the inherent difficulties of including fledgling solar technologies in the static model, which concentrated on mature energy technologies, an active attempt was made to include the impacts of the solar program in the Project Independence analysis. Although solar was not included in the original PIES modeling, a special task force was set up to assess the potential impacts of solar energy: *Project Independence Blueprint, Interagency Task Force Report—Solar Energy, Final Report* (FEA, 1974). The task force did not use any formal projection models but did use extensive technical data and opinion by over a hundred experts who actively participated in preparing the report.

Subsequent studies using PIES included solar impacts in the formal analysis structure. The supply model had explicit activities to represent additional capacity for electric generation. Supply curves were generated for solar to compete with fossil, nuclear, and geothermal energy. This was accomplished by making proposed solar electric systems resemble an equivalent conventional power plant.

In practice, generation load varies by time of day, by day of week, and by season. Thus, utilities typically construct a mixture of plant types with different economic and operating characteristics to meet the varying load requirements. To capture this variation, the model separates electricity into peak, intermediate, and base load categories and requires each kWh of electricity to be produced in prespecified proportions from each category of generation.... Each unit of generation capacity may be operated in convex combinations of the three generation modes. Electricity generation facility types include coal, oil, and gas-fired steam turbines, simple and combined cycle gas turbines, hydro and nuclear plants, plus a variety of exotic technologies including solar, geothermal, and fuel gas plants. (FEA, 1974)

The cost functions were represented as increasing piecewise linear approximations to supply curves for that period. The supply costs for new activities included the variable cost of operations and maintenance as well as the amortized capital costs. In a period of high inflation and interest rates, this placed a heavy emphasis on technologies that had well-understood capital requirements and stable fuel costs, for example, coal. It was not surprising that none of the solar technologies were able to compete in the pre-1990 estimates with the alternative electric options in the PIES model.

The decentralized solar technologies, such as solar heating and cooling, needed to be included in the demand model. Because the PIES model consists of econometrically based sector models, it only projects final demands based on the *current price of supplies*. Individual conservation and technologies are not explicitly represented in this type of model. To develop the likely rates of substitution that might occur from increased energy prices promoting the use of solar and conservation, several sector models were used. Initially the proprietary model of Arthur D. Little, Inc., was used, and later the Oak Ridge model developed for the conservation program was used for side studies.

Brookhaven National Laboratory Model. The team at Brookhaven National Laboratory (BNL) developed national energy and economic scenarios using a set of economic and technology models. The major components of their modeling approach are shown in figure 4.2.

The Brookhaven Time-Stepped Energy System Optimization Model (TESOM) is a detailed technology model that balances energy supply, conversion, and end-use demand. It is formulated as a mathematical programming model that minimizes the cost of meeting a given set of demands using a range of feasible supply and conversion options. The model deals with aggregate technologies (for example, air conditioning and process heat) and has no regional breakout reflecting climate or transportation factors. TESOM, however, is stepped through time to approximate the movement of prices, technology development, and changing demand. TESOM enhances the original static BESOM model used in BNL's initial scenarios.

The Dale Jorgenson Associates (DJA) Long-Term Interindustry Transactions Model (LITM) is a national econometric model that produces a detailed representation of sectoral production and final demand. An input-output model jointly developed by BNL and the University of Illinois is used to map the LITM sectoral outputs and consumption levels

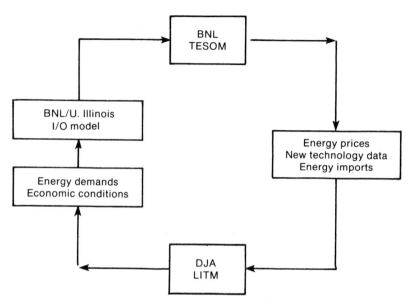

Figure 4.2
Brookhaven National Laboratory model (Groncki, Goettle, and Hudson, 1980).

into a set of demands and energy services consistent with the BNL TESOM model.

Solar technologies are included as conversion technologies in the electric utility and fuels paths, and as end-conversion devices such as solar heating and cooling, which would be compared to heat pumps. The level of aggregation and lack of regional solar performance make this model hard to use for more than the mirror results that were estimated with more detailed modeling techniques. The BNL system is important because it has provided basic price-demand scenarios to many of the major federal policy studies. To the extent that the detailed and macro scenarios are widely different, it makes the integration and eventual creditability of the study difficult.

FOSSIL2 **Model.** The FOSSIL2 model grew out of work done at Dartmouth College under the leadership of R. Naill, who later moved to DOE. FOSSIL2 differs radically from the preceding modeling systems. It uses a dynamic systems modeling process originated by J. Forrester at MIT in the early 1960s. The process combines feedback control theory, organizational behavior, and computer simulation to analyze social systems. The basic modeling process creates a nonlinear flow model reflecting the flow of energy resources, capital, and other variables in the U.S. energy system. Furthermore, FOSSIL2 is a dynamic disequilibrium model that is not based on an optimal set of prices or on a perfect balancing of supply and demand at every point in time. Rather, the model has a set of decision rules *provided by the modeler* governing the flow of investment, goods, resources, and other model flows. It is intended to be used as a policy tool for analyzing potential energy problems and variants used to mitigate these problems.

Although FOSSIL2 is another aggregated model, it does possess many of the features that are important in assessing the eventual path of solar development. Time paths are dynamically developed in a consistent manner, time lags are explicitly included, and nonlinear relationships can be used to model market behavior for new technologies.

Applying FOSSIL2 to analyze NEP-II generated results consistent with the upgraded PIES system, MEFS. NEP-II (DOE, 1979) compares these two models. It provides a good example of an attempt to use widely disparate tools to assess the energy problem in a rational and consistent manner; however, it does not document the modeling methods for the central and decentralized solar technologies.

4.3.2   National Solar Market Penetration Models

Only two national solar models are discussed in this section: the SPURR model developed at the MITRE Corporation and the solar working group model developed by SRI. These two were sufficiently well developed and documented to provide discussions of the methods used to develop models for the planning of federal programs in renewable resources.

System for Projecting the Utilization of Renewable Resources. The MITRE model, A System for Projecting the Utilization of Renewable Resources (SPURR), resulted from the MITRE Corporation's assistance to the Solar Energy Task Force for Project Independence Blueprint. Although the task force had a wealth of valuable technical data, no methodology existed for integrating and coordinating the emphasis on individual projects, matching projects to commercial applications, and developing energy projections. SPURR was intended to fill the gap between the aggregate national energy models and the uncoordinated analyses of individual applications of solar technologies. Its structure and level of de-

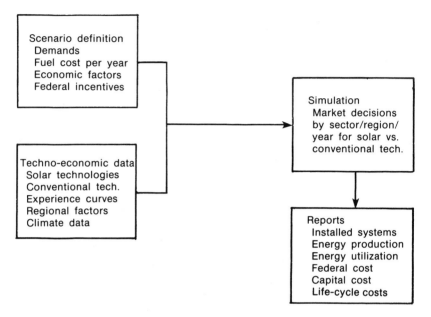

Figure 4.3
SPURR methodology (Bennington et al., 1977).

tail permitted analysis of the key issues affecting the solar program. These issues were the dynamic growth characteristics of emerging energy technologies, the technoeconomic description of a wide range of conventional and new systems, cost reduction from mass production and accumulated experience, the interaction between the utility and consumer sectors, and market behavior.

SPURR consists of a computer simulation model, a technoeconomic database, and scenario definitions. These are combined as shown in figure 4.3 to produce energy projections. The scenario definitions consist of inputs from national energy and economic scenarios, regional climate and energy data, and federal and state demonstrations and incentives. SPURR was to act as a specialized tool for solar analysis and as an adjunct to broader and necessarily more aggregate national studies. As such, it used baseline energy and economic studies to provide year-by-year estimates of energy demand by primary fuel and use, average fuel prices for residential, commercial, industrial, and electric utilities, and real gross national product and inflation rates.

The early SPURR studies took scenarios from BNL's models; the SPURR projections for the National Plan for Accelerated Commercialization (NPAC) of Solar Energy used baseline scenarios for the Domestic Policy Review (DPR) of solar energy. The national estimates were then broken down into estimates from each census region. When these estimates were combined with regional data on solar insolation, wind speeds, and heating and cooling loads, it was possible to develop regional parameters for determining the cost and performance of the solar and conventional alternatives.

Finally, the federal and state solar demonstrations and tax incentives were defined. These programs affected several parts of the model. Tax incentives directly affected the purchase price and life-cycle cost, both of which have a major impact on the market share for a new technology. Demonstration programs resulted in a more rapid user acceptance of the technology, as users were able to see the product in real applications and at a lower cost because the early production was paid for by the demonstration program. Information programs were also assumed to speed acceptance through reducing the buyer's fear of new technologies.

Although the simulation results from SPURR attracted the most attention, the work to develop systems descriptions and engineering costs for the solar and related technologies was possibly the most valuable part of the entire SPURR activity. In cooperation with the federal program managers and their contractors, an extensive technical database was amassed. It presented in a consistent manner the design parameters intended to represent the commercial results of the various development programs. These designs were documented in nine volumes and several thousand pages of material (MITRE Corporation, 1977). Excerpts from one of the fact sheets for solar thermal are shown in table 4.14 to illustrate the type of data included.

A basic assumption about many of the solar technologies was that costs could be significantly reduced by a combination of research and mass production. This assumption was modeled in SPURR by using "experience curves" for major components, such as flat-plate solar collectors, photovoltaic cells, heliostats, and rotors for wind machines. T. P. Wright pioneered the use of learning curves to reflect a manufacturer's ability to reduce the unit cost of an item as total production volume accumulated. The original work on learning curves applied to the labor content of a product. This concept has since been generalized to reflect

Table 4.14
Generic design for solar thermal central receiver fuel saver

First year of operation	1990
Demand type	Fuel saver
Nameplate rating	100 MW$_{th}$
Capacity factor	0.25 to 0.362
Capital cost not subject to experience	274 $/kW ($+40\%$ -20%)
Capital cost subject to experience	585 $/kW

Components	Number built	Projected cost if mass produced ($/kW)
Heliostats	15,000	390
Receiver	2	41
Tower	2	91
Control	600	63
O&M costs		$12/kW/yr
Conventional fuel type		None
Construction time		2 years
Lifetime		30 years
Land use		0.00001 mi^2/kW
Water use		None
Labor		0.010 person-years/kW
Location for generic design		Census regions 7, 8, 9
Factor determining regional variation		Avail. direct sunlight

Source: MITRE Corporation (1977), vol. 5.

the effects of all production factors into an experience curve, which describes the cost reduction achieved for each doubling of accumulated production volume. MITRE reviewed the cost trends for approximately seventy cases as a basis for comparing likely experience curves for solar design components. Figure 4.4 shows a representative case.

The convention for experience curves is to plot cumulative production against unit cost on logarithmic paper to produce a straight line. The experience factor is then described as the slope of the line. For example, the heliostats in the previous generic design were assumed to have a learning curve slope of 80% over their initial production; that is, the cost of heliostats would be reduced by 20% as the *total cumulative production* doubles from 2,000 units to 4,000 units. Another 4,000 units would be required to obtain the next 20% cost decrease, 8,000 more units for the next 20% decrease, and so on. Experience curves do not reflect any specific cause and effect relations but rather are aggregate statistical models of the produc-

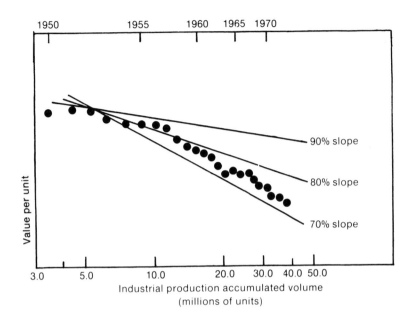

Figure 4.4
Experience curve for nonelectric, forced warm air furnaces (MITRE Corporation, 1977, appendix to vol. 1).

tion process. Their correct use requires careful use and judgment because the cost reduction is extremely nonlinear and not additive. Selecting the starting volume is critical, as the early doublings happen quickly, with an eventual leveling off as the production level grows. Second, splitting a product into two equivalent products, for example, collectors for hot water and for space heating, slows the total effect as compared to producing them together. Last, artificially defining a new technology so that it begins at the bottom of the experience curve with no production allows for several rapid doublings in production, which can understate the cost of the product.

For these reasons SPURR used experience curves for the major components of the generic designs, for example, heliostats. Thus, using one application, such as a solar thermal "fuel saver," could reduce the heliostat costs used for estimating the cost of a solar thermal plant with storage. Because much of a solar design consists of existing items, such as concrete footings, valves, and piping, the portion of the design not subject to experience was explicitly calculated from the original design studies. In the previous example $274 of the mass-produced cost of $859/kW was not assumed to be subject to cost reduction. Regional experience curves were used for such technologies as flat-plate collectors to reflect the expected regionality of the industry. For such components as a solar thermal central receiver, a single experience curve used the combined production of all of the regions.

The market segmentation in SPURR was selected to mirror the introduction into the traditional energy applications as they appear today. For each region the market was segmented into the following sectors or applications:

Residential: hot water, heating, heating and cooling, on-site electricity

Commercial: hot water, heating, heating and cooling, on-site electricity

Industrial: low temperature [to 100°C (212°F)], medium temperature [100–300°C (212°–572°F)], high temperature [over 300°C (572°F)], on-site electricity

Utility: base, intermediate, semipeaking, peaking, fuel savers

Synthetic fuels: methanol, synthetic crude oil, ammonia, synthetic natural gas

For each of these market sectors the market was further segmented by fuel type and new and retrofit demand as appropriate. Thus an example of an individual market segment calculation might be the comparison in 1985 of retrofitting an electric hot water heater in a single-family home located in census region 5 (Alabama, Mississippi, Tennessee, and Kentucky) with one of several solar hot water heater designs. The residential and commercial heating and cooling analysis consisted of over 2,000 individual market sectors for a single year.

Different market decision algorithms were used for the distributed and centralized components of SPURR to compute the market shares allocated to competing technologies within each market segment. For the distributed applications (residential, commercial, and industrial) an S-shaped market share function as shown in figure 4.5 was used. The market share depends on a "figure of merit" (FOM) of the solar versus the conventional system and the years since the technology first became generally available.

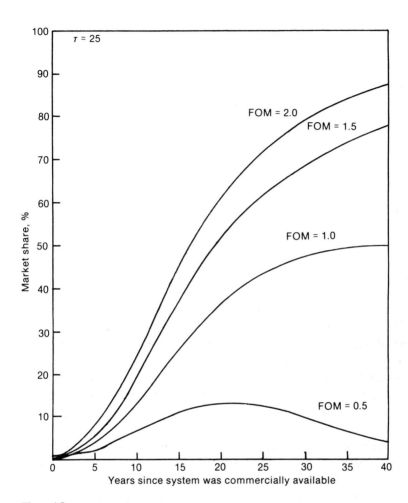

Figure 4.5
SPURR market share function (Bennington et al., 1977).

This market share function has the following characteristics:

Explicit specification of the value of the new technology relative to its conventional competitor.
Technological substitution proceeding exponentially in the early years, finally following an S-shaped curve, similar to that characterized by Fisher and Pry.
The incorporation of a "boom or bust" phenomenon in which the new technology is allowed to enter the market even though its relative value is less than its conventional competitor. However, if the relative value does not increase over time, the technology will taper off and cease to compete. (Bennington et al., 1977)

The FOM used for the residential market was

$$\text{FOM} = C1[(X1/X2) + C2](X1 - C3)X2,$$

where $X1$ is the fuel saving in the first year as a result of solar and $X2$ is the initial solar cost minus the initial conventional cost. The constants $C1$, $C2$, and $C3$ were fit to available market data and attitudinal studies. The FOM for the other two sectors used a ratio of the life-cycle costs of the conventional technology to the solar technology.

The central components of spurr used a market penetration algorithm design to satisfy the given demand according to the probability of least cost. A stochastic allocation was used to handle the wide amount of uncertainty and risk associated with the larger-scale technologies and the uncertainties in the fuel cost. This procedure also had the added advantage of being able to distribute the market share among several alternatives.

For each technology three energy cost estimates were made: expected cost, maximum cost, and minimum cost. These three estimates are used to define a triangular probability distribution for each competing technology, as shown in figure 4.6. A random sampling was then performed, and the three systems with the lowest energy costs were considered for construction. The market is allocated in proportion to the sampled probabilities of the cheapest technology. In figure 4.6 the potential range of energy costs for a coal plant was compared to two wind energy conversion systems (WECSs) and a solar thermal electric plant. If the random sample indicated a 40% chance that the WECS-1 design would be the cheapest for this case, it would be allocated 40% of the new capacity availability for that period.

S-shaped maximum market share constraints were also input to keep a new technology from being accepted too rapidly. If some of the allocated

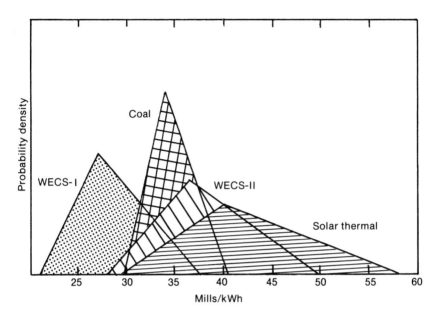

Figure 4.6
Probability distribution for utility market allocation (Bennington et al., 1977).

construction could not be built because of the constraints, the market allocation process was repeated with the residual demand and the technologies that remained unconstrained. Once the market allocations had been completed for the year, the projections for the year were accumulated for each region and the nation for installed systems, energy production by solar, energy utilization, federal cost for incentives, total capital cost for solar installations, life-cycle costs by technology, and labor and materials requirements for solar. The experience curves were updated to reflect the new production levels and the simulation moved to the next period.

SRI Solar Model. SRI developed the SRI solar penetration model in 1976 to assist the DOE Solar Working Group in assessing the priorities for the solar energy program (SRI, 1977). In contrast to SPURR, the SRI model was designed to provide a low-cost analysis tool for studying the implications of the future use of seven major solar energy technologies. The SRI model uses national economic and energy demands derived from a proprietary SRI National Energy Model. Three scenarios were consid-

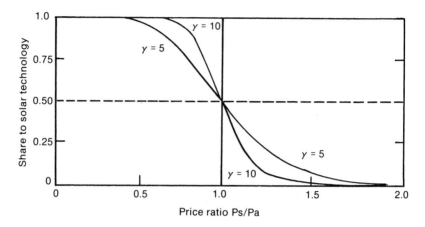

Figure 4.7
Equilibrium market share from SRI solar market penetration model (Schiffel et al., 1978).

ered: a reference case; a solar emphasis case, which assumes solar costs necessary to the 45-quad goal set in ERDA 49 (ERDA, 1975a); and a low-demand case to examine the consequences of minimizing energy demand.

The SRI model divides the country into nine regions and examines seven energy markets—water heating, rural space heat, urban space heat, industrial and agricultural heating, biomass, intermediate electricity, and base electricity. The fourteen solar processes and fuel combinations examined were solar plus gas water heating, solar plus electricity water heating, solar plus fuel oil water heating, solar plus gas space heat, solar plus electric space heat, solar plus fuel oil space heat, solar industrial heating, biomass gas, biomass methanol, biomass electricity, photovoltaic, solar thermal electricity, near-shore OTEC, and tropical OTEC.

By using conventional fuel prices and the technical data on the solar systems, a cost-per-million-Btu can be calculated for each system. An "equilibrium market share" is calculated based on the ratio of the solar and conventional energy prices. The market share function shown in figure 4.7 is an S-shaped curve with a "gamma parameter," which accounts for market conditions such as technical suitability and price variations. The gamma parameter is subjectively assigned for each technology. The market share function is symmetric with regard to a ratio of 1 and has a market share of 0.5 for equal energy prices. This curve represents the long-term market share that would arise based on energy price considerations.

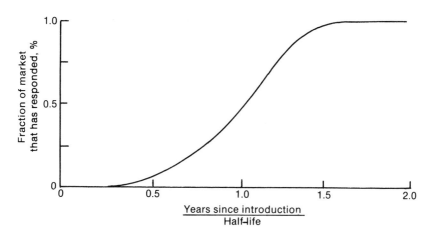

Figure 4.8
Dynamic market response from SRI solar market penetration model (Schiffel et al., 1978).

After obtaining the equilibrium market share for a technology in an application, another S-shaped function, shown in figure 4.8, is used to account for the time required for a new technology to receive widespread acceptance. The behavior lag function depends on the years the technology has been in use and the half-life of the market, that is, the time taken for the curve to reach a value of 0.5. The half-life for each market is assigned subjectively (Schiffel et al., 1978). The product of the equilibrium market share and the behavior lag function determine the solar market share for that application and year. This process is repeated for each application and technology for each region to generate the year's total. The method of determining solar costs is not documented but is assumed to be an input parameter based on the scenario. The model uses five-year time steps through 2020.

SRI's final report, *Solar Energy in America's Future* (SRI, 1977), makes a well-balanced assessment of how to use the market projections of a model such as the SRI model (and most others).

In this study energy projections are not used either for prediction or for planning; instead, they are used to explore a number of alternative futures.... The energy projections of this study represent different energy futures based on alternative assumptions of energy demand levels and energy supply economics. Thus, one or more of these scenarios is likely to appear wildly improbable to different observers. By presenting the results in a parametric fashion, however, we enable each reader to enter the report with his or her own estimates of future energy costs and future energy demand levels and emerge with insights into the implications of his or her own assumptions and the assumptions of others.

4.3.3   Market Segment and Technology Models

I discuss briefly the models developed by Arthur D. Little, Inc., and Energy and Environmental Analysis, Inc., to illustrate alternative approaches to estimating the eventual use of new energy technologies.

Arthur D. Little, Inc. (ADL), contributed the solar heating and cooling estimates for the first Project Independence study and conducted several widely subscribed multiclient studies on the likely use of solar heating and cooling. The ADL model used for the Federal Energy Administration uses ten regions and includes ten market segments. The projections covered between 1977 and 1990. The FOM used is the payback period, that is, the initial capital cost divided by the net annual cost savings. As shown in figure 4.9, the ADL model contains a family of penetration curves starting with an "initial curve" and eventually moving to a "final curve" as the technology matures. For example, a payback of two years would be required to achieve an "initial" market share of 50%, whereas

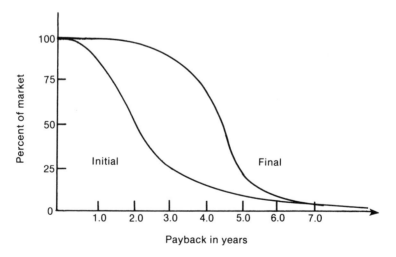

Figure 4.9
Penetration curves used in Arthur D. Little, Inc., market penetration model (Arthur D. Little, Inc., 1977).

this market share could be achieved with a payback of four to five years after the technology matures. The calibration of the parameters defining the family of penetration curves is not given.

Energy and Environmental Analysis, Inc. (EEA), developed a multi-technology model designed to look at all technologies in the industrial sector. The model was characterized by a high level of market segmentation (over 2,000 segments) to ensure that the energy-use patterns and technology matches were correct. No regional data were used. The model develops a price distribution for each technology and market sector based on their complexity, the specifics of their application, and other factors. These distributions were typically based on subjective estimates. The market penetration was performed in three steps. First, "nominal market shares" were determined as the proportion of the uses in a market segment for which the technology would be the cheapest alternative. Second, the nominal market shares were determined by a random sample from the price distributions. An S-shaped curve was used as a multiplier of the nominal market share to approximate the lags in accepting new technologies. Third, the residual unused market share, resulting from the use of the S-shaped curve, was assigned to the incumbent technology, which in this case was assumed to be coal technologies.

4.4   Summary and Conclusions

I have reviewed solar market assessments from twelve major federal studies. The effort expended in attempting to understand the likely impact of solar energy was impressive. Although several of the early studies were performed by a small group in a few months, the later studies were conducted over several years with over 100 person-years and at a cost of several million dollars. Market potential studies have their value, but a major failing of the assessments is in the use of these assessments. There seems to be an irresistible temptation to use studies as predictions of the future. Changing the predictions to "goals" or "technical limits" did little to change the public's or the media's interpretation of these studies. This is not surprising because we all wanted to know how the results of the solar program would turn out.

The effort spent on these assessments accomplished many things. First, the managers of the federal R&D programs were compelled to realize that it would be the participation in the marketplace by the manufacturers, banks, and users that would eventually change the nation's energy usage patterns. Second, basic input data gathered for the assessments provided a uniform framework for comparing the cost/performance profiles for a large number of alternatives. For example, technology descriptions developed for ERDA to support SPURR consisted of a nine-volume summary of solar technologies based on all the available designs that had been thoroughly reviewed by program managers and their contractors. For each of the several hundred basic descriptions, a consistent basis was developed for assessing the likely impact on materials and labor, land use, job creation, and environmental impacts. An equally valuable database was developed by Oak Ridge National Laboratories for assessing the impacts of conservation and solar in buildings.

All the models suffered greatly from lack of market data from which to calibrate the market share calculations. Many investigators tried to calibrate early attitudinal surveys and the early solar hot water sales data. Unfortunately, small changes in assuming market behavior could completely change the outcome of most of the models. If models of this type are to be used again, they should be recalibrated using up-to-date market data and should be extended to include the noneconomic factors that have an impact on solar adoption. In particular, the SERI project was the first effort that tried to predict market penetration based on behavioral theory (Farhar-Pilgrim and Unseld, 1982). Its use in federal policy and solar market assessments was minimal because of the timing of the work. Commercialization of solar was no longer emphasized when this work was published. The four-tiered model of homeowners developed in this study could provide an excellent basis for improved segmentation for modeling as well as constructive insights for direct action.

The use of market potential assessments as predictions was futile. Nonetheless, the use of rigorous accounting structures for energy flows, system counts by market segment, and investment provided the opportunity to assess the implications of achieving different goals. For example, I included the following assessment of what it would mean to achieve a 20% goal for solar as part of a congressional hearing on solar energy futures:

Twenty-two quads of our total energy demand of 115 quads would mean a solar contribution of 20 percent of the nation's energy needs. This would require a capital investment of solar equipment of $1 trillion by the turn of the century. The capital required would be equal to 20 to 30 percent of the nation's investment over the next 20 years.... One out of two existing buildings in the year 2000 would be using some type of solar energy.... Solar sales would amount to over $50 billion per year, a greater than 800-fold increase of current sales. (Bennington, 1979)

These implications were good news or bad, depending on each person's value system. Using solar impacts to understand the magnitude of the effort involved, the benefits, or even whether some goals make sense appears to be the most productive use of assessment tools such as those discussed in this chapter.

References

Arthur D. Little, Inc. 1977. *Solar Heating and Cooling of Buildings (SHACOB) Commercialization Report*, Part B, Vols. I–III. Cambridge, MA: Arthur D. Little, Inc.

Bennington, G. 1979. *Solar Energy Futures*. Testimony to the House Subcommittee on Energy Development and Applications, Richard L. Ottinger, Chairman, and House Subcommittee on Energy and Power, John D. Dingell, Chairman. Washington, DC.

Bennington, G., K. Rebibo, and R. Vitray. 1976. *Preliminary Projections for Centralized Usage of Solar Energy Systems in 1980, 1985, and 1990.* MTR-7174. McLean, VA: MITRE Corporation.

Bennington, G., R. Bezdek, M. Bohannon, and P. Spewak. 1976. An *Economic Analysis of Solar Hot Water and Space Heating*. DSE-2322-1. Washington, DC: Division of Solar Energy, ERDA.

Bennington, G., P. Curto, G. Miller, K. Rebibo, and P. Spewak. 1978. *Solar Energy: A Comparative Analysis to the Year 2020*. MTR-7579. McLean, VA: MITRE Corporation.

Bennington, G., K. Rebibo, P. Curto, P. Spewak, and R. Vitray. 1977. *A System for Projecting the Utilization of Renewable Resources: SPURR Methodology*. MTR-7570. McLean, VA: MITRE Corporation.

Bennington, G., M. Bohannon, R. Gerstein, R. Hartzler, N. Kannon, G. Miller, K. Rebibo, M. Shulman, P. Spewak, and J. Taul. 1980. *Toward a National Plan for the Accelerated Commercialization of Solar Energy: The Implications of a National Commitment*. MTR-79W00004. McLean, VA: MITRE Corporation.

Committee on Nuclear and Alternative Energy Systems (CONAES). 1979. *Energy in Transition 1985–2010*. Washington, DC: National Research Council and National Academy of Sciences.

Council on Environmental Quality (CEQ). 1978. *Solar Energy: Progress and Promise*. GPO Stock 041-011-00036-0. Washington, DC: Council on Environmental Quality.

ERDA. 1975a. *Definition Report: National Solar Energy Research, Development and Demonstration Program*. ERDA-49. Washington, DC: Energy Research and Development Administration.

ERDA. 1975b. *National Program for Solar Heating and Cooling (Residential and Commercial Applications)*. ERDA-23A. Washington, DC: Energy Research and Development Administration.

ERDA. 1976. *A National Plan for Energy Research, Development and Demonstration: Creating Energy Choices For The Future.* ERDA 76-1. Washington, DC: Energy Research and Development Administration.

ERDA. 1977. *A National Plan for Energy Research, Development and Demonstration.* ERDA 77-1. Washington, DC: Energy Research and Development Administration.

Executive Office of the President. 1977. *The National Energy Plan.* NP-21948. Washington, DC: Government Printing Office.

Farhar-Pilgrim, B., and C. T. Unseld. 1982. *America's Solar Potential: A National Consumer Study.* New York: Praeger.

FEA. 1974. *Project Independence Blueprint, Task Force Report; Solar Energy, Final Report.* FEA/N-75/548. Washington, DC: Federal Energy Administration.

FEA. 1976. *National Energy Outlook.* FEA-N-75/713. Washington, DC: Federal Energy Administration.

Groncki, P. J., R. J. Goettle, IV, and E. A. Hudson. 1980. *An Economic Assessment of Alternative Energy Policies.* BNL-27721. Upton, NY: Brookhaven National Laboratory.

Klein, S., W. Beckman, and J. Duffie. 1975. "A design for solar heating systems." Solar Energy 18:113–127.

Lovins, A. 1977. *Soft Energy Paths: Towards a Durable Peace.* Cambridge, MA: Ballinger.

MITRE Corporation. 1973. *Systems Analysis of Solar Energy Programs.* NSF/RA/N-73-III. McLean, VA: MITRE Corporation.

MITRE Corporation. 1977. *Systems Descriptions and Engineering Costs of Solar-Related Technologies.* 9 vols. McLean, VA: MITRE Corporation.

Office of Technology Assessment (OTA). June 1978. *Application of Solar Technology to Todays Energy Needs.* GPO Stock 052-003-00539-5. Washington, DC: Government Printing Office.

Schiffel, D., D. Costello, D. Posner, and R. Witholder. 1978. *The Market Penetration of Solar Energy: A Model Review Workshop Summary.* SERI-16. Golden, CO: Solar Energy Research Institute.

SERI. 1981. *Report on Building a Sustainable Future.* Print 97-K & L. Committee on Energy and Commerce. U.S. House of Representatives, Washington, DC.

SRI. 1977. *Solar Energy in America's Future.* SRI Project URU-4996, Division of Solar Energy. Washington, DC: Energy Research and Development Administration.

Stobaugh, R., and D. Yergin. 1980. *Energy Future: The Report of the Harvard Business School Energy Project.* New York: Ballantine.

U.S. Department of Energy (DOE). 1978. *Domestic Policy Review of Solar Energy.* TID-28835/1. Washington, DC: Department of Energy.

U.S. Department of Energy (DOE). 1979. *National Energy Plan II.* TIC-10203. Washington, DC: Department of Energy.

5 Analyzing the Effect of Economic Policy on Solar Markets

Peter C. Spewak

In response to the oil embargo of 1973, the federal government embarked on a monumental effort to develop new energy resources quickly. Solar energy became a prime candidate as an alternative energy resource. The government felt that solar could provide a significant portion of U.S. energy requirements by the turn of the century, given the successes in other technological areas. These successes seemed to indicate that the considerable technological issues related to solar energy could be solved in this time frame.

However, the federal government realized that the economic and institutional issues would be as important if not more important than the technological issues. To deal with these technical, economic, and institutional issues expeditiously, a significant federal program evolved. Authorized by public law, a series of initiatives were enacted to develop and promote the use of solar energy. In some cases these initiatives were modeled after activities that federal policy had enacted in other energy and nonenergy programs. In other instances new government roles were defined and enacted.

Here I review the incentives proposed and/or enacted for solar energy from a historical perspective. In section 5.1 I discuss the rationale for solar incentives and subsidies, concentrating heavily on the Cone studies, which examine the role of the federal government in energy resource development. I also reference the National Plan for Accelerated Commercialization of Solar Energy (NPAC) material and focus on the potential benefits that can be derived to offset the cost of incentives.

In section 5.2 I discuss the design considerations of specific solar incentives. First, I discuss the forms of incentives in light of what barriers they were designed to remove. Next I turn to the scheduling of incentives, the determination of appropriate level of support, and regional variations of incentives. This information is based on NPAC and subsequent survey reports of incentives (Spewak and Bohannon, 1981).

In section 5.3 I discuss the type, level, and impact of solar incentives enacted. Current solar sales data are extracted from the Energy Information Administration (EIA) solar collector survey (EIA, 1984).

5.1   The Rationale for Federal Incentives and Subsidies

Since the 1920s the federal government has played an active role in the development of energy resources in the United States. Through 1977 over $346 billion (in 1984 dollars) was committed by the federal government to energy production in expenditures and foregone tax revenue (Cone, Brenchley, and Brix, 1980). Historically this sizable investment has been justified for two reasons: (1) financial incentives help underwrite the risk associated with investment in new technologies; (2) financial incentives reflect the value of the resources to society or the nation (national value) when the value perceived by the private sector (or the marketplace) is less than the public national or societal value.

These reasons are not unique to energy technologies. Even today, in the midst of increasing pressures to reduce federal spending, there is a strong emphasis on government support of high technology. Ever since the 1950s the federal government, through the National Aeronautics and Space Administration (NASA), has supported the high-risk high-payoff technology associated with space exploration. Only in the last two years has there been any serious private sector capital investment in space activities. Federally supported research in medicine and health supports national value and allows for health care and treatment that would probably be prohibitively costly if research costs were fully recovered. Loan guarantees from the Department of Housing and Urban Development (HUD) provide national value through financing for adequate housing. Federal- and state-funded agricultural research has been a major factor in the high productivity of our nation's croplands. This research and the related dissemination of information, along with other direct market incentives, help to ensure the stability of U.S. food sources. These are only a few of the many nonenergy related examples.

There are many examples of incentives for energy development and production. Foreign tax credits and oil depletion allowances represent sizable incentives for oil production. The production, sale, and transmission of electric power by federally sponsored regional power authorities represent a significant energy impact. Energy research, development, and demonstration have resulted in commercializing energy technologies, such as nuclear, which would probably have been too risky for development by industry alone. Nuclear continues to receive other substantial incentives, both direct (subsidization of nuclear refinement and enrichment) and indirect (limitations on culpability for nuclear accidents).

National value and parity are two issues that need to be considered in discussing the appropriateness of incentives for solar and conservation. Related to national value are the externalities or, more specifically, the internalizing of the external socioeconomic and political benefits of solar. Related to parity is economic efficiency. Parity is not solely an issue of equity. Economic efficiency might be achieved more successfully by removing an existing incentive that provides a comparative advantage to a conventional energy source over solar rather than by providing solar with an incentive to put it on a par with the conventional energy source.

5.1.1   Total National Value

The potential national value of solar is large. National value benefits fall into several categories: energy savings, industrial development, environmental and health benefits, other indirect sociopolitical benefits, and national defense. The magnitude of these benefits, of course, varies considerably, given the level of conservation and solar use, because the level of use varies as a function of the magnitude of incentives.

Energy Savings Benefits. Estimates are that solar can account for as much as 17.1 quad/yr (10^{24} Btu $= 1.055 \times 10^6$ TJ), roughly 15% of the nation's energy use, as shown in table 5.1 (Spewak and Bohannon, 1981). However, the estimates in table 5.1 depict conventional price-demand scenarios that were felt to be realistic in 1980. These scenarios also represented assumptions of relatively stable economic growth through the 1980s. In hindsight, most of these assumptions were inaccurate, and the resulting projected impacts are considerably higher than current experience demonstrates. Even the most conservative estimate is high. To be "on track" with the 7.2-quad scenario, 1985 annual sales of solar collectors should have been approximately 10^8 ft^2 (10^6 m^2) (Parikh et al., 1980). Based on the results of the EIA collector survey for 1983 (EIA, 1984), actual sales were probably in the range of 15–20×10^6 ft^2 (1–2×10^6 m^2). Subsequent analyses have not been carried out to revise solar use projections based on current scenarios and our expanded understanding of the solar marketplace.

Industrial Development Benefits. A viable U.S. solar industry would provide substantial national benefits. One benefit occurs in the marketplace.

Table 5.1
Solar potential, quads/yr, year 2000 (quads/yr = 1.055 EJ/yr)

Potential	Solar objective document[a]	DPR[b]	NPAC[c]	Solar strategy document[d]
Industrial				
Solar thermal	2.2	2.4	2.2	—
Biomass	3.3	2.2	3.5	—
Wind, PV, hydro	0.5	0.9	0.5	—
Total	6	5.5	6.2	3
Utility				
Solar thermal	0.4	0.4	1.6	—
Ocean thermal	0.1	0.1	0.2	—
Wind	1.2	1.5	2.3	—
Biomass	0.1	0.1	0.2	—
Photovoltaics	0.2	0.3	0.4	—
Hydro	2	2.4	4.7	—
Total	4	4.8	9.4	2
Buildings				
Active	2.4	2	3.1	—
Passive		1	0.4	—
Wind	0.1	0.2	0.6	—
PV	0.4	0.6	0.4	—
Biomass	0.6	0.4	0.8	—
Total	3.5	4.2	5.3	2
Transportation	3	2.2	0.9	0.2
Total Solar	14.5	14.3	17.1	7.2

Source: Spewak (1981).
a. Solar Objectives Document, DOE Division of Solar Energy (1980).
b. Domestic Policy Review (DPR) of Solar Energy (1979).
c. National Plan for the Accelerated Commercialization (NPAC) of Solar Energy (MITRE, 1979).
d. Solar Strategy Document, DOE Office of Conservation and Solar Policy (1980).

Today the solar industry exists primarily in hydroelectric (hardware manufacturers and power producers), industrial and residential biomass, flat-plate collectors, and photovoltaics. Sales of hydroelectric components account for approximately $1 billion per year (about 1.3 billion in 1985 dollars). Because the biomass industry is fragmented, its size is hard to estimate; however, in 1980 approximately 100,000 wood stoves (a component of the biomass industry) were sold, accounting for about $50 million in sales (about 65 million in 1985 dollars). The flat-plate collector industry, including pool heaters, accounted for approximately $784 million (845 million in 1985 dollars) in annual sales (EIA, 1984). Photovoltaics, still used primarily for small-scale remote applications, accounts for about $70 million (75 million in 1985 dollars) in annual sales (EIA, 1984).

Another benefit of the solar industry is its contribution to the job market. If 10 quads of energy are supplied by solar sources in the year 2000, approximately one million new jobs will be created. These jobs include those directly involved in solar manufacture, installation, maintenance, and operation, as well as those indirectly involved in solar manufacture (for example, steel workers). Approximately 330,000 jobs would be directly involved in solar (Bennington et al., 1979).

Environmental and Health Benefits. The environmental and health benefits from solar are similar to the financial benefits to the solar consumer in that initial investments or "front-end" costs are paid back over time with the benefits derived. This analogy also holds in discussing total national value. The front-end costs associated with early developmental incentives are counterbalanced against later benefits.

The manufacture of a solar hot water heater, for example, produces several indirect residuals. The indirect construction residuals associated with the same capacity of an alternative energy form—for example, a coal-fired electric power plant—are noticeably less than those associated with the solar hot water heater. So in the first year, through the manufacture of the materials and components of the solar hot water heater, there is actually an increase in pollution. Specifically the manufacturing process results in particulates [2 lb (1 kg)] and sulfur dioxide [5 lb (2 kg)]. Other pollutants are decreased slightly. In following years, however, the solar hot water heater results in no further residuals in providing hot water, whereas the coal-fired power plant does. Thus after three years the solar hot water heater has caused no net increase in emissions and continues for the rest of the system life benefiting the environment (Bennington et al., 1979).

Under the 10-quad scenario discussed previously, solar provides substantial environmental benefits by the year 2000. Although in the accelerated case solar may be a net producer of pollution at the turn of the century because of its continuing high rate of growth, overall reductions in pollution through time may result in $10–$15 billion in health benefits per additional quad of solar produced (Bennington et al., 1979). These estimates do not take into account the additional benefits derived from reducing the threats associated with carbon dioxide buildup, nuclear waste, thermal pollution, and sludge from coal scrubbers.

A 1,000-MW commercial nuclear power plant produces about 30 metric tons of waste containing approximately 10 kg (4.5 lb) of transuranium elements when discharged. Each additional quad of solar over the basic 10-quad scenario can result in one less 1,000-MW nuclear power plant (based on a reasonable set of assumptions concerning national fuel mix) (Bennington et al., 1979).

Other Indirect Sociopolitical Benefits. The other benefits of solar and conservation that may be used to demonstrate total national value and thus justify the use of incentives are hard to quantify but nevertheless real. United States dependence on imported fossil fuels results in a continual threat of international extortion by means of a fuel embargo. The consequences of this threat are both economic and political. Oil embargoes cause the cost of energy to rise and can cause serious setbacks to gross national product (GNP) growth. After the 1973 embargo the cumulative loss to GNP during the period 1974 through 1976 was $822 billion (in 1985 dollars) (Bennington et al., 1979).

National Defense Benefits. The threat of an oil embargo greatly increased awareness in the United States of the need for energy independence. Developing such domestic energy sources as solar provides greater flexibility in structuring foreign policy and ensuring national security. A decrease of fossil fuel use by the United States could help lessen world pressures on energy resources. Additionally, U.S. development and export of solar and conservation technology would further decrease competition for foreign oil. Solar may be particularly appropriate in developing countries where large centralized energy systems have not yet been established. Without alternative energy resource development, increased energy demand by developing countries could result in further international instability.

5.1.2 Parity

Because other energy technologies have received sizable financial incentives, the issue of parity is central to the argument for providing incentives for solar technology. However, it is not enough to argue for solar incentives purely on the basis of precedent. Economic policies and sociopolitical environments existing at the time historical incentives were instituted distort such an argument. It is appropriate to analyze the continuing impacts of incentives received by conventional energy sources vis-à-vis energy cost. The argument is that, if current energy costs are artificially lowered relative to solar, then incentives for solar (or the removal of incentives for conventional energy sources) may be appropriate if the nation benefits from the resulting parity in energy costs.

In 1980 Battelle Memorial Institute (Cone, Brenchley, and Brix, 1980) studied federal incentives for conventional energy sources. This study followed a historical analysis of subsidies, which analyzed energy subsidies through time versus the energy produced (Cone, Brenchley, and Brix, 1978b). The study found that tax incentives increased oil and gas production by 23 quads and lowered prices from 2.2¢ to 57.5¢/10^6 Btu (per 1.055 GJ) (2.9 to 75¢/10^6 Btu in 1985 dollars). The incentives, which were described as "market activities" (defined in following sections), increased electricity production from hydro and fossil fuels by about 18 quads and lowered the price of electricity from 49.1¢ to $3.23/$10^6$ Btu (per 1.055 GJ) (64¢ to $4.21/$10^6$ Btu in 1985 dollars). Regulatory measures were estimated to increase oil and gas production by about 9 quads and increase prices by about 18.5¢/10^6 Btu (per 1.055 GJ) (24.1¢/10^6 Btu in 1985 dollars). These findings are summarized by energy type in table 5.2. All energy is expressed in thermal units. Incentives applied to utilities are prorated to the primary energy type used to generate the electricity.

The subsidies applicable to each energy source are somewhat different if taken on an annual "snapshot" basis. As seen in table 5.3, the incentives connected with coal, oil, gas, and nuclear in 1977 were substantially different from the historical averages on a dollar per 10^6 Btu basis. These values represent the current level of subsidization, that is, the level at which incentives were decreasing the price of these energy sources.

Table 5.2
Energy production and incentives by energy type (1933–1977)

Energy type	Period	Cumulative incentives[a] ($ × 10^9)	Cumulative production (quads)	Cumulative incentives ($/10^6 Btu = $/1.055 GJ)
Hydro	1933–1977	44	73	0.60
Coal	1950–1977	56	378	0.15
Oil	1950–1977	164	454	0.36
Gas	1950–1977	42	458	0.09
Nuclear	1950–1977	32	10	3.20
Total		338	1,373	
Average incentive				0.25

Source: Cone, Brenchley, and Brix (1978b).
a. GNP deflator applied; values are in 1985 dollars (Department of Commerce, 1985).

Table 5.3
Energy production and incentives by energy type (1977)

Energy type	Annual incentives[a] (1974 $ × 10^9)	Annual production (quads)	Incentives[a] ($/10^6 Btu = $/1.055 GJ/yr)
Hydro	1.5	2.4	0.63
Coal	6.1	16.7	0.37
Oil	21.3	17.3	1.23
Gas	−0.07[b]	20.60	−0.003
Nuclear	3.4	2.7	1.26
Total	32.23	59.7	
Average incentive			0.54

Source: Cone, Brenchley, and Brix (1978b).
a. GNP deflator applied; values are in 1985 dollars (Department of Commerce, 1985).
b. For the period 1955–1977 a net negative incentive results from an increase in intrastate gas prices over regulated interstate gas prices.

5.1.3   The Need for Incentives

The state of energy technologies, the potential benefits of solar, and the need for support in the early years of technological and commercial development made solar a prime candidate for incentives. Many arguments were made and several analyses were carried out to support the value of solar technology development to the nation. Some of the more prominent studies include Bennington et al. (1976, 1979), A Domestic Policy Review of Solar Energy (1978), Lovins (1977), and Council on Environmental Quality (1978). A common theme of all these studies is the true value of solar to the nation in the long run, which may not be reflected in the relative current market price for the technology. Incentives are then required to cause the national value of solar to be reflected in its relative market price.

Just as with conventional energy sources, the need for and therefore the magnitude of incentives decrease as an energy source matures and the technology becomes economically viable. This philosophy is apparent in the structure of incentives originally adopted for solar: heavy initial incentives in RD&D, financial incentives, and institutional programs are phased out as the technology matures and becomes more cost-effective. Ideally producer and consumer confidence increases as the technologies mature and become more familiar, and changes are made in the federal, state, and local legislation, codes, and ordinances to remove institutional barriers.

In the following sections I discuss these issues in more detail. I describe the types of incentives (past, current, or proposed) applicable to conservation and solar in terms of the specific barriers they were meant to overcome. I set these incentives in the context of similar incentives provided for conventional energy sources.

5.2   Incentive Taxonomy

To carry out their historical analysis of incentives for conventional energy sources, Cone, Brenchley, and Brix (1978a,b) developed a taxonomy for analytical purposes. This taxonomy has been widely used in solar incentives analysis. It facilitates comparing proposed approaches to solar incentives with approaches that have been followed by the federal government in the past.

The incentives taxonomy includes seven groups. The category of "fees" was not included because it is used so little. A short summary of each of the categories follows. The incentive categories are creation or prohibition of organizations, taxation, disbursements, requirements, traditional government services, nontraditional government services, and market activity.

The creation or prohibition of organizations includes government action in the creation or prohibition of organizations that perform or carry out some of the remaining types of actions. The government can create or prohibit organizations in federal, other governmental (for example, state), and nongovernmental (for example, cartels) areas. The creation of the Synthetic Fuels Corporation to provide subsidies for synthetic fuels is an example of this type of governmental action. The creation of the Solar Energy Research Institute (SERI) and the now defunct Regional Solar Energy Centers (RSECs) represent solar examples of this incentive type.

Taxation includes the levying of a tax or the exemption or reduction of one that is levied in other situations. These taxes may be levied within or outside the production-consumption cycle. The solar tax credits provided by the National Energy Act are an example of this type of incentive.

Disbursements include actions in which the federal government gives out money without receiving anything in return directly or immediately, such as grants in aid and subsidies. This category includes promises to disburse under certain circumstances as well as actual disbursements. For example, funds provided by the Solar Energy and Energy Conservation Bank to financial institutions to reduce the principal obligation or the interest rate on financing of solar or conservation expenditures are included in this category. Grants provided under the solar hot water program are also included.

Requirements include government demands that are backed by criminal or civil sanctions. The requirement imposed may be judicial, legislative, or administrative. The requirement imposed by the Powerplant and Industrial Fuel Use Act on major fuel-burning installations to reduce or eliminate their use of natural gas and petroleum is an example of this type of incentive.

Traditional government services include all the symbolic or tangible goods or services that are traditional to government and provided to nongovernmental organizations. This category encompasses assistance or benefits provided by the government to a nongovernmental entity without direct charge. The supervision by the Office of Management and Budget of government spending on solar energy is included in this category. Various educational programs, such as the Energy Extension Service, would also be included in this category, as would data collection and dissemination (for example, solar resource data).

The government also provides nontraditional services in addition to providing symbolic or tangible goods and services traditional to government. Funds provided by the government for solar RD&D technology are an example of this type of incentive in that the government has assumed industry's risk for the front-end development of technology, which will primarily benefit industry. RD&D for a technology to be consumed by the federal government (for example, space and military applications) would be considered traditional government services.

Market activity includes government involvement in a market under conditions similar to those faced by nongovernmental producers and consumers. Loan guarantees for ocean thermal energy conversion facilities provided under the Maritime Administration are included in this category. Federal purchase of solar technology for use by government would also fall in this category (for example, the Schools and Hospitals Program and the Solar Energy in Government Buildings Program).

5.3 Solar Incentives

Because of the diversity in the nature and uses of solar energy, the types of incentives cover a broad range. The major differences between technologies and in the financial and tax characteristics of prospective users indicate different incentive policies.

5.3.1 Technology and Application Categories

Potentially, over thirty-five solar technology and application categories exist; however, nineteen of these categories in three market sectors are sufficient for differentiation (Spewak and Bohannan, 1981):

1. Residential: passive, active hot water, active space, wind, photovoltaics, wood

2. Industrial/Commercial: process heat, wood, wind, photovoltaics, active, passive

3. Utilities: solar thermal, OTEC, wind, photovoltaics, wood and bio-mass, current fuels technology (fuels 1), new fuels technology (fuels 2)

Certain special cases are not explicitly treated by these categories. Institutional and governmental markets (for example, schools, hospitals, federal, state, and local government-owned buildings, Federal Power Administrations, municipal utilities, co-ops, and community-owned energy systems) are not specifically listed. In many instances these special cases represent potential federal subsidy mechanisms, in that market development may be stimulated through programmed purchases for these applications. These special cases may represent a significant end-use market in and of themselves. In such cases they may be addressed as a special case of one of the listed technology and application categories (for example, municipal utility and solar thermal applications). In addition, a special market may be integral to a federal support activity, in which case it is addressed as a federal incentive aimed at accelerating market development in one of the listed technology and application categories (for example, federal purchases under the Federal Buildings Program).

5.3.2   Identifying and Setting Priorities for Barriers

Incentives are necessary to remove the barriers that impede solar technology development and full commercialization. To understand the purpose of specific incentives, we need to identify and understand the relative importance of the barriers.

Tables 5.4 through 5.7 summarize the barriers that impede technology development and utilization in each of the four market sectors: residential, industrial, utility, and transportation. Four types of barrier are considered: technological, economic and financial, institutional, and buyer behavior. The barriers have been accorded relative weights on a simple high (H), medium (M), and low (L) scale to reflect the relative importance of each barrier as an impediment. The barriers were derived from data on the primary concerns of decision makers who affect technology utilization in each of the market sectors. Barrier types, in many cases, are interrelated. For example, by treating the economic and financial barrier of high initial cost, the buyer behavior barriers of "first-cost orientation" and "high inferred discount rate" may be positively affected. Similarly, if through federal actions the consumers' "first-cost orientation" could be modified, the high initial cost would be less critical.

Table 5.4
Barriers to solar technology implementation in the residential market sector

Barrier	Technology[a]					
	Passive	Active hot water	Active space conditioning	Wind	Photo-voltaics	Wood
Technological						
Component lifetime	—	H	H	H	M	—
Reliability	L	M	M	H	H	—
Application load match	L	L	L	M	M	—
Environmental	—	—	—	—	—	M
Seasonal/diurnal variability	L	L	L	L	L	—
Supply/resource constraints	—	—	—	—	—	M
Installation quality	L	M	M	L	L	L
Obsolescence	L	L	M	M	H	—
Economic/financial						
Property tax	M	L	M	M	M	—
Resale	M	M	M	M	M	—
Conventional energy subsidies	M	M	M	M	M	M
Average pricing policies	M	M	M	H	H	M
High initial cost	M	H	H	H	H	L
Lack of venture capital	—	M	M	M	M	—
Uncertainty concerning buy-back	—	—	—	H	H	—
Availability of financing	L	—	L	M	M	—
Institutional						
Underdeveloped distribution network	M	L	M	H	H	—
Fragmentation of building industry	M	L	M	L	M	—
Lack of skilled labor	M	M	H	M	H	—
Unresponsive building codes	H	L	M	H	H	—
Lack of standards	—	—	—	M	M	—
Insurance guidelines	L	L	L	M	M	M
Solar access	M	M	M	M	M	—

Table 5.4 (continued)

Barrier	Technology[a]					
	Passive	Active hot water	Active space conditioning	Wind	Photo-voltaics	Wood
Buyer behavior						
Inaccurate perceptions	M	M	M	M	M	L
Unfamiliarity with benefits	M	M	M	M	M	—
Lack of vendor information	H	M	H	H	H	—
First-cost orientation	M	H	H	H	H	—
High inferred discount rate	M	H	H	H	H	—
Uncertainty over conventional fuel cost	M	H	H	H	H	L
General unfamiliarity, newness	M	M	M	H	H	—
Lack of consumer protection	M	M	M	H	H	—
Aesthetics	L	L	L	M	L	—
Renter occupancy	H	M	H	H	H	M

Source: Spewak and Bohannon (1981).
a. Importance of barrier: L = low, M = medium, H = high, — = not applicable.

Table 5.5
Barriers to solar technology implementation in the industrial/commercial market sector

	Technology[a]					
Barrier	Process heat	Wood	Wind	Photovoltaics	Active	Passive
Technological						
Component lifetime	H	L	H	M	H	—
Reliability	M	L	H	H	M	L
Application load match	L	—	M	M	L	L
Environmental	—	M	—	—	—	—
Seasonal/diurnal variability	L	—	L	L	L	L
Supply/resource constraints	—	M	—	—	—	—
Installation quality	M	L	M	M	M	L
Obsolescence	M	L	M	M	M	L
Engineering interface/compatibility	M	—	L	L	M	L
Production/output capacity	M	L	M	M	M	L
Economic/financial						
Property tax	M	L	M	M	M	M
Conventional energy subsidies	M	M	M	M	M	M
Average pricing policies	M	M	H	H	M	M
High initial cost	H	L	H	H	M	L
Lack of venture capital	M	—	M	M	M	—
Uncertainty concerning buy-back	H	—	H	H	—	—
Availability of financing	L	—	L	L	L	L
Institutional						
Underdeveloped distribution network	M	L	M	H	M	L
Lack of skilled labor	H	—	M	H	H	M
Lack of standards	—	—	M	M	—	—
Insurance guidelines	L	L	M	M	L	L
Solar access	M	—	M	M	M	M

Table 5.5 (continued)

Barrier	Technology[a]					
	Process heat	Wood	Wind	Photovoltaics	Active	Passive
Buyer behavior						
Inaccurate perceptions	M	L	M	M	M	M
Unfamiliarity with benefits	M	M	M	M	M	M
Lack of vendor information	H	L	H	H	M	M
First-cost orientation	H	L	H	H	H	M
High inferred discount rate	H	L	H	H	H	M
Uncertainty over conventional fuel cost	M	L	M	M	M	M
General unfamiliarity, newness	M	L	L	L	L	L
Renter occupancy	—	—	—	—	M	M

Source: Spewak and Bohannon (1981).
a. Importance of barrier: L = low, M = medium, H = high, — = not applicable.

Table 5.6
Barriers to solar technology implementation in the utility market sector

Barrier	Technology[a]				
	Solar thermal	OTEC	Wind	Photovoltaics	Wood/biomass
Technological					
Component lifetime	M	M	H	L	L
Reliability	M	M	M	M	L
Application load match	L	L	M	M	—
Environmental	—	L	—	—	M
Seasonal/diurnal variability	M	L	M	M	—
Supply/resource constraints	—	L	—	—	—
Installation quality	L	L	L	L	L
Production/output capacity	M	L	M	M	M
Lack of operating data	H	H	H	H	L
Economic/financial					
Conventional energy subsidies	L	L	L	L	L
High initial cost	M	M	M	M	L
Lack of venture capital	M	M	M	M	L
Availability of financing	H	H	H	H	L
Construction delays	M	H	M	M	M
Institutional					
Lack of skilled labor	L	L	L	L	L
Lack of regulatory homogeneity	M	M	M	M	L
Buyer behavior					
Inaccurate perceptions	H	H	M	H	—
Unfamiliarity with benefits	M	M	M	M	L
First cost orientation	M	M	M	M	L
Uncertainty over conventional fuel cost	M	M	M	M	M
General unfamiliarity, newness	H	H	M	H	—

Source: Spewak and Bohannon (1981).
a. Importance of barrier: L = low, M = medium, H = high, — = not applicable.

Table 5.7
Barriers to solar technology implementation in the transportation market sector

	Technology[a]	
Barrier	Fuels 1[b]	Fuels 2[c]
Technological		
Energy balance	L	M
Reliability and maintenance	L	H
Environmental	M	M
Supply/resource constraints	M	M
Engineering capability	L	L
Production/output capacity	M	H
Lack of operating data	L	H
Economic/financial		
Uncertainty over future fuel prices	L	M
Uncertainty over future feedstock prices	M	L
Conventional energy subsidies	M	M
High initial cost	M	M
Availability of financing	L	M
Construction delays	M	M
Institutional		
Compatibility with existing distribution networks	M	M
Availability of gasoline for blends	M	M

Source: Spewak and Bohannon (1981).
a. Importance of barrier: L = low, M = medium, H = high.
b. Fuels 1 includes current commercial processes.
c. Fuels 2 includes processes currently being refined such as acid hydrolysis and enzymatic hydrolysis.

The difference between "real" and "perceived" problems or issues resulting in the various types of barriers is important. Because of this difference, it may not be sufficient to treat a real barrier, such as high initial cost, and assume that the related buyer behavior barriers will take care of themselves. A good example is the early experiences related to the federal solar tax credits. Although the tax credits and hot water grants significantly reduced the first consumer cost of solar water heaters, as late as spring 1980 most consumers were unaware of the programs or did not understand the possible impacts on system cost. In a national survey conducted during February and March of 1980, over half of the existing homeowners surveyed (53.4%) were unaware of any government incentives (Lilien and Johnston, 1980). Of those aware of the incentives, only 21% correctly identified the incentive as a tax rebate, and of those only 20.3% gave a reasonable estimate of the size of the rebate. Thus, of the

Table 5.8
Summary of most important barriers to solar technology

| Technology | Barrier | | | |
	Technological	Economic/financial	Institutional	Buyer behavior
Residential				
Passive	L	M	M	H
Active hot water	M	M	L	H
Active space	M	M	M	H
Wind	M	M	H	H
Photovoltaics	H	H	H	H
Wood	L	L	L	L
Industrial commercial				
Process heat	M	H	M	H
Wood	L	L	L	L
Wind	M	H	M	H
Photovoltaics	H	H	H	H
Active	M	M	M	M
Utilities				
Passive	L	M	M	M
Solar thermal	M	H	L	M
OTEC	M	M	L	M
Wind	M	H	L	M
Transportation				
Photovoltaics	M	H	L	M
Wood/biomass	L	L	L	L
Fuels 1	L	L	M	—
Fuels 2	M	M	M	—

Source: Spewak and Bohannon (1981).
a. Importance of barriers: L = low, M = medium, H = high, — = not applicable.

homeowners surveyed nationally, only about 2% were adequately informed about solar tax credits (the sample size was 880). In a sample of 744 new-home buyers about 4% were adequately informed.

The experience related to the federal tax credits indicates a need to treat the different types of barriers separately even though they are interrelated. A program that effectively lowers the initial cost of a solar system, such as the federal tax credits, will have little impact unless it is complemented with another program that makes the cost reductions known to potential consumers, thereby reducing the first cost of orientation barrier.

Table 5.8 summarizes the information in tables 5.4 through 5.7. This summary provides insight into the most important barriers by technology in each market sector. Each type of incentive aids in removing one or more of the barriers to technology development and full commercialization. A summary of this "coverage" of barriers is presented by incentive category in table 5.9.

5.3.3 Scheduling Incentives

Scheduling incentives is a case-specific activity based on the facts of each situation. Although it would be nice to be able to generalize from past experience, it is impractical for two reasons. First, the needs differ depending on the technology to which the incentives are applied. The level of RD&D funds applied to date may vary considerably from technology to technology. In addition, the comparative economics of solar technologies versus conventional technologies may be more or less disparate depending on the solar technology considered. Finally, the private sector participation varies widely, depending on the perceived commercialization potential of a particular solar technology. The second reason is that the sociopolitical climate may vary significantly from year to year, resulting in differing perspectives and rationales for energy incentives. Incentives for nonsolar energy technologies may have been put into place for reasons that may not have applied when solar incentives were formulated. Therefore it may be necessary to consider the causes of government action when determining why incentives for other technologies were implemented. The causes for government action fall into four general categories:

1. **Economic.** The federal government acts because it wants to effect a change in market outcome, such as the relationship between production and price or between consumption and price.

2. **Political.** Action occurs as a result of bargaining between individuals, groups, and organizational participants, each of whom is seeking an independent goal. Successful participants include those with a high intensity of performance and high political power.

3. **Organizational.** The government acts by responding to decision problems created by external events or by the actions of nongovernment organizations.

4. **Legal.** The government acts because a body with authority to make a law does so in response to parties appearing before it. Interested parties have requested an authoritative body to clarify an issue or declare a change.

Table 5.9
Coverage of barriers by incentive category

Technology	Incentive category						
	Creation or prohibition of organizations	Taxation	Disbursements	Requirements	Traditional government services	Nontraditional government services	Market activity
Residential							
Technological							
Economic/ financial		×	×	×		×	×
Institutional			×	×			×
Buyer behavior	×		×			×	
Industrial/ commercial							
Technological						×	
Economic/ financial	×	×		×			
Institutional			×	×		×	×
Buyer behavior			×		×	×	×
Utilities							
Technological						×	
Economic/ financial	×	×		×			×
Institutional				×			
Buyer behavior						×	
Transportation							
Technological							
Economic/ financial	×	×	×	×		×	×
Institutional							
Buyer behavior						×	

Source: Spewak and Bohannon (1981).

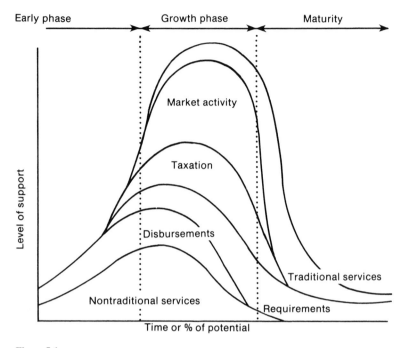

Figure 5.1
Idealized application of incentives (Spewak and Bohannon, 1981).

The difficulties in generalizing from past experience in applying incentives can be easily illustrated. Figure 5.1 illustrates the ideal application of incentives during the development of a particular technology. During the early phases of technology development, nontraditional services, such as funds for basic and applied RD&D, would play a primary role. During a technology's growth phase a different type of incentive is needed to overcome private sector perceptions of commercialization potential or to provide comparative cost-effectiveness. Such government actions as loan guarantees (market activity), tax credits (taxation), grants in aid, and subsidies (disbursements) predominate at this stage. Once the technology has reached full commercialization potential, private sector market forces can be expected to dominate. Traditional government services and requirements are all that may be needed or expected during this mature phase. However, as illustrated in figures 5.2 and 5.3, applying incentives to energy technologies does not necessarily depend on an ideal standard

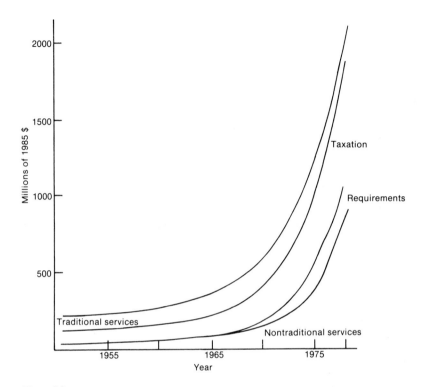

Figure 5.2
Coal incentives (Cone et al., 1978a, b).

but may be applied in response to economic, political, organizational, or legal pressures.

Figure 5.2 presents the principal coal incentives and their magnitude over time in 1985 dollars. Coal was the most important fuel in the United States until the end of World War II. The loss of two markets, steam locomotives and space heating, caused a decline in the industry. In the 1970s RD&D in alternative uses of coal (nontraditional services) and regulation and monitoring of environmental aspects (requirements) resulted in increased incentives despite decreasing production.

Figure 5.3 illustrates federal incentives to hydroelectric energy production. The incentives consist primarily of market activity (buying and selling electricity) and taxation (tax-free bonds). A minor share of the total cost of incentives to hydroelectric energy consists of requirements.

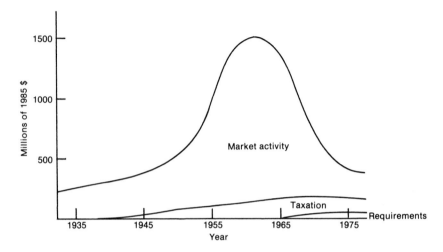

Figure 5.3
Hydro incentives (Cone et al., 1978a,b).

When solar incentives were first designed, the timing of incentives to be applied to solar technologies was determined by considering technology development schedules, comparative cost "crossovers," expected market potential through time, and the economic, political, organizational, and legal environments in which the incentives would be structured.

Technology development schedules were based on the timetables for the federally funded RD&D programs and private industry activity in bringing the technology to the marketplace. Comparative cost crossovers are the projected points in time where the levelized cost ($/10^6 Btu or ¢/kWh) of a solar technology becomes equal to or less than the levelized cost of a targeted conventional energy source. This crossover can be directly related to the RD&D schedule, the effects of volume production on cost (economies of scale and learning curve impacts), and the direct effects of incentives. Ideally it was hoped that through RD&D costs could be decreased to a level at which the technology could be marketed with the aid of financial incentives, which could temporarily lower the cost of the energy produced on an artificial basis. It was thought that, as the demand for the technology increased, the volume of the market would bring the cost of the technology down. As the cost of the technology decreased, the financial incentives could be removed. The final result would be that an adequate market would be established to support a mature industry without financial incentives. This is why market potential through time was an important aspect in the design of solar incentives.

The economic, political, organizational, and legal environments played key roles in defining the incentives as they were established. Equity, total national value, and energy independence defined the real or perceived need for solar incentives. These forces provided the impetus to write and enact legislation mandating solar incentives. These forces were also instrumental in the demise of solar incentives in the first half of the 1980s. Supply-side economics provided an alternative to demand-oriented residential tax credits. Fiscal conservatism resulted in a reduced government role in technology development. And probably most important of all, decreased energy prices brought on by an overabundance of oil lessened the immediacy of economic and national security arguments for alternative energy sources.

5.3.4   The Level of Incentives

The degree of federal commitment to incentives depends on several factors previously discussed and analyzed. The level of incentives for a particular technology depends on technology development requirements, the level of federal support and commitment, market response and expected market potential through time, and the underlying causes for government action. Table 5.10 summarizes the characteristics that affect the level of incentive type.

5.3.5   Regional Aspects of Incentives

Certain incentives require uniformity across the nation to hold down administrative costs or prevent "inefficient" competition. However, in some cases, varying the level of support, type of support, and time phasing of support may be indicated. These variations result from regional variations in comparative solar and conventional cost and performance. They may also result from regional differences in attitudes and market and infrastructure environments.

Regional variations in the barriers to solar energy must be taken into account when assessing an incentive's effectiveness, application, and implementation. Otherwise windfall gains will result in regions where barriers are low, and little impact will be felt where barriers are high. For example, an incentive program in the Southwest (a region characterized by summer

Table 5.10
Characteristics that affect the level of incentive type

Characteristic	Incentive						
	Creation or prohibition of organization	Taxation	Disbursements	Nontraditional services	Market activities	Traditional services	Requirements
Technology development requirements	×		×	×			
Level of federal support/commitment	×	×	×	×	×	×	×
Market response and expected market potential		×			×	×	×
Underlying causes for expected government action							
Economic	×	×	×		×		×
Political	×	×	×	×	×	×	×
Organizational	×		×	×	×	×	×
Legal							×

peaking utilities, rapid growth, state laws already enacted to spur solar development, a healthy construction industry, active capital markets, high insolation, and high dependence on natural gas) may differ substantially from one designed to overcome the Northeast region's current barriers (a region characterized by winter peaking utilities, dependence on oil and coal, comparatively slow economic growth, lower insolation, and relatively stagnant capital markets).

Many barriers are specific to the jurisdictions themselves. For example, restrictive utility commission policies, building codes, land-use requirements, property taxes, and labor union jurisdiction problems differ widely in their degree of complexity and possible constraints on solar energy development. The framework should be capable of adjusting to these particular local conditions. A second reason for regional flexibility is that some states and regions have acted on their own to provide incentives to encourage solar energy. It is important that federal incentives complement rather than substitute for the incentives already enacted. In addition, because the cost of conventional fuel and the degree of insolation differ widely by region, the size of the subsidy necessary to achieve a given market penetration is a strong function of the regional characteristics. Incentives must also be compatible with the highly regional characteristics of the construction industry.

5.4   Solar Incentives to Date

The oil embargo of 1973 stimulated concern over energy supplies. The opportunities and advantages of solar energy were considered as U.S. policymakers sought energy self-sufficiency. This resulted in the creation of federal incentives to increase the national use of solar energy. These incentives are discussed in what follows.

From 1973 to 1984 the federal government committed $6.6 billion (in current dollars), through expenditures or foregone revenues, to solar energy development. The incentive programs designed to remove the technical barriers and influence buyer behavior achieved the most attention. Incentives aimed at reducing the high initial cost of solar systems have received increased attention in recent years.

Table 5.11 summarizes the incentives for solar to date. The first category, creation of organizations, comprises the budget for SERI and the RSECs. The RSECs stopped receiving federal funds in 1982. Market ac-

Table 5.11
Incentives for solar to date (1984 $ × 10)

Incentive category	Year							
	1974–1978	1979	1980	1981	1982	1983	1984	Total
Creation of organizations	10	36	60	56	24	18	16	220
Market activity	262	66	55	89	24	0	0	496
Nontraditional services	1652	636	682	655	271	211	187	4294
Disbursements	0	7	0	0	0	0	0	7
Taxation	57	66	128	229	269	363	468	1580
Total	1981	811	925	1029	588	592	671	6597

Sources: The Budget of the United States (1979–1985); IRS Statistics of Income (1979–1984); The Solar & Conservation Fact Book (1980); Sonia Conley, IRS Bureau of Statistics.

tivity, the second category, comprises the Solar Federal Building Program, the Federal Photovoltaics Utilization Program, market development activities, training, and international trade development activities. It also includes the Solar Energy and Conservation Bank. Nontraditional activities include the RD&D programs initially funded by the National Science Foundation (NSF), then by the Energy Research and Development Administration (ERDA), and finally by the U.S. Department of Energy (DOE). The only true disbursements offered as solar incentives were the HUD solar hot water grants offered in 1979. The incentives in the taxation category include the residential tax credit for solar (active solar heating, cooling, and hot water; wind; and photovoltaics), and the investment tax credit (solar thermal and wind applications for commercial and industrial users).

5.4.1 Solar Incentive Experience

Numerous federal actions have affected the development and commercialization of solar technologies. In addition, other federal actions were considered but not adopted. These federal actions are summarized by incentive category for the residential, industrial and commercial, and utilities market sectors in tables 5.12 through 5.17. Table 5.18 gives information on solar energy savings versus federal incentives. Let us now look at some of the major incentives in each incentive category.

The primary examples of organizations created on behalf of solar and conservation are SERI, the RSECs, the Energy Extension Service (EES), the Synthetic Fuels Corporation, the National Solar Heating and Cool-

Table 5.12
Federal actions by incentive category in the residential market sector

Incentive category	Federal action
Creation or prohibition of organizations	Energy extension services
Taxation	Residential tax credit
Disbursements	Solar hot water grants
Requirements	Residential conservation service
	Standards
	Building information
	Performance standards
Traditional government services	Solar program administration
Nontraditional government services	Active systems development
	Passive systems development
	Photovoltaic systems development
	Wind energy systems development
	Residential demonstration program
	Photovoltaic demonstration program
	Active solar information dissemination
	Passive solar information dissemination
	Residential wood information dissemination
	Active and passive design method development
	Resource data collection, maintenance, and dissemination
	Commercialization planning
	Market analysis
Market activity	Federal Buildings Program
	Overseas marketing assistance
	Solar Energy and Conservation Bank

Source: Spewak and Bohannon (1981).

ing Information Center, and the Interagency Panel on Terrestrial Applications of Solar Energy (IPTASE). Those organizations that are not specific to a market sector include SERI, the RSECs, the National Solar Heating and Cooling Information Center, and IPTASE. SERI was authorized under public law in 1974 and became the center of solar technical, economic, and policy research in the United States. The RSECs were established in response to the need for regional centers of research and information outreach. The Energy Extension Service was established to provide technical information outreach through the state governments in much the same manner that the Agricultural Extension Service provides information through state land grant colleges and through state agricultural schools and organizations. The National Solar Heating and Cooling

Table 5.13
Proposed federal actions by incentive category in the residential market sector (not adopted)

Incentive category	Federal action
Creation or prohibition of organizations	None
Taxation	Passive tax credits Tax deductions Tax incentives for resale
Disbursements	Passive design awards
Requirements	Consumer protection guidelines
Traditional government services	None
Nontraditional government services	Residential woodlot demonstrations
Market activity	Interaction with developers Producer loans Priority treatment for solar in government-supported or insured mortgages, community development block grands, Small Business Administration loans, Federal Photovoltaics Utilization Program

Table 5.14
Federal actions by incentive category in the industrial/commercial market sector

Incentive category	Federal action
Creation or prohibition of organizations	None
Taxation	Investment tax credit Alternative fuels production credit Excise tax exemption
Disbursements	Schools and Hospitals Program
Requirements	Public Utility Regulatory Policies Act Powerplant and Industrial Fuel Use Act Standards Buildings Energy Performance Standards
Traditional government services	None
Nontraditional government services	Technology research, development, and demonstration Technology information dissemination Engineering, financial, and scientific consultation
Market activity	None

Table 5.15
Proposed federal actions by incentive category in the industrial/commercial market sector (not adopted)

Incentive category	Federal action
Creation or prohibition of organizations	Industrial Energy Productivity Corporation
Taxation	Accelerated depreciation Removal of tax credits for conventional fuels
Disbursements	None
Requirements	Broker environmental nonattainment exemptions Generic environmental impact statement Expediting licensing, permitting, and zoning
Traditional government services	None
Nontraditional government services	Industrial wood supply demonstrations Economic demonstrations
Market activity	Federal Site Bank Use of federal lands Supply side loans and loan guarantees Priority treatment for solar in Small Business Administration loans

Table 5.16
Federal actions by incentive category in the utilities market sector

Incentive category	Federal action
Creation or prohibition of organizations	None
Taxation	Investment tax credit
Disbursements	None
Requirements	Residential Conservation Service Powerplant and Industrial Fuel Use Act Public Utility Regulatory Policies Act
Traditional government services	Use of solar by power marketing authorities
Nontraditional government services	Technology research, development, and demonstration Technology information dissemination Engineering, financial, and scientific consultation
Market activity	None

Table 5.17
Proposed federal actions by incentive category in the utilities market sector (not adopted)

Incentive category	Federal action
Creation or prohibition of organizations	Offshore Power Authority Solar Power Marketing Authorities Federal Financing or Leasing Corporation
Taxation	Accelerated depreciation Removal of tax deductions for conventional fuels
Disbursements	None
Requirements	Mandating marginal cost pricing Broker environmental nonattainment exemptions Generic environmental impact statement Expediting licensing and permitting
Traditional government services	None
Nontraditional government services	None
Market activity	Federal Site Bank Use of federal lands Supply side loans Loan guarantees for ocean thermal energy conversion

Table 5.18
Solar energy savings versus federal incentives

Saving or incentive	1980	1981	1982	1983	1984	1985
Solar energy savings (10^{12} Btu/yr)						
Low-temperature plate	39.5	48.2	55.7	60.6	62.6	64.6
Med-temperature plate	43.5	66.5	88.7	112.7	138.7	164.7
Photovoltaics	0.0	0.0	0.0	0.4	0.8	1.3
Wind	0.0	0.0	0.0	0.0	0.3	0.5
Biomass	1500.0	1500.0	1500.0	1500.0	1500.0	1500.0
Total	1583.0	1614.7	1644.4	1673.7	1702.4	1731.1
Solar energy incentives (1985 \$ $\times 10^6$)						
Annual	924	1028	588	591	671	N/A
Cumulative	3714	4742	5330	5921	6594	N/A
Federal cost per 10^6 Btu/yr (10^6 Btu = 1.055 GJ)						
Annual	0.58	0.64	0.36	0.35	0.39	N/A
Cumulative	2.35	2.94	3.24	3.54	3.87	N/A

Information Center was established to provide general information about solar heating and cooling. IPTASE was established to provide a federal interagency forum for identifying research and policy priorities associated with solar.

The primary examples of taxation are the residential tax credits, the investment tax credit (ITC), the alternative fuels production credit, and the excise tax exemption. The residential tax credit, originally established as 30% of the first $2,000 and 20% thereafter for a maximum of $2,200, was replaced by a straight 40% tax credit (maximum of $4,000) in 1980 (through 1985). Residential conservation is eligible for a 15% tax credit (maximum of $300). The investment tax credit, which applies to commercial and industrial users of solar, wind, and geothermal, was increased in 1980 from 10% to 15% (through 1985). An 11% ITC applies to small-scale hydroelectric, and a 10% ITC applies to nongas or oil cogeneration and for certain gasohol and biomass equipment.

The primary disbursement incentives are the solar hot water (SHW) grants and the Schools and Hospitals Program. Under both of these programs the federal government provided users with grants to install solar systems. The SHW grants, sponsored by HUD, offered approximately 11,000 grants of $400. The grants, which were administered by the states involved, were established to give momentum to the solar industry. Because of a lack of publicity and problems encountered in certifying systems, many grants were unclaimed. In New England only 1,362 of the allocated 8,117 grants were used. The Schools and Hospitals Program, also administered through the states, achieved moderate success in installing several large-scale systems.

Included in the requirements are the Residential Conservation Service (RCS), Solar Standards, Building Energy Performance Standards (BEPS), the Public Utility Regulatory Policies Act (PURPA), and the Powerplant and Industrial Fuels Use Act (PIFUA). The RCS required public utilities to provide energy audit and conservation consulting to their residential customers. The standards activity associated with solar were voluntary consensus standards developed through industry participation. Interim standards were, however, produced by the National Bureau of Standards (NBS) and were used in the qualification of systems for federal programs. PURPA required utilities to establish provisions for buying power back from their customers to enhance the economics of alternative power producers that generated excess power (wind, photo-

voltaics, low head hydro, cogeneration). PIFUA requires major fuel-
burning installations to reduce or eliminate their use of natural gas or
petroleum. Major fuel-burning installations include all industrial burners
greater than 10^8 Btu h (30 MW$_t$) or industrial sites using several boilers
between 50 and 10^8 Btu h (15 and 30 MW$_t$) that have a total output of
greater than 250×10^6 Btu h (75 MW$_e$). Several miscellaneous activities
might fall into the requirements area because they deal indirectly with requirements. Legal issues were addressed through the publication of the
Solar Law Reporter, a journal that specializes in zoning, access rights,
consumer rights, and regulatory issues.

The only example of a traditional government service incentive was the
use of solar by federal power marketing authorities (Bonneville Power
Administration, Tennessee Valley Authority).

Numerous solar incentives fall into the category of nontraditional services. The RD&D efforts in all solar and conservation technologies fall
under this category, as well as additional consultation, information dissemination, and resource data collection and dissemination activities.
For a more extensive but by no means complete list, see tables 5.13
through 5.17.

Included in the market activity category are the Federal Buildings Program and the Solar Energy and Conservation Bank. Under the Federal
Buildings Program, the government purchased quantities of solar hardware for its own use in the hope of supporting the fledgling industry
and possibly assisting in initial steps toward mass production of solar
components.

5.5   Summary of the Impacts of Incentives to Date

It is difficult to determine conclusively whether the incentives received by
solar are on a par with the incentives received by other energy sources.
No comprehensive analysis has quantified the impacts of solar energy to
date. However, some estimate of solar impacts can be made based on the
data currently available.

5.5.1   The Equity of Solar Relative to Other Energy Sources

Table 5.18 presents some estimates of solar energy savings versus federal
incentives from 1980 through 1985. The energy impacts for the flat-plate
collectors and photovoltaics are based on installed capacity estimates de-

rived from the EIA Solar Collector Manufacturers Survey (EIA, 1984). Based on these figures, EIA estimated that approximately $64 \times 10^6 \, \text{ft}^2$ ($6 \times 10^6 \, \text{m}^2$) of low-temperature collector, $82 \times 10^6 \, \text{ft}^2$ ($8 \times 10^6 \, \text{m}^2$) of medium-temperature collector, and 34 MW of photovoltaics were installed by the end of 1984. EIA conservatively estimates approximately 12 MW of installed wind capacity based on approximately 1,000 5-kW small wind machines, three 500- to 1,000-kW machines, and a 5-MW wind energy farm at Altamont Pass, California. These values were converted to energy savings based on the following factors:

low-temperature flat plate $= 100,000 \, \text{Btu/ft}^2 \, \text{yr} \, (1.1 \, \text{GJ/m}^2 \, \text{yr})$,

medium-temperature flat plate $= 200,000 \, \text{Btu/ft}^2 \, \text{yr} \, (2.2 \, \text{GJ/m}^2 \, \text{yr})$,

photovoltaics $= 37.5 \times 10^9 \, \text{Btu/MW}_e \, \text{yr} \, (40 \, \text{TJ/MW}_e \, \text{yr})$,

wind $= 37.5 \times 10^9 \, \text{Btu/MW}_e \, \text{yr} \, (40 \, \text{TJ/MW}_e \, \text{yr})$.

These conversion factors were derived from the energy projections developed by Oak Ridge National Laboratories and MITRE for the 1980 solar strategy paper (Parikh et al., 1980). It is also conservatively assumed that the level of biomass use for home heating and industrial heat has not increased since 1980, when it was estimated to be 1.5 quads by DOE (Gill, Majors, and Wang, 1980).

Based on these estimates solar energy provided approximately 1.7 quads (1.79 EJ) of energy in 1985. When compared with the conventional energy incentives depicted in table 5.3, solar, which ranged from \$0.35 to \$0.58/10^6 Btu between 1980 and 1984, was well within the range of annual incentives received by conventional energy sources in 1977.

5.5.2 Incentives as a Reflection of the Total National Value of Solar

All things considered, total national value is a relatively subjective measure. The MITRE NPAC analysis (Bennington et al., 1979) estimated a total national value of \$1.11–\$1.43/10^6 Btu (adjusted to 1985 dollars) for solar energy. These values, however, assumed the costs and impacts of a mature solar industry in the year 2000. Although it is still too early to compare the cumulative costs and benefits against this benchmark, the annual estimated cost per 106 Btu compares favorably with the expected benefit.

Another view of the cost/benefit balance may be to compare the federal incentives with the current size of the solar industry. The Rocky

Mountain Institute estimates the solar industry to be worth approximately $3 billion per year in 1984 (Lamar, 1986). Thus the industry's annual sales volume is worth a little less than half what was invested over the previous twelve years. This also seems to indicate a reasonable benefit for the investment.

5.5.3   The Effectiveness of Incentives in Promoting the Use of Solar

Federal incentives have not been as successful as originally planned in promoting the use of solar energy. Table 5.19 shows the actual impacts versus the impacts projected in the solar strategy document of 1980 (Parikh et al., 1980) for the "legislative mandate" scenario—the scenario that best fits the solar incentive policy adopted. As can be seen from this table, actual impacts fell well short of what was expected for wind and for solar heating and hot water. Photovoltaics generated more energy savings than expected.

The three reasons why incentives fell short of the mark were that energy costs did not increase at the expected rate, a general slowdown in economic growth occurred, and the solar industry pricing model was inconsistent with early assumptions.

Conservative energy pricing scenarios of the late 1970s and early 1980s assumed 1985 oil prices of $35–$45 per barrel. The 1980 solar strategy assumed energy escalation at about 2% over inflation, when in fact it has decreased in real terms. This decrease has had an impact on solar penetration by detracting from the economics of solar relative to conventional alternatives. The decreased savings from solar made the initial investment hard to rationalize. The deteriorated economic benefits from solar in some instances prevented some solar options from even getting to the marketplace. Solar cooling and residential photovoltaics never really emerged from the demonstration phase into commercialization.

The general slowdown in the economy witnessed during 1981 through

Table 5.19
Actual and projected solar impacts of wind, photovoltaics, and solar heat and hot water (10^{24} Btu = 1.055 TJ)

Technology	Actual impact	Projected impact
Wind	0.00045	0.005
Photovoltaics	0.001	0
Solar heat and hot water	0.165	0.228

1983 had some severe impacts on the growth of the solar energy industry. The downturn in housing starts removed an essential market for residential solar heating and hot water. The new market was initially perceived as an important market for solar heating and hot water because of the opportunity to finance the incremental solar cost over a thirty-year mortgage versus a short-term loan associated with a solar retrofit application. The slowdown in industrial growth also limited the market for solar industrial process heat applications. Finally, the slowdown in the growth of demand for electrical power limited the early markets originally perceived for wind and solar thermal fuel savers. Fuel savers are typically smaller-scale systems (less than 10 MW_e) that provide power only when solar or wind resources are available; they do not include storage.

It was originally thought that financial incentives for solar technology could enhance the demand for solar so that mass production could be achieved. Economies of scale and learning curve effects would reduce the cost of solar. The reduced cost would be reflected in reduced price, thus making solar more competitive with conventional energy sources. Although no definitive study exists as to what exactly occurred, it appears that, at best, the price of solar remained constant. In some instances, it appears that the price of solar increased. The incentives allowed the solar dealers to charge more while the consumers, through the tax incentives, paid less. If this is in fact what occurred, the consumer did not reap the entire benefit of the tax credit and the market was not stimulated to the extent originally planned.

In summary, solar incentives do not appear to have been as effective as originally planned. The strongest conclusion that can be drawn from this experience is that the political and economic environment used in designing the initial solar incentives has changed greatly over the past seven years. Because the environment has changed, the actual results have not matched the predicted results.

References

Bennington, G., M. Bohannon, R. Gerstein, R. Hartzler, N. Kannan, G. Miller, K. Rebibo, P. Spewak, and J. Taul. 1976. *A Comparative Analysis of Solar Energy*. McLean, VA: MITRE Corporation.

Bennington, G., M. Bohannon, R. Gerstein, R. Hartzler, N. Kannan, G. Miller, K. Rebibo, P. Spewak, and J. Taul. 1979. *Toward a National Plan for Accelerated Commercialization of Solar Energy*. McLean, VA: MITRE Corporation.

Cone, B. W., D. L. Brenchley, and V. L. Brix. March 1978a. *An Analysis of Federal Incentives Used to Stimulate Energy Production.* PNL-2410. Richland, WA: Battelle Pacific Northwest Laboratories.

Cone, B. W., D. L. Brenchley, and V. L. Brix. December 1978b. *An Analysis of Federal Incentives Used to Stimulate Energy Production.* PNL-2410 (rev.). Richland, WA: Battelle Pacific Northwest Laboratories.

Cone, B., D. L. Brenchley, and V. L. Brix. 1980. *An Analysis of Federal Incentives Used to Stimulate Energy Production.* PNL-2410 (Rev.2). Richland, WA: Battelle Pacific Northwest Laboratories.

Council on Environmental Quality. 1978. *Solar Energy: Progress and Promise.* Washington, DC: Council on Environmental Quality.

A Domestic Policy Review of Solar Energy. 1978. TID-28834. Washington, DC: Solar Domestic Policy Review, Interagency Task Force.

EIA. 1984. *Solar Collector Manufacturing Activity through June 1983.* Washington, DC: Department of Energy, Energy Information Administration, Emerging Energy Statistics Branch.

Gill, G., H. Majors, and H. Wang. 1980. *The Solar and Conservation Fact Book.* Office of Policy Planning and Evaluation, Assistant Secretary for Solar and Conservation, U.S. Department of Energy.

IRS Bureau of Statistics. 1980–1985. *IRS Statistics of Income, 1979–1984.* Washington, DC: Government Printing Office.

Lamar, J. 1986. "Gone with the wind." *Time* 127(3):23.

Lilien, G. L., and P. E. Johnston. 1980. *A Market Assessment for Active Solar Heating and Cooling Products, Category B: A Survey of Decision Makers in the HVAC Market Place.* Final Report DOE/CS/30209-T2. Washington, DC: U.S. Department of Energy.

Lovins, A. 1977. *Soft Energy Paths: Towards a Durable Peace.* London: Friends of the Earth.

Parikh, S. C., R. Barnes, and J. Blue. 1980. *Energy Savings Impacts of DOE's Conservation and Solar Programs.* Oak Ridge, TN: Oak Ridge National Laboratories.

Spewak, P., and M. Bohannon. 1981. *Identify Specific Regimens of Federal Support for Solar Technologies.* Reston, VA: Bennington Enterprises Limited.

U.S. Department of Commerce. 1985. *Statistical Abstract of the United States, 1982–1983.* Washington, DC: Bureau of the Census.

U.S. Department of Treasury. 1979–1985. *The Budget of the United States.* Washington, DC: Government Printing Office.

6 End-Use Matching and Applications Analysis Methodologies

Kenneth C. Brown

Increasing attention to the variety of energy resources available to industrial societies led to studies of energy applications and end-uses and their relationship to the performance of energy technologies. The purpose of these studies was to explore new and potentially more productive uses of available resources through selectively using energy technologies. The guiding philosophy of these efforts was that a productive society with scarce energy resources should appropriately match energy resources to energy end-uses.

For solar thermal technology, matching energy resources to energy end-uses included considering the temperature and form of heat required by the end-use in addition to other common technical and economic parameters. The technology's ability to deliver heat over a wide range of temperatures and at significantly varied efficiencies within those temperature ranges made the task of appropriate end-use matching somewhat unique (see figure 6.1). This uniqueness, and the advantages and disadvantages it implies, was a major factor in the introduction and development of end-use matching concepts for solar thermal technology.

End-use matching and applications analysis were recognized as important elements of the federal research and development (R&D) program for solar thermal technology. As stated in a summary of such programs by the Solar Energy Research Institute (SERI),

The effective use of alternative energy technology (in industry) will depend, to a large extent, on the successful recognition, development and implementation of appropriate applications for alternative energy supply. Because solar energy competes with many other suitable, and perhaps more flexible, energy delivery systems, it is important that applications be chosen to maximize competitive advantages. This requires that issues of location, temperature needs, schedule, fuel price, fuel availability, and environmental constraints be fully explored to identify applications for solar energy. It also necessitates that system design and engineering initially satisfy the requirements of the most appropriate applications and that systems development continue to respond to market needs as these needs develop. (Brown, 1980)

End-use matching and applications analysis, applied in studies of solar thermal technology, emphasized specific studies of particular application needs. These studies focus on technical and economic criteria of the energy user.

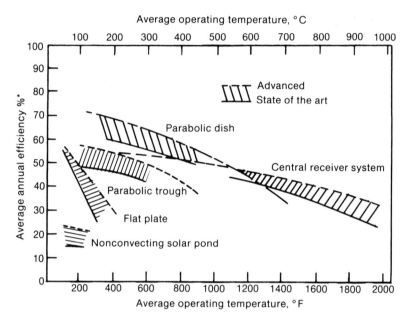

Figure 6.1
Annual efficiency of solar thermal IPH systems versus average operating temperature
(includes typical field thermal transport losses, but not process heat exchange).

In this chapter I describe the development of methodologies and the results of studies from 1972 to 1982 involving applications analysis and
end-use matching of solar thermal technologies in U.S. industry. The
methodologies I describe could have been used to evaluate solar thermal
applications in other economic sectors; however, their primary application was in industry, particularly for industrial process heat (IPH). I also
elaborate on the theory and scope of the studies and their relationship to
technology assessment, market analysis, and system development.

This chapter is divided into four parts. Section 6.1 defines the general
scope and background of applications analysis and its relationship to
other forms of requirements analysis. Section 6.2 summarizes major program thrusts in which end-use matching and applications analysis methodologies were developed and used to support federal solar thermal energy programs. Section 6.3 describes activities in major program elements
in some detail, including applications in solar space heating and cooling,

industrial and agricultural process heat, small power systems, total energy systems, and large power systems. Finally, in section 6.4 I summarize the accomplishments of a decade of applied research and discuss the general contributions to the federal R&D program for solar thermal technology.

6.1   Scope of Applications Analysis

We must define end-use matching and application analysis before we can establish a basis for discussing them in federally supported research, although a precise definition of their scope is difficult. For example, applications analyses are similar to market analyses of technologies within major economic sectors, as both types of study rate markets and technologies by market suitability. However, although market analyses rely heavily on macroscopic characterization of the suitability of markets, an applications analysis generally focuses on a more specific test of the match between technologies and markets. Also, market analysis often considers the dynamics of market penetration and the global impact of economic factors, whereas applications analysis does not. Likewise, applications analysis is similar to specific system development and design studies because site-specific criteria are important to both. System design studies, however tend to be less comprehensive than the typical end-use matching analysis. Applications analysis and end-use matching fit somewhere in between market analysis and system design studies in this continuum of system design detail and comprehensive market coverage.

The intermediate position of applications analysis and end-use matching is of some value in an R&D program. Often such studies confirm the actual technical and economic feasibility of broadly characterized markets because of their attention to site-specific constraints and opportunities. Applications analysis may also provide the basis for market studies of a broader nature by identifying previously unrecognized opportunities.

Applications analysis is supported by two primary facets of research: requirements analysis and end-use matching. Requirements analysis is the investigation, analysis, and documentation of selected aspects of an energy service market. It incorporates features of the application of new technology that are not always directly measurable in economic terms. Requirements analysis focuses on the end-user's technical needs and seeks to characterize important variables in adapting and implementing solar energy resources to the end-use. For example, a requirements analysis of the industrial sector for application of solar thermal process heat would focus on collecting data on (among other things) location and climatology, land availability, process scheduling, peak-to-average heat requirements, required temperature, reliability requirements, current energy costs, capital recovery factors, and atmospheric contaminants. These data would then be documented, organized, and analyzed to indicate the technology's suitability to specific applications.

End-use matching consists of testing, selecting, and refining the application of specific solar technology systems to end-use requirements. For example, for industrial process heat the appropriate selection of the concentration ratio in the collector allows a match of the temperature level of the heat supply to that required in the process. A number of choices of collector may be available for a given end-use, because a variety of solar thermal collectors with different concentration ratios are available. The objective of end-use matching (in general) is to select the collector that most economically meets such requirements as the supply temperature. Such selection ensures that solar thermal energy competes effectively with other conventional forms of energy.

Therefore the basic scope of applications analysis differs from market analysis and system design studies in the following ways:

1. Applications analysis is based on requirements of various end-use applications that are defined with regard to the set of technologies to be analyzed.

2. Applications analysis is characterized by matching technologies and system configurations to end-use requirements rather than by approximating needs by generally defined economic criteria. Applications analysis does not, however, involve the extensive evaluation of a single application to the exclusion of broader markets as does a design study.

3. Applications analysis methods may vary in form and scope depending on the specificity desired, the number of technologies being evaluated, and the breadth of the market being considered. In general, however, such methods must be amenable to the evaluation of multiple technologies over multiple markets and be able to distinguish acceptable applications on the basis of technical as well as economic criteria.

To avoid confusion, let us also distinguish technology assessment from applications analysis. Technology assessment is a long-standing element of the federal energy research program. Although many technology assessment studies are referred to in the discussion of the development of applications analyses for the solar thermal energy program, the two terms are not synonymous. Technology assessment is defined as an area of study that "systematically examines the effects on society that may occur when a technology is introduced, extended, or delayed" (Division of Solar Energy, 1977). As such, technology assessment embodies a number of activities not normally included as part of applications analysis, including the assessment of socioeconomic and environmental impacts, macroeconomic trends, and future energy supply mixes. Applications analysis, which normally follows technology assessments and programmatic mission analyses, focuses more on the private market decisions that will result in the implementation of technologies assumed in the former studies.

6.2 Survey of the Federal Program

Federally sponsored applied research in applications analysis began in 1975. Before that time some studies were conducted to identify major application markets and long-term prospects for solar energy and other energy resources [for example, Project Independence (FEA, 1974)]. Certain industrial energy requirements analyses were conducted as part of the Federal Energy Administration's (FEA's) (Hall et al., 1975) charter in evaluating energy conservation opportunities. Otherwise, FEA maintained only a limited effort in defining market incentives for solar heating and cooling (DOE, 1978). The Energy Research and Development Administration (ERDA) carried out the principal responsibilities in solar research, development, and demonstration (RD&D). ERDA was responsible for examining the total economic and noneconomic "costs" of energy sources and studying major policy issues relating to solar energy. Although these policies hinted at the need for applications analysis, no such initiatives were undertaken until 1975 and 1976.

Total funding for solar energy RD&D also increased dramatically in these years, from only $14.8 million in 1974 to $151.6 million in 1976. Before 1977 the federal government sponsored six major solar energy strategy studies, some of which are reviewed in more detail in section 6.2.1.

These studies included NSF/NASA (1972), FEA (1974), Council on Environmental Quality (1978), ERDA (1975, 1977), and Division of Solar Energy (1977).

These studies, along with the extensive work of the Office of Technology Assessment (OTA) on selected applications and the MITRE Corporation on market penetration modeling, contributed to the understanding of solar technology and its application to energy service sectors. It was not until after 1977, however, that many specific criteria for the market penetration and application suitability of solar thermal technology were evaluated. The basis for applications analysis, which was developed or applied by many of these broad "mission" or strategy studies, is discussed in section 6.2.1.

In fiscal year (FY) 1977 and more extensively in FY 1978, DOE began to support research that was more specific in its coverage of end-use sectors and types of solar thermal technology. Within the Office of the Assistant Secretary for Energy Technology, DOE continued to support overall missions analyses of dispersed power applications (power, total energy, and process heat) for solar thermal systems under the guidance of the Aerospace Corporation. Applications analyses of specific markets (industrial, institutional, commercial, residential) were expanded under this program, as were the Aerospace studies of small power system applications. In the Office of Conservation and Solar Technology, two studies of the industrial market for process heat systems (by InterTechnology and Battelle) during 1977 catalyzed a rapid expansion of research and development and end-use matching studies in solar thermal process heat. Key results of some of these studies are discussed in the section 6.2.2.

These efforts, as well as the development of techniques for the evaluation of specific markets, continued in FY 1978 and FY 1979. For example, end-use matching for industrial applications was initiated at SERI (Lameiro and Brown, 1978) and a study of utility prospects for repowering power plants with central receiver technology was sponsored by the Public Service Company of New Mexico (Public Service Co. of New Mexico, 1978). New techniques for evaluating alternative small power systems were developed at the Jet Propulsion Laboratories and applied at SERI and Battelle Pacific Northwest Laboratories. Other contracts awarded in FY 1979 that relate to market and applications analysis are summarized in table 6.1. The impact of these studies on applications analysis methodology is discussed by major application sector in later sections.

Table 6.1
Market and applications analysis programs funded in FY 1978 and FY 1979

Program	Funding ($ thousands)[a]	Funding source
Effect of system factors on the economics of demand for small solar thermal power systems (utility and industry)	$100	General Electric
Comparative ranking of 0.1 to 10 MW$_e$ small solar thermal power systems	$330	Solar Energy Research Institute Battelle Pacific Northwest Lab Jet Propulsion Laboratory
Solar thermal plant impact analysis and requirements definition study	$255	Science Applications, Inc.
Small solar power systems mission analysis	$781	Oak Ridge National Laboratory
Conceptual system evaluations for solar central receiver hybrid power systems	$629	Bechtel National
Conceptual system evaluations for solar central receiver hybrid power systems	$634	Rockwell International
Solar thermal electric plants in a hydroelectric grid	$100	Bureau of Reclamation
Economic assessment of advanced solar thermal plants	$221	Westinghouse
Solar thermal repowering assessment	$468	Public Service Co. of New Mexico

Source: DOE (1979).
a. Current year dollars.

During FY 1979 the president's Domestic Policy Review established goals for displacing conventional energy resources with solar and hydroelectric technology that amounted to approximately 20% of the nation's projected energy needs in the year 2000 (Easterling, Grace, and Kettle, 1980). The targeted displacement was equivalent to 18.5 quads (1 quad = 10^{15} Btu = 1.05 EJ), of which 3 quads was due to solar thermal systems (PRC Energy Analysis Co., 1980). To accomplish these goals, the Solar Thermal Program (under the Assistant Secretary for Energy Technology) established a program strategy that included subsystem design and testing, applications analysis, advanced subsystem development, and commercialization. To direct research toward marketable applications, DOE established goals for end-use matching of solar technologies to markets. These are reflected in figure 6.2. These goals were based on numerous applications studies, but not on overall end-use matching assessments.

The Advanced Technology Subprogram, which was within the Solar Thermal Program, was responsible for identifying candidate components, subsystems, systems, and processes for solar thermal applications (PRC Energy Analysis Co., 1980). This identification process was supported by analytical studies in applications of fuels and chemicals, high-temperature [$> 1000°F$ ($540°C$)] process heat, and advanced electric power.

DOE was reorganized during FY 1980 to place all solar R&D programs under a single assistant secretary for conservation and solar energy. Under the deputy assistant secretary for solar energy, four program offices were established: Solar Applications for Buildings, Solar Applications for Industry, Solar Power Applications, and Alcohol Fuels. This organization shifted emphasis toward the appropriate application of solar technology and encouraged cooperation and cross-technology comparisons within end-use sectors. Solar thermal heating and cooling systems were assigned to Applications for Buildings. Other solar thermal programs (including power) were assigned to Applications for Industry. All three applications offices were responsible for market analysis, testing, and development within their respective market sectors. One of the first thrusts of these new offices was to commission market strategy analyses of their respective sectors.

In general, solar thermal applications analyses from FY 1981 and FY 1982 aimed at recapitulating and reformulating market thrusts for solar thermal technology. Review efforts during this period are summarized in section 6.2.3.

Applications \ Technologies	Distributed receiver: Parabolic trough	Distributed receiver: Fixed bowl	Distributed receiver: Parabolic dish	Central receiver
Low temperature	▲ Heating and cooling ◆ Low-temp IPH			
Industrial process heat	▲ Retrofit ◆ Mid-temp	▲ Small	◆ Small high-temp	▲ Retrofit
Mechanical power		▲ Small irrigation ◆ Large irrigation		
Total energy/ cogeneration			▲ Industrial ◆ Residential	▲ Small cogeneration
Small electric	▲ Community scale	◆ Community scale	▲ Remote ▲ Community scale	
Utilities			◆ Bulk electric	▲ Repowering ◆ Bulk electric
Fuels and chemicals processing			◆ Chemicals processing	◆ Large-scale fuels production

▲ Early, near-term ◆ Ultimate market application

Figure 6.2
Solar thermal power systems technology and applications matching (PRC Energy Analysis Co., 1980).

In FY 1981 solar research budgets were restricted, and thus difficult decisions on the future application of limited funds were necessary. Earlier studies on comparative ranking of technologies for various applications and cost goals were used to select the relative program funding needs and emphases. For the purpose of multiyear program planning efforts, DOE focused on the following applications for major technology groups:

1. Central receiver systems: electric utility applications
2. Parabolic trough systems: industrial process heat (IPH) applications
3. Parabolic dish systems: small community and on-site electric applications
4. Hemispherical bowl systems: small utility steam electric applications
5. Solar ponds: electricity and low-temperature heat at favorable sites

Overall program emphasis was also refocused on R&D technology and on specific application demonstrations. No new thrusts in end-use matching or applications analysis were proposed. However, $2.6 million (out of a FY 1981 budget of $141.8 million) was allocated to planning and assessment studies geared toward identifying market opportunities and constraints (DOE, 1981). Applications and requirements analyses initiated in FY 1980 were completed during 1981 and included an extensive review of industrial energy requirements, user attitudes, and application suitability for industrial process heat.

6.2.1 Early Developments

ERDA makes an early reference to the broad concept of energy service analysis in their 1975 national policy plan (ERDA, 1975). ERDA used the reference energy system developed by Brookhaven National Laboratory to analyze broad energy technology development scenarios in view of their ability to meet a given level and mix of end-use energy services. Energy sources were linked to end-uses by various processing, conversion, and distribution stages specified by the reference energy system. Among the conclusions of their study was the argument that improvements or modifications of end-use technologies are crucial to the management of the nation's liquid fuels shortage. However, ERDA did not identify specific applications analysis or end-use matching as a high priority for RD&D programs in either solar electric technologies or solar heating and cooling (ERDA, 1975).

The reference energy system used in ERDA's planning allowed for economic optimization of energy resource and energy service matching on a macroscopic level. Energy demand and supply projections were developed independently, and the reference energy system then solved for the least-cost set of energy supply sources that met energy demand projections based on exogenously specified unit costs (ERDA, 1975). The least-cost solution was defined macroscopically; that is, a set of energy resource technologies was selected based on overall costs of production, processing, conversion, and distribution (but not utilization). In this way the ERDA approach indicated a macroeconomic optimization of the supply-demand balance but did not reflect specific near-term competition among energy technologies in selected applications. Because the reference energy system did not account for specific end-user financing, site factors, or other constraints, its results were useful to indicate trends in but not specific utilization of new energy technologies.

The Stanford Research Institute (SRI) used a similar macroscopic energy system model to determine the most economical means of meeting energy needs in exogenously defined scenarios in its research for ERDA during FY 1976 (Division of Solar Energy, 1977). The primary purpose of this study was to perform a technology assessment of solar energy implementation in the future, not to provide specific market penetration estimates or technology rankings for selected applications. However, the principles that guided this analysis, as with the previous ERDA study, assumed the validity of the energy resource and energy service matching paradigm in long-term energy policy analysis. The SRI model assumed twenty-two different types of energy end-use demands in nine geographic regions. Although this was not an extensive or detailed universe of energy end-uses, it was nonetheless suitable for a broad energy policy comparison of multiple resource, processing, and distribution alternatives (for example, fourteen different solar process combinations were examined). Solar energy use in various end-uses was based on relative cost (using discounted present value of costs) of a generic solar system compared with other alternatives. In their study SRI noted the difficulty in developing "generic" solar system performance and cost estimates for these comparisons. They also noted the failure of their model to reflect various site-specific factors, conventional energy supply prices, and institutional barriers that have an impact on the actual use of solar technologies.

A later study by SRI investigated the relative market penetration potential of eleven solar technologies in fifteen end-use markets in nine geographic regions. The purpose of the study was to contribute to policy decisions by DOE's advisory board, the Solar Working Group. The method used by SRI in this study did not specifically address end-use matching but rather relied on a traditional market penetration model (solar penetration model, or SPM) to define the share of a given end-use market allocated to specific technologies based on relative price, energy demand, and market response parameters (Solar Working Group, 1978).

ERDA also sponsored a limited utilization of end-use matching in selecting residential solar heating and cooling demonstration sites in 1976. Residential, nonfederal demonstrations in the Department of Housing and Urban Development (HUD) program were selected on the basis of solar energy systems matched with specific geographic locations to provide unique research and market opportunities (Division of Solar Energy, 1976). In doing this, ERDA used a selection matrix that mapped location constraints (housing types, building codes, zoning, climate, energy demand, and fuel costs) against solar system characteristics.

Although ERDA believed that this approach to site selection was useful in the first round of residential demonstrations, it was not utilized in later demonstration program cycles. ERDA adopted a policy of evaluating system-site combinations on their individual merit in subsequent program cycles, with no attempt to influence selections through preanalysis. The decision to abandon the site-system selection methodology is reflected in R&D program decisions as well; among the market development and economic analysis initiatives proposed by ERDA in their 1976 program plan, there is no mention of a need for broad-based application analyses.

One of the first serious attempts to recognize and deal with the site-specific, application-specific nature of solar energy usage was the extensive study by the Office of Technology Assessment (OTA), which concluded in June 1978 (OTA, 1978). This study notes the potential importance of time (in terms of energy-use scheduling) and temperature in defining the value of solar energy, variables that were not a part of energy value assessments for conventional fuel technologies. As noted by OTA, "It is important to compare competing technologies on the basis of their ability to perform a specific set of tasks in a specific location—generalizations and simple 'measures of merit' can be very misleading" (OTA, 1978). OTA evaluated the relative merit of on-site solar technologies that met specific energy needs (single-family residence, apartment, shopping mall, mixed community, and selected industrial processes) in four locations (Albuquerque, Boston, Fort Worth, and Omaha). Obviously, by selecting the specific set of site and application scenarios, OTA was restricted from providing broad market estimates of future energy supply mixes or solar market potential. Their analysis allowed one to judge relative "competitiveness" of certain types of on-site solar systems but not comprehensive applicability of those systems. Furthermore, their analysis could not assess the competitiveness of solar equipment on the basis of an "optimal" system design for the selected site and end-use. However, other important aspects of applications analyses were addressed, such as comparison of alternatives based on their ability to meet the same end-use demand, life-cycle costs, economically "rational" end-user decision criteria, and integrated energy systems. This approach required, of course, the assembly of data regarding the cost of alternative and backup energy sources, the cost of solar equipment, specific aspects of end-use demands, and local site climatological factors.

ERDA contracted with the MITRE Corporation (METREK Division) in 1975 to study the potential utilization of solar energy to the year 2020 (Bennington et al., 1978). Like the previous studies, the MITRE work was comprehensive and broad in its coverage, but was more specific than others in its treatment of market penetration factors and specific application requirements. Much attention was dedicated to evaluating market penetration methodologies in the development of MITRE's own model, called SPURR (System for Projecting the Utilization of Renewable Resources). MITRE also developed considerable analytical background on the cost and performance of typical solar thermal systems. This and other market penetration studies are covered by Bennington in chapter 4.

MITRE considered seven basic technology groups in its study: wind energy conversion systems, fuels from biomass, solar thermal systems for hot water, heating, or cooling of buildings, solar thermal process heat systems, solar thermal electric utility systems, photovoltaic central power systems, and ocean thermal energy conversion systems.

The MITRE study included fifteen technology options in solar thermal alone and was clearly more detailed in its scope and more consistent with applications analysis methodology than any of the previously discussed strategic studies. Two energy scenarios were tested in MITRE's analysis: a recent trends scenario, based on ERDA's baseline data developed in the reference energy system studies, and the national energy plan scenario, based on proposed incentives and regulations proposed in legislation under the same name.

As stated in MITRE's description of their methodology, the SPURR system was "intended to fill the gap between the aggregate national energy models and the uncoordinated analyses of individual applications of solar technologies" (Rebibo et al., 1977). SPURR was intended to permit the analysis of a wide range of technologies on a regional basis, with explicit accounting for the dynamics of market penetration, changes in cost resulting from experience in production, and changes in prices and energy policy over time. Production experience between applications was shared in the model, as was intersectoral demand.

The MITRE study showed significant potential market penetration for solar thermal hot water, heating, and cooling systems (between 0.9 and 1.6 quads by 2000) and approximately the same potential for solar thermal process heat. The study also indicated generic application types and general regional response but not specific applications. The MITRE study was a major step toward applications analysis and end-use matching, but it stopped short of developing the specificity required to utilize these methods fully.

6.2.2 Targeted Market Studies

Almost coincident with several of the strategic studies noted, ERDA and DOE sponsored several analyses of specific market sectors to define long-term potential for solar thermal technology. Several parallel studies of solar thermal total energy systems were conducted for industrial, residential, institutional, and commercial applications of solar total energy under the overall direction of the Aerospace Corporation (Aerospace Corp., 1978b). These studies evaluated overall markets and selected several "test" generic application requirements for solar system evaluation. The McDonnell-Douglas Corporation undertook a comprehensive analysis in its study of the industrial sector (Rogan et al., 1975).

The objective of the McDonnell-Douglas study was "to define solar energy systems that are technically and economically feasible and can satisfy all or part of selected industry demands, and to determine the market potential of such systems" (Rogan et al., 1975). Unlike the aggregate market studies of the early phase of the ERDA program, the McDonnell-Douglas study first selected those industrial applications with the highest energy demand and then ranked those applications according to the economic viability of a solar thermal total energy system. A review of industrial energy demand data was followed by phone surveys of 1,000 plants to determine actual energy demand schedules, process parameters, etc. Forty applications were selected for first-level system designs, and then five conceptual designs were completed for each of six geographic locations. This study can be considered the first true attempt at applications analysis in the solar thermal program. It demonstrates both the advantage of evaluating technologies at true end-use levels and the limited market perspective allowed by such detail.

In 1976 the InterTechnology Corporation (Fraser, 1977) and Battelle Laboratories (Hall, 1977) assessed the potential for solar thermal applications for process heat. Each contractor was asked to identify and characterize industrial energy requirements and specify state-of-the-art solar systems related to these requirements. The contractors were also asked to perform preliminary assessments of relevant nontechnical issues surrounding the adoption of solar thermal systems by industry. The conclusions of the two studies differed significantly because of several factors, but the net result of the studies was to increase the attention given to solar industrial process heat applications.

Both contractors pursued an approach similar to those used in later attempts at end-use matching. InterTechnology, for example, evaluated the performance of generic solar pond, flat-plate, evacuated tube, parabolic trough, and paraboloid collector systems (with storage), using simulations at specific sites. These simulation results were adjusted for temperature and location meteorology and applied against actual industrial demands in each state. For the first time in the solar thermal program, a database of industrial energy requirements was developed that included not only process, industry, and demand but also the temperature and form of heat required. Over seventy-eight industry groups accounting for 59% of total industrial energy consumption were included in the Inter-Technology database. The Battelle study, on the other hand, concentrated on only six major energy consuming industries (accounting for 80% of total manufacturing consumption) and on actual applications requiring heat below 350°F (180°C).

The Aerospace Corporation and the Jet Propulsion Laboratory evaluated the dispersed small power systems market between 1978 and 1980. Their work included several surveys of both utility and industrial use of small power systems and a review of applications of solar thermal technology to these needs (Aerospace Corp., 1979b; Aerospace Corp., 1980). The Jet Propulsion Laboratory directed work on technical system analysis of point-focus distributed collector systems for power applications and also the pioneering work in system concept "ranking" using detailed decision analysis in 1979 (Feinberg, Kuehn, and Miles, 1979). This ranking system was later utilized in an extensive study of small solar thermal power system technologies (Thornton et al., 1980). Although this study concentrated on the evaluation of system performance and economics under "generic" operating conditions, the ranking criteria method with which it evaluated small power system concepts was a useful contribution to the understanding of utility applications.

6.2.3 Review and Reformulation

Following DOE's reorganization into applications-oriented departments in 1980, two major efforts were sponsored to assemble market development strategies for all solar technologies in power applications and industrial applications (DeAngelis, Edesses, and Wilson, 1980; DOE, 1980). These studies provide a good general review of previous work on end-use matching, applications analysis, and market penetration studies. In addition, new information was provided on a current review of end-user attitudes and requirements. This review resulted in an overall strategy formulation for solar thermal as well as other solar technologies. Chiefly these strategies called for targeting specific near-term markets for solar thermal energy based on its relative strength in major application markets as compared to other renewable technology alternatives and conventional fuels. An example of the preliminary matching of solar thermal technologies to industrial markets is shown in table 6.2.

The end-use matching and location mapping techniques discussed here were used by SERI in a broad study of the potential for solar energy and conservation in U.S. industry during 1980 and 1981 (SERI, 1981). Part of this evaluation involved using the computer programs developed as part of the SERI end-use matching analysis to identify energy displacement potentials by region and industry in the year 2000. The availability of land or roof area was identified as a crucial issue in the SERI study, but data regarding such constraints were not available.

Following the completion of these market and application strategy studies, DOE sponsored little more research in end-use matching and applications analysis for solar thermal energy. However, at least one significant effort was undertaken to apply these same methodologies, with much improved application data, to other energy supply technologies. Under the Assistant Secretary for Conservation and Renewable Energy

Table 6.2
Matching of solar thermal technologies to industrial markets

Industrial market	Technology type[a] Passive	Flat-plate/ Evacuated tubes	Concentrators Line focus	Point focus Thermal	Point focus Electric	Central receiver Thermal	Central receiver Electric
Agriculture							
Crops production (01)	1	1	×	4		×	×
Livestock production (02)	1	1	×	4		×	×
Agricultural Services (07)	1	1	4	4		×	×
Forestry (08)	1	1	×	4		×	×
Fishing (09)	3	3	×	4		×	×
Manufacturing industry							
Food (20)	2	2	3	4	4	4	4
Tobacco (21)	1	3					
Textile mill products (22)	2	3	4	4	4	4	4
Apparel (23)							
Lumber and wood (24)	1	2	4	4	4	4	4
Furniture and fixtures (25)	1	2	4	4	4	4	4
Paper (26)	4	4	4	4	4	4	4
Printing and publishing (27)							
Chemicals (28)	×	×	×	4	4	4	4
Petroleum and coal products (29)	×	×	4	4	4	4	4
Rubber and plastics (30)							
Leather (31)							
Stone, clay, and glass (32)	×	3	3	4	4	4	4
Primary metals (33)	×	4	4	4	4	4	4
Fabricated metals (34)	3	3					
Nonelectric machinery (35)	3	3					

Table 6.2 (continued)

Industrial market	Technology type[a]		Concentrators				
	Passive	Flat-plate/ Evacuated tubes	Line focus	Point focus		Central receiver	
				Thermal	Electric	Thermal	Electric
Electric machinery (36)	3	3					
Transportation equip. (37)	3	3					
Instruments and related products (38)							
Miscellaneous manufacturers (39)							
Mining							
Metals (10)	×	3	3	4	4	4	4
Anthracite (11)	×	×	×	×	×	×	×
Bituminous coal and lignite (12)	×	×	×	×	×	×	×
Oil and gas (13)	×	4	3	3	3	3	3
Nonmetallic minerals (14)	×	3	3	4	4	4	4
Construction							
Buildings (15)							
Heavy (16)							
Trade contractors (17)							

Source: DOE (1980, p. 57).
a. 1, very competitive now; 2, competitive soon, some applications now; 3, competitive soon; 4, applicable but marginally competitive within ten years; ×, not applicable; blank, insufficient data.

(Office of Industrial Programs), General Energy Associates (GEA) evaluated the plant-specific potential for sixteen different cogeneration fuel-technology combinations in U.S. industry (GEA, 1983). GEA used the database it had compiled as part of a previous DOE contract concerning waste heat. This database, comprising 10,000 U.S. industrial plants, undoubtedly the most complete and detailed database of industrial process energy requirements ever assembled, was instrumental in the effective application of end-use matching techniques by GEA.

As part of this study, cogeneration fuel-technology combinations were matched to plant-specific energy requirements on the basis of total plant heat requirements. An appropriately sized cogeneration system was then evaluated on the basis of return on investment (considering both fuel and power savings as well as excess power sales revenue), and all combinations were then screened to identify those exceeding a minimum return criterion. The GEA study identified 3,131 plant sites (over 30% of the total) that exceeded the minimum 7% real return on investment criterion. The study also used forecasts for each industry and region to predict potential energy savings resulting from industrial cogeneration through the year 2000.

The GEA study was completed in 1983. No other extensive applications of end-use matching methodologies under DOE sponsorship are documented from 1982 to 1984.

6.3   Major Program Elements and Development of Methodologies

The results of some applications analysis programs briefly mentioned earlier deserve special attention in the context of discussing the technical requirements review of solar thermal technologies. These program elements constitute examples of particularly important methodological advances in applications analysis for solar thermal energy. The following subsections discuss these advances in more detail, including application analysis of large power systems and the resultant repowering strategy analysis, industrial process heat end-use matching, industrial requirements analysis, case studies, and geographic suitability analysis.

6.3.1   Applications Analysis for Large Power Systems

Often the specific goal of a technology-related market study is to define application characteristics of a portion of the market best suited to a

solar technology and thereby to identify additional technology development needed to increase penetration of that market. Studies of this sort
were an important part of the Solar Central Receiver Program of DOE,
managed by Sandia National Laboratories (Livermore). Initial directions
in this program emphasized applications to the utility market, where
well-defined and consistent market requirements could be identified.

Interest in potential application of central receiver technology to industrial process heat (IPH) applications began in FY 1979. Aerospace performed basic research on high-temperature IPH applications (Aerospace
Corp., 1978a; Aerospace Corp., 1979a) while SERI (Wright and Dougherty, 1979) informally reviewed the feasibility of certain moderately high-
temperature applications. These studies and others were reviewed by
Sandia (Fish, 1980) to establish a basis for applicable industrial markets
in the DOE program. Sandia's report notes the applicability of central receiver equipment to two classes of IPH requirements: direct production
of process steam and preheating of combustion furnace air or heat transfer fluids. In particular, the Sandia review suggests central receiver systems for providing process steam [below 550°F (290°C)] and heat transfer
fluid preheating [< 1,200°F (650°C)] to oil refineries and air preheating
[up to 1,000°F (540°C)] to lime calcining. The report does not address a
specific matching methodology but does note the need to address technological problems of storage, receiver cost versus exit temperature, and
general economic barriers.

The effect of receiver exit temperature on the cost of the overall system
led Sandia Laboratories to question the actual distribution of temperature requirements and on-site energy needs at various sites. Analyses
by Sandia Laboratories (Iannucci, 1981) used data from the Census of
Manufacturers and other sources to estimate the distribution of IPH
usage by size and end-use temperature. Data were extracted from the
comprehensive work by Energy and Environmental Analysis, Inc. (1979),
which compiled information on the thirty-nine largest energy-consuming
industries in the United States from U.S. Census, FEA Major Fuel Burning Installation, and Bureau of Mines surveys. Data used in establishing
the energy consumption distributions were adjusted for feedstock fuel
consumption in iron and steel and in ethylene production but were not
adjusted for electrical energy use (estimated by Iannucci to be only 10%
of energy use) or for other non-IPH usage. The study considered actual
data representing about 67% of total manufacturing energy consumption

(Iannucci, 1981). Iannucci's estimates of plant energy demand were not based on actual individual facility data. Instead, the energy demand rating was calculated on the basis of an "average" industrial plant assumed to operate for 8,760 hours per year at a constant energy demand rate.

From this study, Iannucci concluded that a significant fraction of the total *number* of industrial energy-using facilities are less than 3 MW_t in size and consume end-use heat at less than 450°F (30°C) (see figure 6.3). On the other hand, a substantially greater portion of energy consumed by industry is utilized at much higher temperatures and in larger facilities (see figure 6.4). More than 70% of the energy consumed is used in facilities between 30 MW_t and 300 MW_t. Furthermore, the study indicates that food processing is generally conducted in smaller facilities with low temperature requirements (except for wet corn milling and sugar refining). Petroleum refining, chemical processing, glass manufacturing, and iron and steel applications are predominantly conducted in larger facilities with higher temperature requirements.

6.3.2   Repowering Program and Applications Analysis

In a technical and economic assessment study completed by the Public Service Co. of New Mexico in 1978 (Public Service Co. of New Mexico, 1978) market analysis was provided that identified significant opportunities in "repowering," the partial substitution of central receiver thermal plants for input heat at existing electric generating stations. Additional analysis of utility and industrial repowering market potential provided by the MITRE Corporation (MITRE Corporation, 1978; and Curto, 1978) and Battelle (Hall, 1977) indicated that the repowering market might exist for industrial installations as well as electric power. Convinced of the overall market potential, DOE began a program of site-specific design studies (which were expected to lead to a limited number of actual cofunded demonstrations) rather than more extensive market analyses. The site-specific studies considered technical and economic applications and can be considered case studies for central receiver technologies. The participants and locations of each of these case studies are shown in tables 6.3 and 6.4.

The original conceptual design studies, including eight utility repowering designs and six industrial retrofit designs completed between 1980 and 1981, were summarized in a report published by Sandia Laboratories

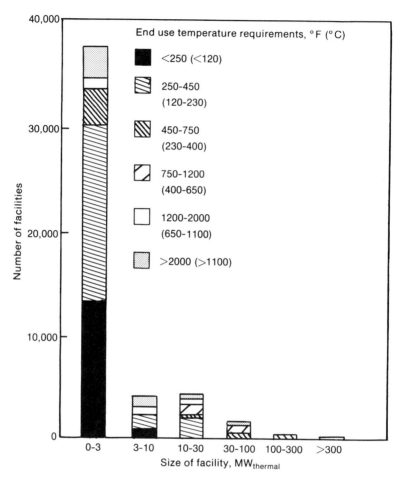

Figure 6.3
Distribution of *number* of facilities by size and temperature (Iannucci, 1981, p. 13).

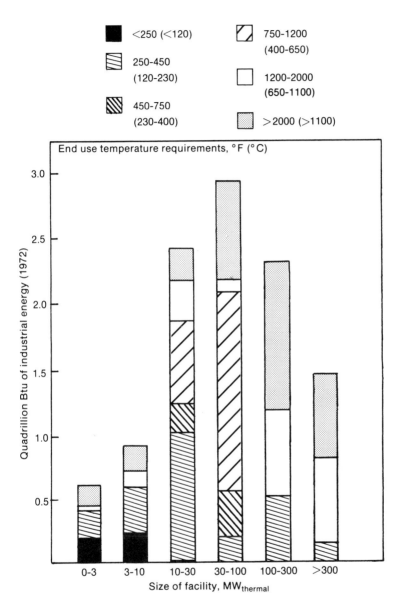

Figure 6.4
Distribution of *energy consumed* at facilities by size and temperature (Iannucci, 1981, p. 14).

Table 6.3
Repowering electric utility applications[a]

Prime contractor	Subcontractor	Site location	Plant data Net output (MW$_e$)	Turbine inlet	Type/date	Receiver technology and output temperature	Receiver output power (MW$_t$)	Number of heliostats
Arizona Public Service (APS) (Eric Weber)	Martin Marietta; Badger Energy; Gibbs and Hill	Saguaro Power Station, Tucson, Arizona	113.2	1000°F (538°C) 1450 psig (10 MPa)	Nonreheat 1954	Molten salt, quad cavity. 1050°F (566°C)	350	10,500/49.05
El Paso Electric Co. (EPE) (Jim Brown)	Stone & Webster; Westinghouse	Newman Station, El Paso, Texas	82	1000°F (538°C) 1450 psig	Reheat 1960	Advanced water/steam, external. 1020°F. Separate reheat recycled	92/13	2776/81.8
Black & Veatch Consulting Engineers (Sheldon Levey)	Public Service Co. of Oklahoma (PSO); Babcock & Wilcox; Baily Controls Co.	Northeastern Station, Oologah, Oklahoma (near Tulsa)	145	1000°F (538°C) 1800 psig (12.4 MPa)	Reheat 1961	Advanced water/steam, external (240° sector). 1012°F (545°C)	73.3	2255/49.05
McDonnell Douglas Astronautics Co. (C. R. Easton)	Sierra Pacific Power, Foster Wheeler Division; Stearns-Roger; Desert Research; University of Houston; Westinghouse	Ft. Churchill Plant, Yerington, Nevada (near Reno)	115	1000°F (538°C) 1800 psig	Reheat 1967	Molten salt, partial cavity. 1050°F (566°C)	330	8411/56.42
General Electric (Jim Elsner)	Southwestern Public Service (SPS); Kaiser Engineers	Plant X, Earth, Texas (near Lubbock)	100	1000°F (538°C) 1450 psig	Reheat 1955	Liquid sodium, cylindrical, external. 1100°F (593°C)	142	4809/49
Rockwell International Energy Systems Group (Tom Springer)	Stearns-Roger; McDonnell-Douglas; University of Houston; Texas Electric Service Co.	Permian Basin Station, Monahans, Texas (near Odessa)	115	1000°F (538°C) 1465 psig (0.02 MPa)	Reheat 1958	Liquid sodium, cylindrical, external. 1100°F (593°C)	158.5	4742/56.42
Rockwell International Energy Systems Group (Tom Springer)	West Texas Utilities; University of Texas; Boeing; Sargent & Lundy	Paint Creek Station, Stanford, Texas (near Abilene)	110	1000°F (538°C) 1800 psig	Reheat 1972	Liquid sodium, cylindrical, external. 1100°F (593°C)	226	7882/49
Public Service of New Mexico (J. D. Maddox)	Stearns-Roger; Westinghouse	Reeves Station, Bernalillo, New Mexico (near Albuquerque)	46	950°F (510°C) 1250 psig (8.7 MPa)	— 1957	Water/steam, Rockedyne once though, based on Barstow. 1050°F (566°C)	111	40,086/38

Prime contractor	Total mirror area (m²)	Land area (acres)	Field configuration	Solar output design point (net MWₑ)	Annual energy produced (GWhₜ)	Storage	BBL of oil (equivalent) saved annually	Capital costs (heliostat unit price)	$/KWₑ	Annual energy/area (MWhₜ/m²)	Cost/MWhₜ fuel displaced	Solar fraction (annual)
Arizona Public Service (APS) (Eric Weber)	515,025	547	Surround	111	719.5	3.8 hr @ 305 MW	493,000	$166.8 M ($230/m²)	523 523	1.397	$199	0.273
El Paso Electric Co. (EPE) (Jim Brown)	227,077	320	160° N	41	206.8	None	133,000	$93.1 M ($230/m²)	1067	0.98	$397	0.60
Black & Veatch Consulting Engineers (Sheldon Levey)	110,617	126	Near React. North	30	115.2	None	88,000	$55.1 M ($260/m²)	757	1.04	$368	0.08
McDonnell Douglas Astronautics Co. (C. R. Easton)	474,549	520	130° N	77	759	6 hr	490,000	$196 M ($224/m²)	593	1.6	$258	0.23
General Electric (Jim Elsner)	235,881	212	Surround	57	290.5	10 min full power	200,000	$116 M ($230/m²)	811	1.23	$399	0.23
Rockwell International Energy Systems Group (Tom Springer)	267,544	322	Surround	50	355.5	1 hr	237,000	$112 M ($260/m²)	869	1.346	$313.9	0.28
Rockwell International Energy Systems Group (Tom Springer)	386,218	480	Surround	60	482.5	4 hr full power	351,000	$145 M ($230/m²)	619	1.249	$234.6	0.32
Public Service of New Mexico (J. D. Maddox)	155,268	—	Surround	25	—	None	—	$125 M	—	—	—	0.5

Source: Gibson (1982, p. 7).
a. All dollar values are in 1979 dollars ($\times 1.421 = 1985$ dollars), except in the last column. Those values are in 1978 dollars ($\times 1.547 = 1985$ dollars).

Table 6.4
Industrial process heat applications[a]

Prime contractor	Subcontractors	Site location	Plant data Process	Process fluid	Temperature	Receiver technology	Output temperature	Receiver output power (MW$_t$)	Number of heliostats/m^2	Total mirror area (m^2)
Northrup Inc. (Roy Henry)	Arco Oil & Gas Co.	North Coles Levee, Plant 8 (22 mi west of Bakersfield, CA)	Natural gas production	Hydrolight Cycle Oil	575°F (302°C)	Oil fluid, once thru cavity	560°F (293°C)	11.8	320/52.6	16,832
Martin Marietta Denver (Dave Gorman)	Exxon Research & Engineering; Foster Wheeler Development Corp.; Black & Veatch	Edison Oil Field, Bakersfield, CA	Enhanced oil recovery	Wet steam	450°–670°F (232°–354°C) 80% quality	Water/steam twin cavity, wet steam	567°F (297°C) 82%	29.32	818/49.05	40,123
McDonnell Douglas Astronautics Co. (L. W. Grover)	Gulf Research & Development Co.; Foster Wheeler; Univ. of Houston	Mt. Taylor Uranium Mill, San Mateo, NM (60 mi west of Albuquerque)	Uranium processing	Saturated steam	366°F (186°C)	Water/steam external, drum type	399°F (204°C)	15.7	383/56.4	21,601
Foster Wheeler Development Corp. (R. Raghaven)	Foster Wheeler; Energy Corp.; Provident Energy Co., Inc.	Provident Energy Refinery, Mobile, AZ (near Phoenix)	Refinery	Steam (separate fossil superheater	700°F (371°C)	Water/steam external, flat panel	520°F (271°C)	43.5	1,174/56.4	66,214
Boeing Engineering and Construction Co. (D. K. Zimmerman)	United States Gypsum; Institute for Gas Technology; Shawinigan Engineering Corp.; North America Turbine	Sweetwater Plant, Sweetwater, TX (40 mi west of Abilene)	Gypsum board drying	Hot air	800°F (424°C)	Air cavity, forced circulation	1,335°F (723°C)	11.92	469/44	20,636
PFR Engineering Systems, Inc. (Tzvi Roseman)	Valley Nitrogen; McDonnell Douglas; Univ. of Houston	El Centro Ammonia Plant, El Centro, CA (100 mi east of San Diego)	Ammonia production	Solar-heated steam/methane refining	1450°F (788°C)	Reradiation, inner cavity	1,450°F (788°C)	34.5	1,040/56.4	58,656

Prime contractor	Land area (acres)	Field configuration	Solar output design point (net)	Annual energy produced (GWh$_t$)	Storage	BBL of oil (equivalent)	Capital cost ($) (heliostat unit price)	$/kWt	Annual energy (MWh$_t$/m^2)	Cost/fuel displaced (per MWh$_t$)	Solar fraction (annual)
Northrup Inc. (Roy Henry)	32	North	9.5 MWt (32.5 × 10^6 Btu/hr)	23.2	None	21,336	$8.34 M ($301/m^2)	741	.34	$368	0.90 (0.24)
Martin Marietta Denver (Dave Gorman)	47	150° North	29.3 MWt (100 × 10^6 Btu/hr)	55.9	None	44,058	$14 M ($230/m^2)	478	1.39	$188	(2.6)
McDonnell Douglas Astronautics Co. (L. W. Grover)	34	North	13.9 MWt (47.4 × 10^6 Btu/hr)	31.8	None	21,135	$12 M ($230/m^2)	1,182	1.47	$332	(0.21)
Foster Wheeler Development Corp. (R. Raghaven)	73	North	43.2 MWt (147.4 × 10^6 Btu/hr)	105	3 min. Press. water	76,724	$27.6 M ($272/m^2)	603	1.59	$212.3	(0.208)
Boeing Engineering and Construction Co. (D. K. Zimmerman)	13	North	10.54 MWt, 1.3 MWe cogenerating electricity	16.8	None	11,734	$9 M ($230/m^2)	632	0.814	$427	0.07 (does not include 0.221 MWe)
PFR Engineering Systems, Inc. (Tzvi Roseman)	60	North	34.0 MWt (115.8 × 10^6 Btu/hr)	84.5	None	Additional 10,500 tons of ammonia produced	$24.9 M ($230/m^2)	721	1.44	$261	0.66 (0.226)

Source: Gibson (1982, p. 11).
a. All dollar values are 1979 dollars (× 1.421 = 1985 dollars).

in June 1982 (Gibson, 1982). DOE did not identify the type of technology, application, or location to be considered; therefore there were significant differences in actual design. All of the selected sites are in the southwestern United States.

The conceptual designs illustrated the wide variety of system technology configurations that would be suitable among the numerous location and application combinations throughout the United States. None of the designs demonstrated compelling economic merit (although some applications were marginally acceptable with moderate government cost subsidies). More important, these studies illustrated the technical and economic factors crucial to eventual commercialization of the technology. These factors can be added to a growing list of important criteria in solar technology end-use matching (see table 6.5).

Two major results of the conceptual design studies emphasize factors that influence the ability to match solar technology to applications. First, the temporal nature of plant operations (including overall system dispatch in the case of power generation) is extremely important in facilitating system compatibility and design. Second, end-users review alternative energy systems with respect to the same economic factors affecting fuel availability, cost, and capital financing. Solar thermal systems are only one of a number of alternatives available. Users indicated that the economic merit of solar retrofit must be compared to other fuel-saving or switching options (such as coal-fired generation or conservation). To be the preferred option, a solar retrofit investment must provide returns greater than these other options, not simply net savings against current fuel-use patterns.

The substantial body of applications analysis and system design work for solar central receiver systems done by Sandia Laboratories during this period was summarized in a design guide (Battleson, 1981). This report addresses site selection and system configuration criteria, including insolation, land, water, and storage requirements. Although Battleson treats insolation and gross land and water requirements generally, he introduces a detailed list of site selection criteria that should be considered in complete application analyses of central receiver systems.

Battleson also discusses general criteria for the minimum economic size of a solar electric application (100–400 MW_e) and the expected number of heliostats and tower height for representative applications by output power. Although not discussed in this report, electric repowering retrofit

Table 6.5
Technical and economic factors influencing the acceptability of central receiver technology in repowering or industrial retrofit applications

Technical factors		Economic factors	
Utilities	Industry	Utilities	Industry
System performance is dependent on overall utility system characteristics	Land availability	Selective dispatch (peak load) capability must be provided to maximize value; implies that storage is necessary	Storage is generally not cost-effective
Installation must be of sufficient size to have an impact on overall system dispatch strategy		Technology risk will have an adverse impact on economic merit	Perceived technical risk
Technology must demonstrate adequate safety and reliability		Impact of repowering on useful life of existing plant should be determined	Additional operating and maintenance staff requirements
Simplicity in design and familiarity of operating materials and systems is desirable	Technical demonstration necessary	Magnitude of initial capital requirements have an impact on ability to finance construction	Impact on plant operation and proven feasibility of full output/control
Atmospheric emissions	Atmospheric emissions	Economic viability of repowering is time dependent because of the increasing age of available plants and the increasing availability of coal-fired capacity	Rapid capital payback often crucial

Source: Gibson (1982).

applications probably have a significantly smaller minimum economic size (approximately 50 MW$_e$) than stand-alone solar thermal power plants. Industrial retrofit applications have an even smaller economic size because substantial thermal storage is generally not required.

6.3.3   End-Use Matching

The market studies of MITRE Corporation, InterTechnology, and Battelle catalyzed interest in the development of solar thermal technologies to supply industrial process heat. However, the approach adopted in these studies did not go far enough in identifying specific near-term opportunities for solar IPH by attempting to match system configurations with plant-specific energy needs on a broad scale. Although it is difficult to predict actual aggregate market penetration, one could estimate prime market applications and thus provide targets for commercial development. This modification of purpose, from one concerned with absolute energy displacement estimates to one concerned only with the "targeting" of primary markets, is a key distinction between market analysis and end-use matching.

During late 1977 analysts at SERI began to formulate a research program plan for evaluating IPH markets by end-use matching. Kreith, Lameiro, and Brown (1977) introduced the end-use matching approach in a paper that demonstrated the need to match solar system designs in a given location with both quantity and quality of energy service demands. This initial research resulted in the design of a program to delineate an end-use matching methodology. As stated in a SERI program summary published in April 1978, the purpose of end-use matching was to "provide a coarse map of selected industrial plant sites with suitable solar process heat systems matched to those plant requirements, and an analysis of their costs." In addition, SERI saw the need to refine further this coarse mapping through case studies of selected sites and used such case study information to help select promising collector technologies and define crucial research needs (Lameiro and Brown, 1978).

Figure 6.5 shows SERI's overall program approach adopted for end-use matching. This program approach emphasized the need for both "industrial acceptance case studies" and for "plant site/system mapping," which later became important aspects of solar IPH application analysis at SERI and elsewhere. The program diagram also points out the importance of data gathering (IPH technical requirements, eco-

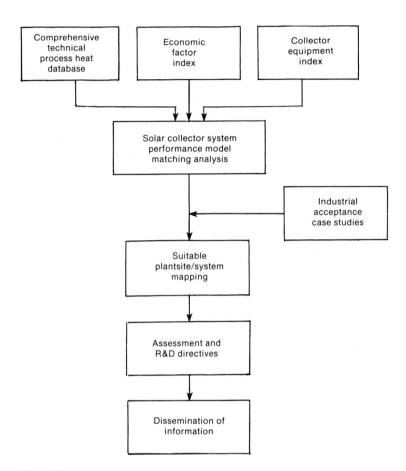

Figure 6.5
End-use matching program approach.

nomic factors, and collector equipment information) to end-use matching. SERI identified the need to establish locations, minimum terminal process temperatures, minimum heat rates, process parameters, operation schedules, and reliability requirements for a large number of standard industrial classified (SIC) industries. These data were part of an industrial process heat database (IPHDB), the basic contents of which are outlined in table 6.6. In addition, SERI established a fuel price index by location and a collector equipment manufacturers' database. A computerized evaluation model for solar system performance was proposed to select the least-cost system satisfying the specific IPH requirements found in the database. This selection process was, by definition, end-use matching.

After considerable development work on a suitable solar system performance model and data collection effort to supply minimum requirements, SERI produced a set of results for six selected locations in early 1980 (Brown et al., 1980). Figure 6.6 illustrates the process that led to these results. Although the results of this study were limited, SERI was successful in developing and implementing a methodology for effective end-use matching. The project identified some viable applications that were normally obscured in other studies and helped to distinguish the feasibility of applications in different locations, even though actual process and system selections were limited to available site-specific data.

SERI identified four factors that governed the technical and economic feasibility of supplying IPH from a solar thermal system: adequate quantity of delivered heat, adequate quality (that is, temperature and medium) of delivered heat, the means to transfer heat from collector source to process, and relative cost (or economic feasibility) of delivered heat using the system (Brown et al., 1980). SERI's method sought to selectively match solar thermal IPH systems to industrial processes in various cities by considering system performance and economics against each of these factors.

To perform such end-use matching, a requirements analysis was performed on elements of industrial process needs (quantity, quality, and schedule), collector delivery potential, and economics (competing fuel cost, system installed cost, etc.). The potential complexity of these specific comparisons and the number of such comparisons that might be required (there are some 200,000 individual manufacturing plants in the United States, well over 1,000 processes that might be encountered, and a multitude of solar collector and system combinations available) argued for use of a computerized approach to such end-use matching.

Table 6.6
Contents of the Industrial Process Heat Data Base (IPHDB) proposed by SERI and used in end-use matching

Label	Description or comment
SIC	Standard Industrial Classification code describing process
ALPHA	Alphanumeric character to distinguish segmented process
NAME	Description of industry
TMP	Required process temperature ($°C$ or $°F$)
HEATR	Required process heat rate (MW or MBtu/h)
FLOWR	Maximum required steam flow rate (kg/s or 10^3 lb/h)
SAE	Standard annual energy use [10^{13} J/yr or 10^{10} Btu/yr]
SYS	Possible solar systems, in order of applicability
	1 Direct hot water
	2 Indirect hot water
	3 Direct hot air
	4 Indirect hot air
	5 Steam flash
	6 Steam generator
BF	Backup fuel
	1 Natural gas
	2 Electricity
	3 Oil
	4 Coal
OP	Operation schedule
	1 Continuous process
	2 Batch process
SOP	Seasonal operation
	0 Continuous
	1 Seasonal
ENERGY	Energy required per unit of industrial output
UNIT	Unit of production for ENERGY

Source: Brown (1980, p. 17).

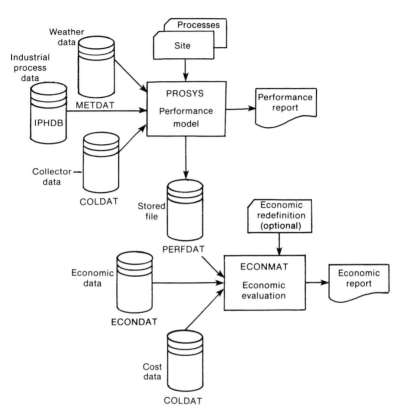

Figure 6.6
End-use matching process: relationships between models and data (Stadjuhar, 1981, p. 3).

The computer approach used consisted of two main programs: a long-term average performance evaluation for solar IPH systems, known as PROSYS, and an economic analysis routine and system matching selector, known as ECONMAT. SERI developed these two programs and their databases between 1978 and 1981. Researchers used the system to provide market data in test data situations with six cities and in response to certain private requests directed to the program. SERI published the computer programs in 1981 as a software package with a user's guide (Stadjuhar, 1981).

This first and most extensive survey of solar IPH feasibility through end-use matching concluded that the accuracy of PROSYS and ECONMAT as a means for end-use matching was directly dependent on the accuracy

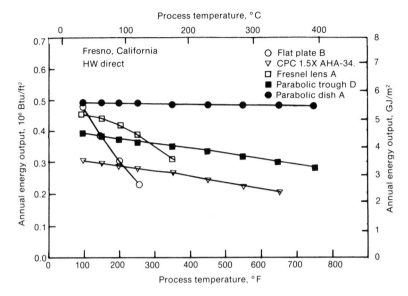

Figure 6.7
Annual energy output of several collector types over a range of temperatures for Fresno, California (Brown et al., 1980, p. 88).

and completeness of input data (particularly with regard to industrial energy use). Furthermore, the limitations of PROSYS as a simulation tool were important in interpreting the results, because PROSYS is neither a dynamic simulation model nor a detailed system heat balance analyzer.

Further refinements of PROSYS resulted in improvement of its accuracy to within 10% of other solar system performance simulation models—a significant accomplishment considering the simplicity of the initial basis of solar performance evaluation (Stadjuhar, 1981). PROSYS is based on a generalized collector performance model, which is based on the "utilizability" concept advanced by Rabl and Collares-Pereira (Rabl and Gordon, 1982; Collares-Pereira and Rabl, 1979). PROSYS assumes a constant, average working temperature for the collector and approximates system heat and parasitic power losses not associated with the collector, but it does not model storage. PROSYS was modified after its initial end-use matching applications to accommodate single-application sensitivity analyses and case studies.

PROSYS and ECONMAT yielded the site-specific performance charts shown in figure 6.7, which were produced when collector-system combinations were tested against a variety of IPH temperature requirements. PROSYS and ECONMAT also yielded a histogram of application SIC codes in a given range of cost (where cost is that of the least-cost matching solar system), which was of more use to those interested in targeting available markets (see figure 6.8). These histograms yield intuitive as well as specific information for market analyses.

In view of the initial successes and limitations of end-use matching for solar energy, SERI recommended more definition of industrial energy requirements, more specific capabilities for PROSYS, more information on solar IPH system costs, and more utilization of case studies in supporting solar thermal research and development. These recommendations were pursued through continuing research programs in IPH applications at SERI and elsewhere from 1981 to 1982.

6.3.4   Industrial Energy Requirements Analysis

FEA, ERDA, and DOE sponsored a number of industrial energy requirements studies between 1974 and 1981 to support the solar IPH, total energy, and solar small power systems programs, as well as ongoing programs in conservation and geothermal energy. Although information in many of these studies was useful, some data necessary for evaluating solar process heat system performance in applications analysis were not readily available. The lack of available data was cited as a constraint to effective end-use matching in the study by SERI (Brown et al., 1980) and a general program need in both the Solar Thermal Power and AIPH programs. Near the end of 1978 SERI coordinated and integrated industrial energy requirements studies with several DOE contractors, resulting in a report issued in October 1979 (Green, 1979). Although the report outlined a program for coordinating industry contacts and sharing data, not all entities actively participated.

Contractors involved in these early discussions on industrial energy use carried out a variety of separate studies with divergent objectives. McDonnell-Douglas, for example, collected fairly detailed data on a small number of industries as part of its study in support of total energy mission analysis (McDonnell Douglas Astronautics, 1977). A subsequent contract in 1978 supported additional industrial demand studies, which investigated "modularization" approaches for total solar energy systems. Insights West studied industrial energy use and industry attitudes among manufacturers in southern California between 1975 and 1979 for the

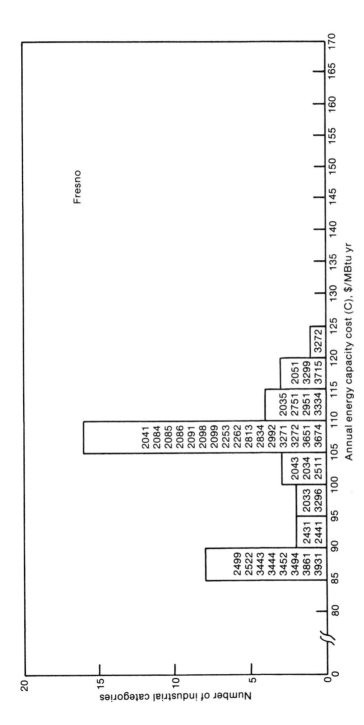

Figure 6.8
Ranking of solar IPH applications in Fresno, California, on the basis of energy capacity cost. The numbers on the bars are SIC codes. 1980 dollars × 1.303 = 1985 dollars. From Brown et al. (1980, p. 94).

Southern California Gas Company and the Gas Research Institute. These studies, which covered industrial plants within a "most likely" category, included field interviews from two hundred selected plants across the nation and the development of a "matching matrix" of available solar hardware to process heat requirements (Insights West, 1980). The Insights West studies were unique in their attention to owners' perceptions of decision factors (economic), attitudes, and identification of suitable applications. Insights West continued this type of fact-finding field interview for SERI after 1980 (Wilson, Williams, and Ball, 1981).

The Jet Propulsion Laboratory (JPL) reviewed potential commercial and industrial applications of solar energy in California through twenty-five field interviews (Barbieri and Pivirotto, 1978). This review was followed by case studies of selected applications (French and Bartera, 1978) and a survey of solar thermal cogeneration applications in twelve locations in California. JPL also cooperated in a study of solar applications for heating buildings conducted between 1972 and 1979 with the Southern California Gas Company (SAGE Project).

The Aerospace Corporation studied both high-temperature process heat and industrial on-site power applications for DOE. Data collection efforts in both studies included extensive telephone surveys, analyses of Federal Power Commission forms, and site surveys. Aerospace surveyed industrial power requirements of 409 companies with small (less than 10 MW_e) generating units and obtained information on peak and average power requirements, scheduling, and fuel costs (Aerospace Corp., 1978a; Aerospace Corp., 1979a). They also conducted field surveys of IPH potential in primary metals, paper and pulp, stone, clay, and glass industries in FYs 1978 and 1979. These surveys were followed by fourteen selected application reviews that analyzed investment criteria, land availability, schedule, and investment attitude issues (Aerospace Corp., 1979a).

Lawrence Livermore Laboratories studied industrial energy requirements (and applications analysis) in 1977, reviewing the suitability of solar process heat technology in ten industrial operations (Casamajor and Wood, 1978). The operations studied were selected from those industries identified as having "excellent" prospects in the InterTechnology (Fraser, 1977) and Battelle (Hall, 1977) studies. The eighteen-month field study revealed the need to check the general findings of these studies to accurately assess their validity for long-range program planning at DOE.

Lawrence Livermore decided to approach this need for additional information on industrial plant sites by industry and region. To initiate this process, Lawrence Livermore was engaged by DOE to survey food-processing plants in a five-state western area (Arizona, California, Hawaii, Oregon, and Washington). The survey concentrated on land and roof availability for solar thermal systems at each of the 1,330 food-processing plants located in the area. The research began by reviewing all plants identified by a commercial plant-site database and estimating the plants' annual energy consumption through correlations with data from the Census of Manufacturers. The estimated solar collector area required to supply at least 25% of the estimated annual energy requirement for each plant was checked against available adjacent land and roof area. Lawrence Livermore concluded (Casamajor, 1980) that the basic potential for displacing conventional energy resources by solar IPH systems (in the food-processing industry) is limited to only 25% of the apparent market because of land or roof area constraints alone. Furthermore, the study notes that site-specific inspections are mandatory for an accurate understanding of industrial energy requirements and constraints. This implies that a monumental data collection effort would be required to fulfill complete applications analysis data needs.

In addition to informally coordinating data collection efforts for industrial energy analysis, SERI also conducted several industrial energy studies under its own management in FY 1978 through FY 1981. The first of these was an overview of data availability and suitability for manufacturing energy analysis (Ketels and Reeves, 1978, 1979). This study used data from the Census of Manufacturers, as well as the results of other solar energy applications studies, to rank manufacturing industries by three- and four-digit SIC codes in terms of energy consumption, location, and temperature regime. These rankings, along with other process-specific data, were used to help select promising industry categories for further applications analysis. For example, Wright and Dougherty (1979) used these data to assess the technical feasibility of key markets for solar IPH in the temperature range of 550° to 1,100°F (290°–590°C). Their study concluded that, despite the large energy consumption in this range, the unsuitability of solar thermal systems to supply petroleum refining energy needs (which accounts for 98% of overall energy requirements) leaves only a small potential market for actual solar applications in that temperature range.

To supplement and extend this work on industrial energy requirments, SERI initiated research in November 1979 to develop reliable data on current and future industrial energy demands, end-uses, and costs, and to synthesize information on future energy needs in the industrial sector by state, four-digit SIC categories, end uses, and temperature ranges. This initial study is summarized in a three-volume report published in 1980 (Krawiec and Limaye, 1980). The study characterized energy demand by fuel type and cost within all four-digit manufacturing categories in fifteen selected states. Fuel-use intensity and expected growth factors were also included in the database. Projected energy requirements were calculated using state energy models developed by the research contractor, Synergic Resources Corporation.

These research activities were continued into FY 1981. The major objectives of this continuing research were to (1) refine industrial energy requirements by two- and four-digit SIC categories, end-uses, and temperature ranges for the fifteen states considered in the FY 1980 study using data from the 1977 Census of Manufacturers; (2) develop industrial energy requirements by two- and four-digit SIC, end-use, and temperature ranges for the thirty-five states not considered in the FY 1980 study using data from the 1977 Census of Manufacturers data; and (3) integrate base year (1977) data and projections (1990) for the industrial mix in each state to obtain the temperature distribution for industrial energy utilization at the state and national level.

The 1980 and 1981 research results were presented in a three-volume report published in 1981 (Krawiec and Limaye, 1981). This report updated state-level energy price forecasts using assumptions incorporated into DOE's 1980 Annual Report to Congress. This report also generated energy end-use profiles for primary temperature intervals of hot water, steam, and hot air end-uses. This extension of the database, which considers end-use profiles, is a major advance in establishing industrial requirements analysis.

6.3.5 Industrial Case Studies

An important conclusion of both end-use matching and other market identification studies is that detailed, site-specific information is necessary in making well-defined selections of technology for industrial applications. Although excessive attention to the details of specific cases defeats the purpose of broader end-use matching, researchers find that well-selected case studies add a knowledge of typical opportunities and constraints that can often be generalized and applied to end-use matching.

SERI researchers studied two local industries as part of the original end-use matching program in 1979 (Brown et al., 1980) and conducted an additional seven case studies the following year (Hooker, May, and West, 1980). Hooker used PROSYS and ECONMAT to determine the appropriate size and annual performance and costs of solar thermal systems for manufacturers in crude oil extraction, containers, wet corn milling, polymeric resins, fluid milk, baking, and meat processing. Hooker found few economical applications for the near term. Economical applications generally resulted from special or unusual circumstances, such as poor existing fuel-use efficiency. However, the results of this study spawned related research in a number of areas. For example, researchers at SERI later published several papers (May and Hooker, 1980; May, 1980a,b) on applying solar energy to certain unit operations in industry. These papers demonstrated the benefits of modifying certain industrial process operations (such as air-drying procedures and high-pressure hot water heat exchangers) to enhance the feasibility of solar applications.

As part of this research, May (1980c) observed that applying solar thermal energy to certain operations, such as the high-temperature catalytic and distillation processes of petroleum refining, does not satisfy industry operational requirements despite the apparent compatibility of temperature regime. Table 6.7 summarizes the factors that favor the application of solar thermal systems to industrial needs. This list of factors, a significant result of the SERI case study program, provides an extended list of criteria applicable to future end-use matching efforts.

SERI was not the only organization to find the case study approach of merit in defining the applicability of solar energy to industry. JPL performed one of the earliest set of case studies in 1978 (French and Bartera, 1978) on behalf of the state of California. As with the SERI studies, four industries were selected for detailed study after a more extensive review of industrial solar applications statewide (Barbieri and Pivirotto, 1978). JPL found that retrofit costs and performance varied widely among applications and that a number of technology developments would be necessary to achieve economic performance. The ability of a plant to fully utilize the process heat provided by a solar system was also found to be a serious issue.

Regional case studies were also conducted by several regional solar energy centers, including an extensive survey of fourteen selected applications by the Mid-America Solar Energy Complex (MASEC) and asso-

Table 6.7
Factors favoring the application of solar IPH systems

Environmental factors	Process factors	Economic factors	Company factors
High insolation levels, either total or direct, depending on solar technology proposed	Low-temperature process so that cheapest type of collector, operating at a high efficiency, can be employed	High and rapidly escalating fuel costs	Industrial plant wants to install a solar system and has an enthusiastic work force from top management down
High ambient temperatures to reduce thermal losses (particularly for nonconcentrating collectors) and to allow the possible use of water as a heat transfer fluid	Continuous, steady operations (24 h/day, 7 day/wk) where exact temperature control is not necessary	Uncertainties regarding fuel, supplies, such as interruptible natural gas contracts	Plant possesses a skilled maintenance and engineering work force, so as to run and maintain the solar system at maximum efficiency
Pollution-free microclimate, so as not to dirty or corrode collector surfaces	Built-in process storage, which serves to even out fluctuations in thermal output of collectors and which can act as a reservoir to store heat (during the weekend or on long summer evenings) produced by the solar system	Industry has, or has access to, sufficient capital to finance investments in a solar energy system	Progressive management, which gives some recognition to the noneconomic but social values of solar energy, such as public relations, security of long-term supply, and reduced air pollution, leading perhaps to the application of less stringent payback criteria to investments in solar systems
Polluted microclimate or area with strict air pollution regulations, where no additional air pollution emissions are allowed and where such controls are a restraint on levels of production	Easy retrofit of solar system, so as to minimize costs	Industry applies long payback periods or demands low rates of return on energy investments	
	Inefficient present fuel usage, not easily rectified, so that energy delivered from the solar system replaces more than the equivalent Btu content of fossil fuel	High federal, state, or local tax incentives for solar investments	
		Industrial operation is energy intensive, and energy costs represent a large fraction of value added	
		Industry has exhausted economic energy conservation measures	
		Cheap land or strong roof is available close to delivery point of the required energy. Salt is available at little or no cost, if a salt pond is a solar option	
		Low labor cost area, because solar installations are labor intensive	
		New plant, allowing a solar system to be incorporated from the beginning with savings on the conventional heating system	

Source: Hooker and May (1980).

ciated regional assessment analyses (Murray, 1980a,b; Mid-America Solar Energy Complex, 1981). The Southern Solar Energy Center also used selected case studies to evaluate the economic feasibility of solar thermal IPH (Montelione, Boyd, and Branz, 1981).

6.3.6 Applications Mapping

One of the last major efforts in developing and implementing methodologies for end-use matching from 1972 to 1982 came in a joint project by SERI and the Colorado School of Mines (Turner and DeAngelis, 1981). This project, although only the first step in this direction, was a logical and ultimate extension of end-use matching to the actual large-scale mapping of promising solar thermal industrial applications using criteria overlay processes.

Using the Generalized Map Analysis Planning System (GMAPS) developed by the Colorado School of Mines, the research team successively overlaid twelve market-matching criteria for the continental United States in a system modeling strategy, shown in figure 6.9. The computerized overlay of separate maps of each criterion resulted in a composite map showing relative attractiveness of various areas of the United States for a given technology (as shown for parabolic trough technology in figure 6.10). The authors concluded that, although such an analysis is useful in identifying prime application markets for solar thermal technology, it must be combined with plant-specific factors to indicate actual suitability. This map overlay system, in conjunction with an end-use matching system using discrete plant data and industrial case studies, could provide an extremely comprehensive and accurate means to fulfilling the ultimate objectives of end-use matching.

6.4   Contributions of Applications Analysis

As mentioned in the beginning of this chapter, end-use matching and applications analysis became extremely important elements of the federal program for solar thermal energy, particularly with respect to industrial applications, because of the ability to match solar output (both quality and quantity) to specific end-use needs. Indeed, this apparent advantage of solar thermal systems over conventionally fueled energy systems became recognized as a crucial factor in the market viability of solar industrial applications. Because of the relatively high capital cost of solar

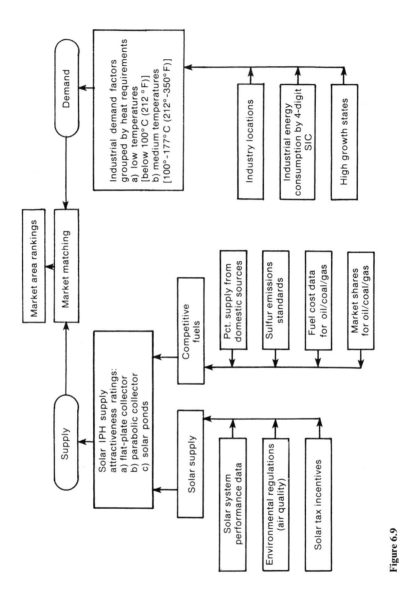

Figure 6.9
Mapping strategy model for solar thermal process heat end-use matching (Turner and DeAngelis, 1981).

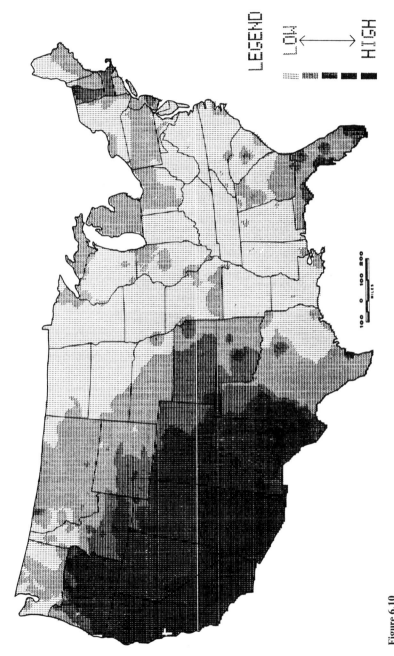

Figure 6.10
Composite "attractiveness index" map for parabolic trough technology applied to industrial process heat (Turner and DeAngelis, 1980).

thermal technology, solar thermal systems must exploit every advantage possible in competing with alternative energy supply systems. Matching solar thermal capabilities to the specific needs of energy end-use markets, in conjunction with collector cost reduction research, became key elements of the federal solar energy strategy by 1980.

The research programs sponsored between 1972 and 1982 in the field of applications analysis and end-use matching were beneficial to the overall solar thermal program in six major respects:

1. They provided valuable data on industrial energy use and operating practices. Such data were useful not only to the solar thermal program but also to other energy technology and conservation programs.

2. They resulted in a better understanding of the factors influencing solar system design, selection, and competitiveness. In most instances this understanding flowed back to technology development programs.

3. They provided a more realistic perspective on the financial, legal, environmental, social, and economic factors affecting solar thermal applications in the marketplace. This perspective resulted in a more realistic appraisal of the prospects for solar technology in the near term.

4. They identified specific application opportunities for solar thermal energy, often generating immediate industry interest. Furthermore, the process of end-use matching analysis often brought together analysts and industry personnel in a way that fostered more interest in the technology.

5. They led to case studies and demonstration programs that provided a means of establishing a more defined link between specific site studies and broader market penetration analyses.

6. They produced logic structures and computer-assisted models that apply not only to solar thermal technology development, but also to development and marketing of many other alternative technologies. Furthermore, computerizing these systems made "market" sensitivity studies of design modifications possible for the first time.

In summary, the development of end-use matching methodologies benefited the formulation of more effective and responsive federal programs for research and development in solar thermal energy. While not contributing directly to the displacement of conventional fuels by solar energy, these studies did point the way toward markets where solar energy stood better chances of successfully competing for market share. Ultimately, the greatest value of the research in this program may be realized in other areas of technology development as the technology/application selection process in the United States becomes increasingly complex.

References

Aerospace Corp. 1978a. *High-Temperature Industrial Process Heat: Technology Assessment and Introduction Rationale*. ATR-78 (7692-03)-2. El Segundo, CA: Aerospace Corporation.

Aerospace Corp. 1978b. *Solar Thermal Dispersed Power Systems Program: STES Project*. ATR-78 (7692-01)-1. El Segundo, CA: Aerospace Corporation.

Aerospace Corp. 1979a. *Field Survey of Solar High-Temperature Industrial Process Heat Application*. ATR-79(4820)-1. El Segundo, CA: Aerospace Corporation.

Aerospace Corp. 1979b. *Solar Thermal Small Power Systems Study*. ATR-79 (7780)-1. El Segundo, CA: Aerospace Corporation.

Aerospace Corp. 1980. *Solar Thermal Power Systems Summary Report*. DOE/CS/21036-01. Washington, DC: Department of Energy.

Barbieri, R. H., and D. S. Pivirotto. 1978. *Process Heat in California: Applications and Potential for Solar Energy in the Industrial, Agricultural, and Commercial Sectors*. JPL78-33. Pasadena, CA: Jet Propulsion Laboratory.

Battleson, K. W. 1981. *Solar Power Tower Design Guide: Solar Thermal Central Receiver Power Systems, a Source of Electricity and/or Process Heat*. SAND81-8005. Livermore, CA: Sandia National Laboratories.

Bennington, G. E., P. Curto, G. Miller, K. K. Rebibo, and P. Spewak. 1978. *Solar Energy: A Comparative Analysis to the Year 2020*. MTR-7579. McLean, VA: MITRE Corporation.

Brown, K. C. 1980. *Applications and Systems Studies for Solar Industrial Process Heat*. SERI/TR-351-481. Golden, CO: Solar Energy Research Institute.

Brown, K. C., D. W. Hooker, A. Rabl, S. A. Stadjuhar, and R. E. West. 1980. *End-Use Matching for Solar Industrial Process Heat*. SERI/TR-34-091. Golden, CO: Solar Energy Research Institute.

Casamajor, A. B. 1980. *The Impact of Land Use on Solar Industrial Process Heat for the Food Processing Industry*. Livermore, CA: Lawrence Livermore National Laboratory.

Casamajor, A. B., and R. Wood. 1978. *Limiting Factors for the Near-Term Potential of Solar Industrial Process Heat*. UCRL-52587. Livermore, CA: Lawrence Livermore Laboratory.

Collares-Pereira, M., and A. Rabl. 1979. "Simple procedure for predicting long-term average performance of nonconcentrating and of concentrating solar collectors." *Solar Energy*: 23(3):235–253.

Council on Environmental Quality. 1978. *Solar Energy: Progress and Promise*. Washington, DC: Council on Environmental Quality.

Curto, P. A. 1978. "Market analysis of high-temperature solar process heat," in *Proceedings of the Solar High-Temperature Industrial Process Workshop*. SERI-0637-4. Golden, CO: Solar Energy Research Institute, 139-148.

DeAngelis, M., M. Edesess, and V. Wilson. 1980. *A Market Development Plan for Solar Thermal Energy Systems in the Industrial and Utilities Sector*. Golden, CO: Solar Energy Research Institute.

Division of Solar Energy. 1976. *National Program for Solar Heating and Cooling of Buildings.* ERDA 76-6. Washington, DC: Energy Research and Development Administration, Division of Solar Energy.

Division of Solar Energy. 1977. *Solar Energy in America's Future: A Preliminary Assessment.* DSE-115/1. Washington, DC: Energy Research and Development Administration, Division of Solar Energy.

Easterling, J., A. Grace, and R. Kettle. 1980. *Summary of the DPR.* SERI/TR-733-394. Golden, CO: Solar Energy Research Institute.

Energy and Environmental Analysis, Inc. 1979. *Data Sources and Methods for Industrial Energy Analysis.* DOE/ET/2344-1. Washington, DC: U.S. Department of Energy.

ERDA. 1975. *A National Plan for Energy Research, Development and Demonstration: Creating Energy Choices for the Future. Volume 1, The Plan.* ERDA-48. Washington, DC: Energy Research and Development Administration.

ERDA. 1977. *The Need for Deployment of Inexhaustible Energy Resource Technologies. Report of Inexhaustible Energy Resources Planning Study (IERPS).* Washington, DC: Energy Research and Development Administration.

FEA. 1974. *Project Independence: Final Report of the Solar Energy Panel.* Washington, DC: Federal Energy Administration.

Feinberg, A., T. J. Kuehn, and R. F. Miles, Jr. 1979. *Decision Analysis for Evaluating and Ranking Small Solar Thermal Power System Technologies.* JPL5103-47. Pasadena, CA: Jet Propulsion Laboratory.

Fish, J. D. 1980. *Solar Industrial Process Heat Markets for Central Receiver Technology.* SAND80-8214. Livermore, CA: Sandia National Laboratories.

Fraser, M. D. 1977. *Analysis of the Economic Potential of Solar Thermal Energy to Provide Industrial Process Heat.* ERDA/InterTechnology Report 0002891-1. Warrenton, VA: Inter-Technology Corporation.

French, R. L., and R. E. Bartera. 1978. *Solar Energy for Process Heat: Design/Cost Studies of Four Industrial Retrofit Applications.* JPL78-25. Pasadena, CA: Jet Propulsion Laboratory.

GEA (General Energy Associates). 1983. *Industrial Cogeneration Potential (1980–2000) Targeting of Opportunities at the Plant Site (TOPS): Final Report,* 5 vols. DOE/CS/40362-1. Washington, DC: U.S. Department of Energy.

Gibson, J. C. 1982. *Solar Repowering Assessment.* SAND81-8015. Livermore, CA: Sandia National Laboratories.

Green, H. J. 1979. *Cooperative Effort for Industrial Energy Data Collection (IEDC).* SERI/RR-333-422. Golden, CO: Solar Energy Research Institute.

Hall, E. H. 1977. *Survey of the Applications of Solar Thermal Energy Systems to Industrial Process Heat.* ERDA TID-27348-1. Columbus, OH: Battelle Columbus Laboratories.

Hall, E. H., W. T. Hanna, L. D. Reed, J. Varga, Jr., and D. N. Williams. 1975. *Evaluation of the Theoretical Potential for Energy Conservation in Seven Basic Industries.* PB-244 772/OST. Washington, DC: Federal Energy Administration and Battelle Columbus Laboratories.

Hooker, D., and E. K. May. 1980. *Industrial Process Heat Case Studies.* SERI/TR-333-323. Golden, CO: Solar Energy Research Institute.

Hooker, D. W., E. K. May, and R. E. West. 1980. *Industrial Process Heat Case Studies.* SERI/TR-733-323. Golden, CO: Solar Energy Research Institute.

Iannucci, J. J. 1981. *Survey of U.S. Industrial Process Heat Usage Distributions.* SAND80-8234. Livermore, CA: Sandia National Laboratories.

Insights West. 1980. *Solar-Augmented Applications in Industry.* Chicago, IL: Gas Research Institute.

Ketels, P. A., and H. R. Reeves. 1978. *Market Characterization of Solar Industrial Process Heat Application: Progress Report.* SERI/PR-52-212. Golden, CO: Solar Energy Research Institute.

Ketels, P. A., and H. R. Reeves. 1979. *Market Characterization of Solar Process Heat Applications: Progress Report Second Quarter 78–79.* SERI/PR-353-212. Golden, CO: Solar Energy Research Institute.

Krawiec, F., and D. R. Limaye. 1980. *Current and Future Industrial Energy Service Characterization,* 3 vols. Golden, CO: Solar Energy Research Institute.

Krawiec, F., and D. R. Limaye. 1981. *Energy End-Use Requirements in Manufacturing,* 3 vols. SERI/TR-790R. Golden, CO: Solar Energy Research Institute.

Kreith, F., G. F. Lameiro, and K. C. Brown. 1977. "Optics applied to solar energy conversion," *Proceedings of the Society of Photo-Optical Instrumentation Engineers,* Vol. 114. Bellingham, WA: Society of Photo-Optical Instrumentation Engineers, 22–34.

Lameiro, G. F., and K. C. Brown. 1978. *Industrial Process End-Use Matching: Technical Planning Report.* SERI-27. Golden, CO: Solar Energy Research Institute.

May, E. K. 1980a. "Industrial process heat case studies: Potential for solar thermal applications." Paper in *Proceedings of the National Conference on Renewable Energy Technologies.* Manoa, HI: Hawaii Natural Energy Institute, 6-46–6-47.

May, E. K. 1980b. "Potential for supplying solar thermal energy to industrial unit operations," *AICHE Symposium Series* 77:61–69.

May, E. K. 1980c. *Solar Energy and the Oil Refining Industry.* SERI/TR-733-562. Golden, CO: Solar Energy Research Institute.

May, E. K., and D. W. Hooker. 1980. *Reducing Fuel Usage through Application of Conservation and Solar Energy.* SERI/TP-733-665. Golden, CO: Solar Energy Research Institute.

McDonnell Douglas Astronautics Co. 1977. *Industrial Applications of Solar Total Energy.* 5 vols. SAN/1132-2/1-5. Washington DC: U.S. Department of Energy.

Mid-America Solar Energy Complex. 1981. *Summary of Some Feasibility Studies for Site Specific Solar Industrial Process Heat.* MASEC-R-81-010. Minneapolis, MN: Mid-America Solar Energy Complex.

MITRE Corporation. 1978. *Solar Thermal Repowering: Utility/Industry Market Potential in the Southwest.* MTR-7919. McLean, VA: MITRE Corporation.

Montelione, A., D. Boyd, and M. Branz. 1981. *Economic Feasibility of Solar-Thermal Industrial Application and Selected Case Studies.* SSEC/TP-31298. Atlanta, GA: Southern Solar Energy Center.

Murray, O. L. 1980a. *Solar Feasibility Study for Site-Specific Industrial-Process-Heat Applications.* MASEC-SCR-80-014. Minneapolis, MN: Mid-America Solar Energy Complex.

Murray, O. L. 1980b. "Solar IPH in mid-America: A MASEC study," in *Proceedings of the Annual Meeting of the American Section of the International Solar Energy Society.* Vol. 3.1, Newark, DE: American Section of the International Solar Energy Society, 39–43.

NSF/NASA. 1972. *An Assessment of Solar Energy as a National Energy Resource.* Washington, DC: National Science Foundation.

OTA. 1978. *Applications of Solar Technology to Today's Energy Needs.* Vol. 1. Washington, DC: Office of Technology Assessment, U.S. Congress.

PRC Energy Analysis Co. 1980. *A Guide to the Solar Thermal Programs Fiscal 1980.* Washington, DC: U.S. Department of Energy.

Public Service Co. of New Mexico. 1978. *Technical and Economic Assessment of Solar Hybrid Repowering (Final Report).* SAN/1608-4-1. Washington, DC: U.S. Department of Energy.

Rabl, A., and J. M. Gordon. 1982. "Design, analysis, and optimization of solar industrial process heat plants without storage." *Solar Energy* 28(6):519–530.

Rebibo, K., G. Bennington, P. Curto, P. Spewak, and R. Vitray. 1977. *A System for Projecting the Utilization of Renewable Resources: SPURR Methodology.* MTR-7570 (ERHQ/2322/77-4). McLean, VA: MITRE Corporation.

SERI. 1981. *Report on Building a Sustainable Future*, Print 97-L, Vol. 2. Washington, DC: Committee on Energy and Commerce, U.S. House of Representatives, 492–518.

Solar Working Group. 1978. *Solar Energy Research and Development Program Balance.* HPC/M2693-01. Washington, DC: U.S. Department of Energy.

Stadjuhar, S. 1981. *PROSYS/ECONMAT User's Guide: Solar Industrial Process Heat Feasibility Evaluations.* SERI/SP-733-724. Golden, CO: Solar Energy Research Institute.

Thornton, J. P., K. C. Brown, J. Finegold, A. Herlevich, T. Kriz, J. Kowalik, and J. Gresham. 1980. *Final Report: A Comparative Ranking of 0.1-10* MW$_e$ *Solar Thermal Electric Power Systems.* SERI/TR-351-461. Golden, CO: Solar Energy Research Institute.

Turner, A. K., and M. DeAngelis. 1981. *A Geographic Market Suitability Analysis for Low-and-Intermediate-Temperature Solar IPH Systems.* SERI/TR-733-1194. Golden, CO: Solar Energy Research Institute.

U.S. Department of Energy. 1978. *Solar Energy: A Status Report.* DOE/ET-0062. Washington, DC: U.S. Department of Energy.

U.S. Department of Energy. 1979. *Solar Thermal Power Systems Program Summary.* DOE/CS-0145. Washington, DC: U.S. Department of Energy.

U.S. Department of Energy. 1980. *Solar Industrial Market Strategy.* Washington, DC: U.S. Department of Energy, Office of Solar Applications for Industry.

U.S. Department of Energy. 1981. *Solar Thermal Energy Systems Multi-Year Program Plan*, Vol. 1. Washington, DC: U.S. Department of Energy.

Wilson, V., J. Williams, and H. Ball. 1981. *Solar Industrial Process Heat: Industrial Applications and Attitudes.* SERI/TR-733-1015. Golden, CO: Solar Energy Research Institute.

Wright, J. D., and D. A. Dougherty. 1979. *Technical Feasibility of Solar Industrial Process Heat in the Temperature Range 550–1100°F.* Internal Task Memorandum 5121.26. Golden, CO: Solar Energy Research Institute.

7 Net Energy Considerations

Robert A. Herendeen

At first glance, net energy analysis (NEA) is an attractive and reasonable concept. Loosely stated, NEA compares the energy required for all inputs in developing a new energy technology (energy itself, materials, and services) with the energy that the technology will eventually produce. In particular, NEA reflects the fear that certain technologies could end up being net energy users rather than producers. Furthermore, some practitioners of net energy analysis claim that standard economic analysis will not always indicate this problem (especially if subsidies are involved) and that therefore net energy analysis should be carried out largely independently of economic analysis (Herendeen, Kary, and Rebitzer, 1979). However, the usual argument is that a net energy analysis is a supplement to, not a substitute for, economic analysis in decision making.

Concern over net energy resulted in 1974 federal legislation requiring net energy analysis of federally supported energy facilities: "The potential for production of net energy by the proposed technology at the state of commercial application shall be analyzed and considered in evaluating proposals" [Public Law 93–577, Sec. 5(a) (5)]. Despite this legislation, NEA remains an elusive concept subject to inherent generic problems that limit its usefulness. Some of the problems that persist represent fundamental ambiguities in NEA that can be removed only through decision. These include the difficulty of specifying the system boundary and the questions of how to compare energy produced and consumed at different times and of different thermodynamic qualities. These questions make NEA harder to perform and interpret (Herendeen, Kary, and Rebitzer, 1979). Some analysts therefore reject *net* energy analysis but support energy analysis (Leach, 1975). Energy analysis considers the output as a good or service rather than as energy. This view is reasonable and useful in determining how much energy is required to keep a room warm or to make a pair of skis, but it is not reasonable and is generally useless in discussing the net energy situation of a coal mine. This fundamental objection attacks the basis of NEA—that the economic and human life-support system can be separated into the "energy system" and the "remainder" and that studying the energy system in isolation is valid. For many applications this separation, although not perfectly defensible, seems to be acceptable.

Before proceeding, we should ask what evidence we have that NEA has actually been used to make decisions. In particular, because economic analysis has already been developed, a natural question is whether NEA provides a unique or different result from an economic analysis. In 1982 the U.S. General Accounting Office (GAO) issued a report criticizing the U.S. Department of Energy (DOE) for not following the 1974 law's requirement for NEA of federally funded energy projects (GAO, 1982). DOE's response, which is included as an appendix to the GAO document, is summarized in what follows: "This is in accord with DOE's view that the benefits of [NEA] are not worth the time and effort involved for general application. This result also belies GAO's assertions that its report demonstrates the feasibility for use of such analysis" (GAO, 1982).

Little has occurred since then to challenge that statement, and in my opinion NEA has seldom, if ever, been used as a critical decision criterion. In the grain belt, although NEA arguments were sometimes quoted regarding ethanol production from grain (Chambers et al., 1979), the sorting out that occurred (small operators using dry milled corn shut down while larger operators using wet milling flourished) could be understood on the basis of the economies of scale and the flexibilities of the two milling processes.

In fairness, a large debt is owed to NEA because it stimulated proper thinking about resources. Predicting resource availability on the basis of past trends has been shown to be fallacious by the example of *domestic* U.S. oil (Cleveland et al., 1984). To the extent that prediction by extrapolation is an economics technique and that prediction by incorporating the physical aspects of the resource is an NEA technique, NEA deserves credit. However, proper economics should incorporate physical realities. Once that is granted, the question of usefulness lies in the details of NEA.

That NEA has not been used is, of course, difficult to prove, as is the thesis that it will never be used. In this chapter, therefore, I review a number of net energy studies of solar technologies and include the results of my own analysis of several others. Given the issues raised previously, it will be necessary to discuss at length the details of the problems of NEA. We can anticipate two specific noncontroversial outcomes. If a technology appears to be a clear energy loser even after all uncertainties of technique and data are accounted for, the result is useful—the program should be scrapped. Similarly, if a technology appears to be an unambiguous energy producer, it is likely that NEA can then be deemphasized and the decision on whether to proceed can be based on other criteria. If the technology is close to the energy break-even point, things are more complicated and NEA seems more appealing. Of course, in this case a higher degree of accuracy in the result and hence in the needed data will be required.

7.1   Basic Definitions and Philosophy

The statement that an energy technology can be an energy winner implies three assumptions. First, because energy is not created or destroyed (mass energy in the case of nuclear energy), it would seem that at best an energy (conversion) technology can only break even. NEA usually assumes that the energy in the ground (or coming from the sun) is not to be counted as an input; that is, it is outside the system boundary. This will be discussed in detail later in this section. Second, it is usually assumed that high-quality energy (in the thermodynamic sense) is desirable, whereas low-quality energy is not. This would be covered properly if "free energy" is used instead of energy. I follow the common practice and use the term "energy" when I actually mean thermodynamic free energy. Third, I have assumed that all material and service inputs to an energy facility can be expressed in energy terms. There is a broad literature on the "energy cost of goods and services" (Bullard and Herendeen, 1975; Bullard, Penner, and Pilati, 1978).

Figure 7.1 shows the basic NEA framework, assuming only one energy type. It is useful to have normalized indicators of the NEA situation, two being

$$\text{Incremental Energy Ratio} = \frac{E_{\text{out, gross}}}{E_{\text{in, support}}},$$

$$\text{Absolute Energy Ratio} = \frac{E_{\text{out, gross}}}{E_{\text{in, gross}} + E_{\text{in, support}}}.$$

Support energy is that energy obtained directly from the rest of the economy or used by the economy to provide materials and service inputs. $E_{\text{in, gross}}$ is the energy from the ground or sun, which is usually referred to as "resource." $E_{\text{out, gross}}$ is the energy supplied to the rest of the economy by the facility.

The incremental energy ratio (IER) parallels standard economic practice in that the "cost" of using a resource is the cost of extraction and includes as inputs only those inputs taken from the rest of the economy.

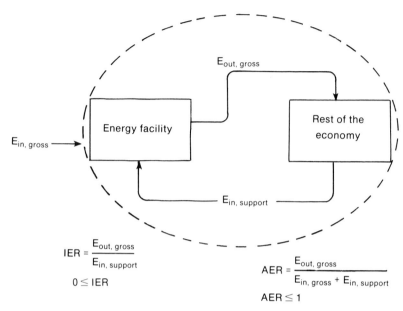

$$\text{IER} = \frac{E_{\text{out, gross}}}{E_{\text{in, support}}}$$

$$0 \leq \text{IER}$$

$$\text{AER} = \frac{E_{\text{out, gross}}}{E_{\text{in, gross}} + E_{\text{in, support}}}$$

$$\text{AER} \leq 1$$

Figure 7.1
Incremental energy ratio (IER) and absolute energy ratio (AER) analysis. Energy flows are embodied energy and therefore include energy in nonenergy goods and services as well as energy itself. The dashed line indicates the system boundary for incremental analysis. For absolute analysis, the boundary expands to include $E_{\text{in, gross}}$. $E_{\text{in, support}}$ is the energy provided by the rest of the economy to develop and operate the energy facility.

The extraction cost can be thought of as "invested" energy. IER is called by some workers "energy return on investment" (Hall and Cleveland, 1981).

The absolute energy ratio (AER) is appropriate to the more global question of the physical efficiency with which a given (say, fossil) energy source can be converted into useful energy. AER (which never exceeds 1) is useful for determining how much of a given stock of energy resource can actually be used by the rest of the economy. It is considerably less useful in the case of a flow resource (for example, sunlight), which generally cannot be stored. The difference between the two ratios is the system boundary. The following discussion will emphasize IER.

To exhibit how energy flows change over a facility's lifetime, we can use a "power curve," shown in figure 7.2. The power curve tells "everything" (assuming that system boundary and other factors are properly stated) but is cumbersome. Some summary indicator of the whole time

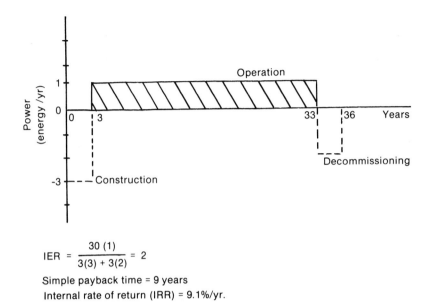

$$IER = \frac{30\ (1)}{3(3) + 3(2)} = 2$$

Simple payback time = 9 years
Internal rate of return (IRR) = 9.1%/yr.

Figure 7.2
Example power curve (for incremental analysis). Only one kind of energy is assumed. Power is obtained by dividing the energy flows in figure 7.1 by the appropriate time unit.

path is desired, such as energy payback time, energy internal rate of return, or energy ratio (already defined). These indicators have economic analogues. The first two have identical titles in economics, and the third, energy ratio, corresponds to benefit-cost ratio. (The economic case is discussed in chapter 2.) All these indicators require summing energy over many years, and it is necessary to decide how to weigh the present against the future. Standard economic practice (chapter 2) is to geometrically discount the future relative to the present. However, discounting is nowhere near universally accepted in energy analysis, and I do not discount here.

Such indicators summarize information, sacrifice detail, and hence introduce ambiguity. For example, consider simple payback time (payback time with no discounting), which is defined as the time at which the energy facility has produced as much energy as it has consumed. This definition implicitly assumes a net power curve that is negative at first and then becomes, and remains, positive during facility operation. This definition would be less useful and even misleading for more complicated power curves, such as the curve for a facility that requires significant energy inputs late in the lifetime, as for repair, maintenance, or decommissioning (see figure 7.2).

The energy ratio is also difficult to interpret. For example, if an electric plant uses some of its own output, should that be added to the inputs or subtracted from the output? This makes no difference to payback time or internal rate of return, but it does affect energy ratio.

If the power curve is typical, that is, negative at first and then positive over the whole lifetime, then the following are equivalent statements of positive net energy production:

IER > 1,

Simple payback time $<$ facility lifetime,

Internal rate of return > 0.

Comparisons in this chapter emphasize IER whenever possible.

7.2 Conceptual and Procedural Difficulties

Conceptual and procedural difficulties with NEA are summarized in table 7.1. The various difficulties overlap considerably, and economic analysis is also subject to several. Here I expand on three: end-use considerations, opportunity cost, and the dynamic problem.

An example of the effect of end-use considerations on NEA is the gasohol miles per gallon question. Gasohol (a mixture of 90% unleaded gasoline and 10% ethanol) gets more miles per gallon than would be expected based on the enthalpy of combustion of the mixture as compared with that of pure gasoline. This increased output can be viewed as extra energy production, that is, an increase in $E_{out, gross}$. For the purpose of NEA, should this increased output be considered to originate wholly in the ethanol-producing process? Or, because the improvement is really due to the existence of the mixture, is it inadmissible to perform NEA of ethanol only? And what if the ethanol goes to an end-use other than gasohol, where this issue is moot? There is no clear answer. Chambers et al. (1979) demonstrate how strongly IER for ethanol depends on how this question is answered.

The gasohol example can also be thought of in terms of opportunity cost, because the increased miles per gallon does not represent actual energy produced. Another example is energy conservation, considered a

Table 7.1
Conceptual and procedural difficulties with net energy analysis (NEA)

Issue	Example, solar	Example, nonsolar or general	Remedied by careful statement of problem?	Comments
1. Specification of system boundary				
a. Spatial	Is sunlight "free"? Is auxiliary heating energy included?	Are CO_2 mitigation effects included?		Differences of opinion persist.
b. Temporal	Is soil quality maintained in biomass production?	Will nuclear plants require significant energy inputs for decommissioning?	Yes	
c. Conceptual	How much of passive solar retrofit is "redecorating"?	Should the energy required to support labor be included?		
2. End-use considerations (related to system boundary)	Is the fact that gasohol gets better mpg than expected allocable as an energy benefit to an NEA of ethanol from grain?	Should an NEA of an electric power plant depend on whether that electricity will be used in heat pumps?	No	Fundamental question is to what extent the energy system can really be separated from the rest of economy.
3. Opportunity cost (also related to system boundary)	Is the energy output ("benefit") of a solar facility just that energy, or is the energy not needed from another source?	If energy is diverted from society at large to build a power plant, what about the energy needed to provide substitute inputs to other industries?	Yes	Difficult to set ground rules that are not too situation specific.
4. Dynamic problem.	Will a program of developing many energy facilities produce positive net energy for the first n years? Occurs when the doubling time for the building program approaches the single facility energy payback time. Always a potential problem even if single facility NEA is favorable.		Yes	Raised regarding nuclear energy in the mid-1970s but proved not to be significant. (One reason is that the projected rapid construction has not occurred).
5. Existence of more than one kind of energy.	How is electric output of a power plant compared with the coal inputs to produce the plant's steel components?		Not completely	A truly vexing problem.
6. Average vs. marginal accounting	Going from 70% to 100% solar hot water heat requires more energy inputs than from 0% to 30%.	New oil has a lower IER than old oil.	Yes	A general problem in resource analysis
7. Dependences of NEA on some economic factors.	Energy requirements for steel production are dependent on the price of energy.		No	Energy analysis, because it is based on real technologies, is simply not free of economic influences.

source of saved energy. A third example is the argument that energy used to construct an energy facility cannot be used in some other industry. That industry responds to lowered energy availability by using more of other inputs, such as metals, which themselves require more energy (Baumol and Wolff, 1981).

The dynamic problem arises from the rapid buildup of an energy technology. Even though a single facility may have an IER > 1, a program of building many facilities may be a net energy sink for many years. (If the technology is growing fast, the average facility is not even finished!) Concern over this problem was stimulated by rapid building programs for nuclear reactors promoted by several countries in the early and mid-1970s (Chapman, 1975). The predicted exponential growth did not occur, and the issue has largely gone away. The issue was also raised for solar heating by Whipple (1980).

7.3   The Type of Net Energy Analysis Used in This Chapter

Net energy results for various solar technologies (and a few nonsolar for reference) are compared here using what is called engineering-based energy analysis, which explicitly excludes the energy requirements of labor or the energy consequences of environmental disruption beyond that covered by on-site pollution control devices. This approach does not account for such off-site problems as CO_2, acid rain, heavy metals, and soil erosion.

Energy for labor is excluded for two reasons. First, it is difficult to agree on a correct reckoning: is it the energy consequences of a worker's whole paycheck or just that portion spent for food and lodging? Second, including labor means that the economy is viewed as producing goods for government consumption, exports, and capital purchases. Consumption by residences is assumed to be fixed in total and, especially, in relative amounts for different consumer goods and services. Other energy research, such as that in conservation, exploits the flexibility of consumption patterns (to purchase more insulation and less home heating fuel, for example). For consistency with this other work I prefer to keep the analysis independent of residential consumption of all goods and services.

I exclude energy related to environmental disruption because the theory and techniques to do so [mostly the work of H. T. Odum and his students, for example, Odum et al. (1981)] are not sufficiently developed to stand the test of self-consistency, completeness, and coherence. Although an objection to this kind of analysis has had to be withdrawn [advanced in Herendeen (1981) and withdrawn in Costanza and Herendeen (1984)], there remain several significant unresolved objections (Herendeen, 1984).

The NEA presented here, therefore, is little changed from that originally proposed in the early years of the field at the International Federation of Institutes for Advanced Study (IFIAS) meetings in Sweden in 1974–75 (IFIAS, 1975), the National Science Foundation (NSF) Workshop at Stanford University (Stanford, 1975), the Institute of Gas Technology (IGT) Symposium in Colorado in 1978 (IGT, 1978), and the several outputs of the Institute for Energy Analysis (Perry, Devine, and Reister, 1977; Perry et al., 1977; Devine, 1979). What has changed since those efforts is an understanding of what problems can and cannot be corrected. For example, today we have a better understanding of the system boundary problem, but it is not much easier to deal with. Our understanding shows just how arbitrary the choice of boundary is and how sensitive the result is to it.

In presenting results, I usually dispense with the power curve and use the IER or payback time, either applied to primary or premium (oil and gas) energy. Calculating the IER for premium energy requires counting oil and gas but not coal inputs and outputs. This is another variation of the system boundary and different fuel-type issues, and has been discussed regarding liquid fuel technologies. For example, an ethanol-from-grain plant (which produces premium fuel) can burn oil, gas, coal, or biomass for the distillation and associated processes. Although the net energy balance in primary terms looks quite precarious, if coal is used, the plant looks much better as a producer of net premium fuel (Chambers et al., 1979).

System inputs are assigned energy values converted to primary terms, that is, coal, crude petroleum, natural gas, and hydro- and nuclear electricity. To do this requires determining the direct and indirect energy of the entire economy to produce the materials, energy, and service inputs needed to construct, operate, maintain, and dispose of the facility. In practice this determination is almost always done using process analysis, input/output analysis, or a hybrid of the two. The following is a short description of these techniques.

Process analysis is a detailed bookkeeping of the production of a given product, in which energy inputs are recorded along the chain of production. Thus, to determine the total energy to produce a solar panel, one would start at the final assembly plant and there record the energy used as well as all material and service inputs from other economic sectors. Then the same data would be collected at the fabrication site of each of these inputs, then at the suppliers of their inputs, and so on. Energy consumed at one stage would be "embodied" in the flows to the next stage according to some adopted convention, such as energy per unit weight of materials. In principle, this bookkeeping could go on forever because, for example, solar panels could be used in producing aluminum, which is one input to solar panels. In practice, the process usually converges mathematically in a few stages; the error from truncation is acceptably small. The energy inputs are traced back to the well or mine, and the sum of all energies recorded is the primary energy required to produce the product, its "energy intensity." Process analysis is potentially accurate, but extremely time consuming. As a result it has been done for only a limited number of products. For a review, see Bullard, Penner, and Pilati (1978).

Input/output (I/O) analysis uses an existing economic database as a substitute for the painstaking data collection involved in process analysis. The monetary flows between economic sectors are used as surrogates for materials flows and the data are manipulated in a way that is equivalent to a process analysis carried out to infinity. The advantage of I/O analysis is the large database that already exists (developed by the U.S. Department of Commerce). In Hannon, Casler, and Blazeck (1985), data are manipulated to yield the energy intensities of 398 commodities spanning the entire U.S. economy.

There are several significant disadvantages to I/O analysis (Bullard and Herendeen, 1975). First, even with 398 sectors aggregation is a problem. For example, all plumbing fixtures are contained in one sector; more detail might be desired for solar applications. Second, the data are typically at least seven years old, reflecting the enormity of the task. Third, the data represent averages. Fourth, monetary flows may not be the appropriate surrogate to allocate energy.

What technique to use depends on the desired accuracy and the available time and effort. Sometimes a preliminary analysis will show that I/O analysis is adequate, or that a hybrid approach is satisfactory. For a hybrid approach process analysis is used for the first several production stages (which are probably not typical of the I/O average data), and I/0 analysis is used for the remaining stages (which may be more typical). The hybrid approach is described in Bullard, Penner, and Pilati (1978).

Table 7.2
Net energy results for nonsolar technologies

Energy type	IER
Coal, U.S. average[a]	37
Eastern surface coal[a]	43
Crude petroleum (delivered to refinery)[a]	7
Natural gas (delivered through gas utility to user)[a]	11
Geothermal electricity (vapor dominated)[b]	13 ± 4
Geothermal electricity (liquid dominated)[b]	4 ± 1
Coal mine–electric power plant[c]	8
Natural gas well–electric power plant[c]	2.3
Uranium mine–light water reactor[c]	5
Conservation: ceiling insulation as energy displacer[a]	136

a. Hannon (1982). U.S. averages, ca. mid-1970s.
b. Herendeen and Plant (1981).
c. Pilati (1977). U.S. averages, ca. mid-1970s.

7.4 Net Energy Analysis of Nonsolar Technologies

For comparison with results of solar technologies, NEA results for several nonsolar technologies are presented in table 7.2. Although these IERs are all significantly greater than unity, note that they represent U.S. averages as of the mid-1970s. Especially for crude petroleum and natural gas, the marginal picture is not so bright. Hall and Cleveland (1981) point out that IER (at the wellhead) for production of oil and gas has dropped from around 50 in the 1940s to about 8 in the late 1970s. A similar problem occurs for conservation-as-energy producer. The high IER (> 100) given in table 7.2 is for a modest amount of ceiling insulation in a poorly insulated residence. The marginal benefit from insulating an already well-insulated house is lower, and hence so is the IER. From table 7.2 we see that traditional fossil energy sources have been characterized by IER of the order of several tens.

7.5 Net Energy Analyses of Solar Thermal Technologies

In this section I review a number of published NEAs of solar thermal technologies and add my own NEA of several others. In all cases the results have been converted to conform with consistent assumptions about system boundary, etc. The output is often low-temperature heat, which is usually assigned an energy value according to the high-quality-energy consumption it displaces. Such a criterion is reasonable in many applications but does not answer the ultimate question of whether a society can use solar energy to fashion its next-generation solar equipment, that is, to be self-sustaining. The comparisons in table 7.3, therefore, incorporate the following assumptions:

1. System output is assigned an energy value that is just the enthalpy of that output, whether it be low-temperature heat or electricity. Thus, for example, 1 kWh of electricity = 3,413 Btu, the equal amount of energy (but not thermal energy required to generate the electricity).

2. System inputs are assigned energy values converted to primary terms, as described in section 7.3.

3. Regarding system boundary, the inputs to a solar technology to be converted to energy terms are those that are considered in a normal economic accounting of the energy facility over the entire lifetime, including original expenses as well as operation and maintenance. Examples are given in Mann and Neenan (1982); these are the basis for several of the NEAs presented in this chapter.

A comment on uncertainty is necessary. The figures in table 7.3 have no indicated error bounds. Such bounds are usually not estimated in the original sources, and the data to do so are often nonexistent. A few attempts at error estimation in energy analysis have been made, but it is beyond the scope of this chapter to perform them where they are absent. [See Herendeen, Kary, and Rebitzer (1979) for an error analysis of NEA of the solar power satellite, or Herendeen and Plant (1981) for a similar analysis of geothermal electricity.] On the other hand, it is necessary to say something about the validity of the results in table 7.3. On the basis of experience I estimate that accuracy is of the order of a factor of 2 for the results in table 7.3.

Essentially every study here is a prospective one, assuming some level of performance and durability for a projected facility or solar system. Retrospective studies using historical data obviously would be superior, but there is insufficient long-term experience to do so.

The entries in table 7.3 can be divided into four general categories: thermal electric energy, biomass energy, residential and commercial space heat and hot water, and high-temperature heat. These are discussed in turn.

Table 7.3
Net energy results for solar thermal technologies

Technology	IER[a] or EPT[b]	References and comments
Thermal electric power		Electric output at 3,413 Btu/kWh
160 MW OTEC	EPT = 4.7–6.2 yr	Carlson and Goss (1978)
	IER = 6.6	Perry et al. (1977)
100 MW power tower	IER = 1,1	Moraw Schneeberger and Szeless (1977)
	= 20	Meyers and Vant-Hull (1978) for Barstow, CA)
	> 6	Baron (1978) (for Barstow, CA)
	= 20–30	Vant-Hull (1985) (Vant-Hull points out that Baron used all the most pessimistic energy requirements for construction) (for Barstow, CA)
	EPT = 3 yr	Curto and Nikodem (1978)
Biomass		
Direct combustion	IER = ~2.7	Oswald and Eisenberg (1980); agricultural residues
	< 15–23	Strauss et al. (1984), poplar; 6–14 dry ton/ha-yr
	= 10–22	Herendeen and Brown (1986), without fertilizer
	= 2.5–3.8	with fertilizer, several species
Liquefaction		Herendeen and Dovring (1984),
	= 2	methanol
	= 5.6	methanol, premium-fuel basis
		Chambers et al. (1979),
	= 1–1.4	ethanol from grain
	= 2.3–3.8	ethanol from grain, premium-fuel basis
Gasification	= 0.8	Mann and Neenan, (1982)[c] (NEA performed by R. Herendeen), manure digestion, gas used to generate electricity
Direct use of heat/space heat		Thomas and Cambel (1982), single
Solar pond	IER = 1.8	single house (for Columbus, OH)
	= 2.7	20 houses (for Columbus, OH)
	= 13	town size in western Massachusetts
Active	= 1.6	Sherwood (1978), 60% solar,
Passive	= 8–16	(for Santa Fe, NM)
Active	= 3–5	Rogers (1980) (for Toronto) 50% solar, "moderately" insulated, ~2 days storage (adjusted by Herendeen)
Active	= 1.7	Bailey (1982), (for Michigan and New
Passive	= 10–25	York) location
Active	= 2.0	Mann and Neenan (1982)[c]
Passive	= 2.6	(for Denver, CO)
Passive	EPT = 1.5–2 yr	Wagner and Turowski (1979) (for Germany)

Table 7.3 (continued)

Technology	IER[a] or EPT[b]	References and comments
Direct/hot water	IER = 5.2	Bailey (1981) (for Michigan)
	= 2.5	Mann and Neenan (1982) (for Albuquerque, NM) (NEA by Herendeen)
	= 2.2	Lund and Kangas (1983) same system as Bailey (1981) (for Finland)
	EPT < 5 yr	Payne and Doyle (1978)
Direct/high temperature	IER = 2.5	Mann and Neenan (1982), 250°C steam, trough collectors (for Madison, WI)
∼ 500°C steam	IER ≳ 60	Based on steam thermal energy for power (Vant-Hull, 1985)

a. IER = incremental energy ratio.
b. EPT = energy payback time, years.
c. This reference supplies energy inputs for several solar technologies.

7.5.1 Thermal Electric Energy

Thermal electric energy includes central receivers ("power towers") and ocean thermal energy conversion (OTEC) applications. Both are net energy producers, with IER greater than 6 or payback time less than three years out of a facility lifetime of twenty years. These returns are sensitive, of course, to locations and the actual capacity factors. The IERs compare favorably with those for fossil fuel electric generation. Pilati (1977) found IERs between 2 and 7 for the full mine-to-bus-bar process for various oil, coal, natural gas, and nuclear electricity generators.

7.5.2 Biomass Energy

For biomass the net energy question has been potentially worrisome, as at least one biomass technology, ethanol from grain, is indeed near the energy break-even point. Small-scale production of ethanol was popular in the late 1970s in the grain belt, but many operators closed down or never materialized (Herendeen and Reidenbach, 1982). On the other hand, large producers are in business, even expanding, and ethanol is now used (usually inconspicuously) as an octane extender in road gasoline, rather than (conspicuously) as a major component of gasohol.

In any case ethanol from grain has an IER of about 1. With modern energy-saving technology, this goes up to about 1.4 (Chambers et al., 1979). Both of these figures are in primary energy terms. However, because the aim of an ethanol program is to take the pressure off liquid fuel supply, it is also of interest to define a premium fuel IER, in which all inputs that can be converted away from oil and gas to nonpremium fuel (for example, coal), are so converted. The major input that cannot be converted is agricultural energy, which depends heavily on premium fuels through farm equipment fuel and chemical inputs. In premium terms IER is then between 2.3 and 3.8, a significant improvement.

There has been relatively little net energy analysis on commercial-scale conversion of biomass to liquid fuels. The only other study given for biomass liquids indicates that producing methanol from woody biomass has a primary IER of about 2; for premium fuels this rises to approximately 6 (Herendeen and Dovring, 1984).

Biomass for direct combustion has been studied intensively and extensively. Use of agricultural residues has some potential, but because collection and handling activities are comparable to those for more intensive biomass production (which has yields of over five times as great), the IER is quite low. Thus in table 7.3 we see that a rough estimate of using residues leads to an IER of 2.7, whereas IERs from biomass plantations are between 10 and 22.

The relatively high IERs for unfertilized biomass production must be questioned on sustainability. Recent studies (Strauss et al., 1984; Herendeen and Brown, 1987) show significantly lower IERs with fertilization, between 2.5 and 1.5. The sustainability issue simply cannot be settled yet, as biomass plantation work is typically no more than five years old.

7.5.3   Residential and Commercial Heat and Hot Water

Table 7.3 indicates that, although some active systems have shown fairly low IERs (a maximum of 5, more typically of order 2), passive residential systems show much higher ones [≥ 8 in the U.S. Southwest (Sherwood, 1978) and ≥ 10 (Bailey, 1982)]. Left unstated here are the degree of solar coverage (percentage of heat supplied by solar versus a backup system), the fraction of a residence's cost allocable to the space conditioning system, and how well insulated the structure is in the first place. This is a prime example of where the question of "total energy needed to provide so much conditioned space" seems more useful than that of "NEA of the solar system." Some authors [for example, Sherwood (1978)] have stated results in the former way.

Hot water applications show IERs in the range 2 to 5. It is interesting to compare regional dependence. Lund and Kangas (1983) state that the flat-plate water heater system of Bailey (1981), which has an IER of 5 in Michigan, would have an IER of 2.2 in Finland.

7.5.4 High-Temperature Heat

The one example, medium-temperature steam [482°F (250°C)] from a 700,000 ft^2 (65,030 m^2) concentrating collector producing 130×10^9 Btu/yr (137 TJ/yr), has an IER of 2.5.

7.5.5 One Technology with IER < 1

In table 7.3 one entry has IER < 1. This is an anaerobic digester whose methane output is burned to produce electricity. Even when the manure input is considered to have zero energy intensity, the entire system has an IER = 0.8. For comparison, an average conventional natural gas well/gas utility/gas-fired electric power plant has an IER of approximately 2.3 (Pilati, 1977). I consider this to be a fairly unlikely solar thermal application.

The dynamic question is mentioned in many analyses. For example, Whipple (1980), in considering the consequences of the Domestic Policy Review of Solar Energy, found that, if the most rapid growth scenario actually occurred (which would involve a doubling of solar heating capacity every three years for the period 1985–2000), the whole program would have an IER (for the period) of 0.5–1.5, depending on the payback time of individual solar units. (He assumed a range of 3–9 years, roughly corresponding to IERs of 2.5 to 7.) In fact, this rapid growth now seems unlikely, which reduces the likelihood of the program's IER being below unity. The dynamic problem has been ridiculed by some analysts, who state that exponential program growth simply cannot be sustained for long. Historically this has turned out to be true, but the potential for a dynamic problem persists.

7.6   Summary and Conclusions

Within the confines of the several significant objections to NEA, a review has been made of NEA studies of solar thermal energy technologies, supplanted by my own analyses. With two exceptions (one a methane digester whose output is burned to produce electricity and the other ethanol from grain), the technologies are net energy producers. Solar thermal electric has IERs that are no lower, and sometimes higher, than fossil-based production. For producing residential and commercial space heating and hot water, low-temperature industrial IERs are as low as 2. Passive residential heat generally has an IER greater than that for active. The highest active system IER is 13 (a town-sized solar pond for residential heat) but, except for that, active systems show IER ≤ 5. A wide range of IERs is seen for biomass, but for fertilized systems IER ≤ 4.

Except for solar thermal electric, these IERs are generally less than those that have charactrized fossil fuel resources until recently. This, plus the observation that IER is decreasing for newly developed fossil fuels, representing depletion, is sobering. Despite such useful broad insights, I have argued that NEA has so many inherent problems that using it to compare and rank different energy technologies is often impractical. Especially in this time of heightened attention to efficiency of energy use, the separation between the "energy-producing system" and the "rest of the system" (the fundamental assumption of NEA) is less justifiable. Energy analysis, without the "net," has more potential.

References

Bailey, R. 1981. "Net energy analysis of eight technologies to provide domestic hot water heat." *Energy* 6:983–997.

Bailey, R. 1982. "Net energy analyses of three passive solar heating systems," in *Proceedings of Progress in Passive Solar Energy Systems*, J. Hayes and C. Winn, eds. New York: American Solar Energy Society, 599–603.

Baron, S. 1978. "Solar energy: Will it conserve our non-renewable resources?" in *Proceedings of the 1978 Annual Meeting of the American Section of the International Solar Energy Society*, Vol 2.2, K. Boer and G. Franta, eds. Newark, DE: International Solar Energy Society, 617–621.

Baumol, W., and E. Wolff. 1981. "Subsidies to new energy sources: Do they add to energy stocks." *Journal of Political Economy* 89:891–913.

Bullard, C., and R. Herendeen. 1975. "Energy costs of goods and services." *Energy Policy* 3:263–278.

Bullard, C., P. Penner, and D. Pilati. 1978. "Net energy analysis handbook for combining process and input-output analysis." *Resources and Energy* 1:267–313.

Carlson, T., and W. Goss. 1978. *Comprehensive Energy Analysis Applied to an Ocean Thermal Energy Conversion System*. Paper 78-TS-6. New York: American Society of Mechanical Engineers.

Chambers, R., R. Herendeen, J. Joyce, and P. Penner. 1979. "Gasohol: Does it or doesn't it produce positive net energy?" *Science* 206:789–795.

Chapman, P. 1975. "Energy analysis of nuclear power stations." *Energy Policy* 3:285–294.

Cleveland, C., R. Costanza, C. Hall, and R. Kaufmann. 1984. "Energy and the U.S. economy: a biophysical perspective." *Science* 225:890–897.

Costanza, R., and R. Herendeen. 1984. "Embodied energy and economic value in the United States economy, 1963, 1967, and 1972." *Resources and Energy* 6:129–164.

Curto, P., and Z. D. Nikodem. 1978. *Solar Thermal Repow*ering. Report MTR-7861. Springfield, VA: National Technical Information Service.

Devine, W., Jr. 1979. *Energy Accounting for Solar and Alternative Energy Sources*. ORAU/ IEA-79-11CR. Oak Ridge, TN: Institute for Energy Analysis.

GAO. 1982. *DOE Funds New Energy Technologies without Estimating Potential Net Energy Yields*. GAO/IPE-82-1. Washington, DC: U.S. General Accounting Office.

Hall, C., and C. Cleveland. 1981. "Petroleum drilling and production in the United States: Yield per effort and net energy analysis." *Science* 211:576–579.

Hannon, B. 1982. "Energy discounting," in *Energetics and Systems*, W. Mitsch, R. Ragade, R. Bosserman, J. Dillon, Jr., eds. Ann Arbor, MI: Ann Arbor Science Publishing, 73–100.

Hannon, B., S. Casler, and T. Blazeck. 1985. "Energy intensities for the U.S. economy— 1977." Available from Bruce Hannon, 1208 W. Union, Champaign, IL.

Herendeen, R. 1981. "Energy intensities in ecological and economic systems." *Journal of Theoretical Biology* 91:607–620.

Herendeen, R. 1984. "Challenges to ecologists and economists for the merging of their work for policy decisions," in *Integration of Economy and Ecology: An Outlook for the Eighties, Proceedings from the Wallenberg Symposium 1–3 September, 1982*, A.-M. Jansson, ed. Saltsjobaden, Sweden: Asko Laboratory, University of Stockholm, 105–110.

Herendeen, R., and S. Brown. 1987. "A comparative analysis of net energy from woody biomass with implications for Illinois." *Energy* 12:75–84.

Herendeen, R., and F. Dovring. 1984. *Liquid Fuel from Illinois Sources*. Report AERR 192. Urbana, IL: Agricultural Experiment Station, College of Agriculture, University of Illinois.

Herendeen, R., and R. Plant. 1981. "Energy analysis of four geothermal technologies." *Energy* 6:73–82.

Herendeen, R., and D. Reidenbach. 1982. *Ethanol from Grain: Economic Balances of Small Scale Production (0.25-2.5 million gal/yr)*. Report 82 E-222. Urbana, IL: Department of Agricultural Economics, University of Illinois.

Herendeen, R., T. Kary, and J. Rebitzer. 1979. "Energy analysis of the solar power satellite." *Science* 205:451–454.

IFIAS. 1974. *Workshop Report: Energy Analysis*. Stockholm, Sweden: International Federation of Institutes for Advanced Study.

IFIAS. 1975. *Workshop Report on Energy Analysis and Economics*. Stockholm, Sweden: International Federation of Institutes for Advanced Study.

IGT. 1978. *Symposium Papers: Energy Modelling and Net Energy Analysis*, Fred Roberts, ed. Chicago, IL: Institute of Gas Technology.

Leach, G. 1975. "Net energy: Is it any use?" *Energy Policy* 2:332–344.

Lund, P., and M. Kangas. 1983. "Net energy analysis of district solar heating with seasonal heat storage." *Energy* 8:813–819.

Mann, G., and B. Neenan. 1982. *Economic Profiles of Selected Solar Energy Technologies for Use in Input-Output Analysis*. LA-9083-TAES. Los Alamos, NM: Los Alamos National Laboratory.

Meyers, A., III, and L. L. Vant-Hull. 1978. "The net energy analysis of the 100 MW$_e$ commercial solar tower," in *Proceedings of the Meeting of the American Section of the International Solar Energy Society*, K. Boer and G. Franta, eds. Newark, DE: International Solar Energy Society, 786–792.

Moraw, G., M. Schneeberger, and A. Szeless. 1977. "Energy investment in nuclear and solar power plants." *Nuclear Technology* 33:175–183.

Odum, H., M. Lavine, F. Wang, M. Miller, J. Alexander, and T. Butler. 1981. *A Manual for Using Energy Analysis for Plant Siting*; Final Report to the Nuclear Regulatory Commission. Gainesville, FL: Energy Analysis Workshop, Center for Wetlands. University of Florida.

Oswald, W., and D. Eisenberg. 1980. "Biomass generation system as an energy resource," in *Proceedings of the Bio-Energy '80 World Congress and Exposition*. Washington, DC: Bio-Energy Council, 123–126.

Payne, P., and D. Doyle. 1978. "Fossil fuel cost of solar heating," in *Proceedings of the Intersociety Energy Conversion Engineering Conference*, New York: Institute of Electrical and Electronics Engineers.

Perry, A., W. Devine, and D. Reister. 1977. *The Energy Cost of Energy—Guidelines for Net Energy Analysis of Energy Supply Systems*. ORAU/IEA(R)-77-14. Oak Ridge, TN. Institute for Energy Analysis.

Perry, A., W. Devine, Jr., A. Cameron, G. Marland, H. Plaza, D. Reister, N. Treat, and C. Whittle. 1977. *Net Energy Analysis of Five Energy Systems*. ORAU/IEA(R)-77-12. Oak Ridge, TN: Institute for Energy Analysis.

Pilati, D. 1977. "Energy analysis of electricity supply and energy conservation options." *Energy* 2:1–7.

Rogers, D. 1980. "Energy resource requirements of a solar heating system." *Energy* 5:75–86.

Sherwood, L. 1978. "Total energy use of home heating systems," in *Symposium Papers: Energy Modelling and Net Energy Analysis*, F. Roberts, ed. Chicago, IL: Institute of Gas Technology, 665–672.

Stanford. 1975. *Report of the NSF-Stanford Workshop on Net Energy Analysis*. 27976-6001-RU-00. Washington, DC: Office of Energy R&D Policy, National Science Foundation.

Strauss, C., P. Blankenhorn, T. Bowersox, and S. Grado. 1984. "Net financial and energy analyses for producing populus under four management strategies," in *Symposium Papers: Energy from Biomass and Wastes VIII*. Chicago, IL: Institute of Gas Technology, 251–271.

Thomas, S., and A. B. Campbel. 1982. "Net energy analysis of residential solar ponds." *Energy* 7:457–463.

Vant-Hull, L. L. 1985. "Solar thermal power generation: The solar tower, progress toward commercialization." *Natural Resources Journal* 25:1099–1117.

Wagner, H., and R. Turowski. 1979. *Energy Analysis of Solar Energy Systems, Heat Pumps and of Improved Insulation of Single-Family Houses*. Report BMFT-FB-T-79-01. Available from National Technical Information Service. In German.

Whipple, C. 1980. "The energy impacts of solar heating." *Science* 208:202–266.

8 Cost Requirements for Active Solar Heating and Cooling

Mashuri L. Warren

The economics of the marketplace plays a dominant role in shaping the future of solar space conditioning and water heating technologies. Several methods have been used to assess the economics of active solar energy systems, including simple payback, life-cycle cost, discounted present value, return on investment analysis, and levelized annual energy cost. All these methods require establishing the solar technology costs compared to conventional technology costs; establishing the system performance to meet the specified heating, cooling, and/or domestic hot water load; and making assumptions about the expected future costs of factors such as fuel, inflation, interest rates, and taxes.

Cost requirements (goals) for active solar heating and cooling or for any new technology are useful but difficult to determine because of the need to predict future economic conditions and technical changes. Thus clear and consistent methods must be used. This chapter presents a methodology developed for establishing cost requirements for active solar heating and cooling systems. The cost requirement analyses of active solar cooling systems illustrate the method.

Market studies indicate a relationship between market penetration (percent of market captured) and payback periods for heating, ventilating, and air conditioning systems. The payback period is related to the expected real return on investment, using fuel escalation and inflation rates. By postulating the commercial introduction of solar cooling systems in a given year with the market share increasing at a set rate, payback and return on investment goals for cooling systems are established as a function of year of purchase.

The return on investment goals are used to calculate the twenty-year present value of the fuel saved by the solar energy system. To be cost-effective, the incremental solar system cost must be equal to or less than the present value of the energy savings. This equality establishes the link between incremental solar system cost and the return on investment goal and determines the system cost goals as a function of year of purchase. Return on investment goals are shown to be equivalent to discounted-cash-flow payback analysis.

To evaluate the cost-effectiveness of systems that have not yet been built requires (1) estimating the load to be served and calculating the contribution of the active solar system to meet that load, (2) calculating the contribution of an efficient conventional system to meet the same load, thereby establishing the annual energy and first-year fuel cost savings attributable to the solar system, (3) estimating the cost of the active solar system with full backup and the cost of the conventional system, and (4) comparing the energy cost savings to the incremental solar system cost with suitable economic assumptions. The first three examples in chapter 2 (sections 2.7.1–2.7.3) are especially pertinent in this regard.

Cost requirements can be determined for a broad range of solar heating and cooling technologies, including domestic hot water, space heating, and space cooling and heating systems. The cost goals are then broken down to the subsystem level. Expected costs for each of the different component subsystems are based on typical present-day costs and reasonable scenarios for reducing those costs by means of mass production economies, design simplifications, and performance improvements resulting from additional research and development. In effect, a cost "budget" is established for each component subsystem and is addressed as component subsystem goals for engineering development effort.

8.1   Market Potential

Figure 8.1 shows the relationship between market potential and payback period for heating and air conditioning products as indicated in market studies (Lilien and Johnston, 1980). Lilien and Johnston point out that there are additional subjective factors that influence the "decision to buy" that must be overcome before market potential is converted into market penetration. Therefore market penetration goals may be converted into an economic performance goal for the solar system that is understandable to the consumer and would influence the decision to buy (a goal such as a payback period or a return on investment). The behavior of solar cooling systems in the marketplace is assumed to be similar to that of other heating and air conditioning products. Details on market penetration studies are found in chapter 5.

To develop cost requirements, a market penetration scenario is postulated. As an example, active cooling market penetration begins with commercial introduction of active solar cooling systems in 1986 and attains a 20% annual market share of the entire national cooling capacity by the year 2000. This postulated market penetration curve increases linearly starting in 1986 and can be compared to the actual penetration

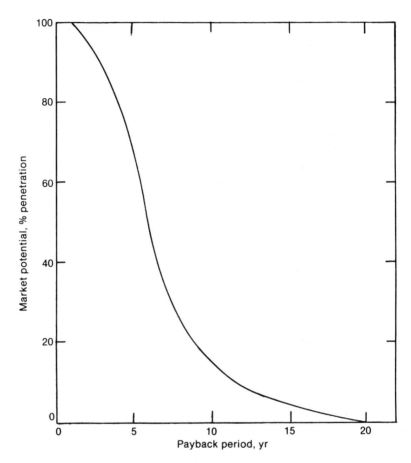

Figure 8.1
Market potential as a function of payback period (Lilien and Johnston, 1980).

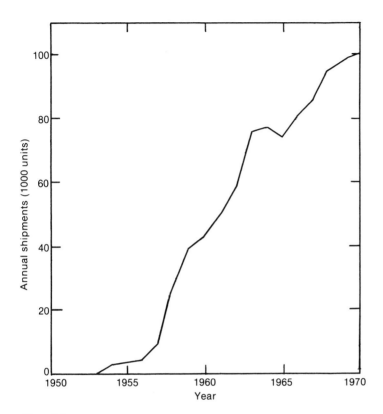

Figure 8.2
The heat pump market 1954–1970 (ARI, 1980).

curve achieved by heat pump sales in the United States from 1953 to 1970, as shown in figure 8.2. This pattern is typical of historic early market penetration achieved by major heating ventilating and air conditioning (HVAC) products. During the same period the price of residential heat pumps dropped significantly (figure 8.3) as the technology evolved and the volume of production increased.

Air conditioning demands are expected to grow significantly over the next twenty years, driven by population shifts to the "sun belt" regions of the country. It is estimated that more than 90% of the new construction in that region will have central air conditioning. The postulated annual penetration goal applies equally to each of the geographic regions considered.

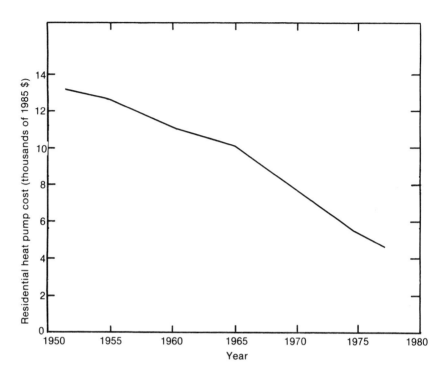

Figure 8.3
Cost of residential heat pump systems normalized to 1985 dollars by the consumers price index shows a 64% price reduction in the 25 years from 1952 to 1977 (Planco, Inc., 1981).

Energy conservation and passive cooling efforts are expected to reduce significantly the sensible cooling and heating loads. However, the substantial latent (that is, humidity) cooling loads in all buildings and internal heat gains in commercial buildings will remain, despite conservation and passive measures, and will require the use of active cooling systems.

The annual energy used for cooling buildings in the year 2000 can be estimated by accounting for the expected mix of building types, regional differences in building stock and cooling loads, and residential and commercial building energy requirements. An estimated annual energy displacement of 0.14 quad (0.15 EJ) of conventional energy sources by solar in the year 2000 (for new residential, new multifamily, retrofit residential, and commercial market sectors) would result from a 20% penetration by solar cooling of the annual projected air conditioning market by the year 2000 (Warren and Wahlig, 1982).

8.2 Economic Goals

I selected the discounted-cash-flow payback period as the economic performance criterion because it is easy to understand and use. The discounted-cash-flow payback period is the number of years it takes for the discounted future fuel cost savings to equal the present incremental cost to produce those savings. The payback period is the number of years required for the present worth of future savings in fuel costs to equal the present worth of the initial investment plus that of the operating costs for the solar system. This payback period is related to the "real" return on investment, taking into account the effect of inflation. A more complete discussion of these and other economic criteria is in chapter 2.

The "discount rate" that is assumed for economic analysis is the subject of much discussion. There are at least three discount rates that can be used:

1. From the consumer's perspective, the time value of money (time rate preference) is the expected return from money placed in another investment, such as a savings account or money market security. Typically this rate is a few percent (2–4%) above the inflation rate. If the inflation rate is 6–8%, then the time value of money is 8–12%. It is close to the mortgage interest rate.

2. The effective cost of capital is the cost the consumer must pay to borrow the money to finance the investment. For instance, a home improvement loan may be 8–10% over the inflation rate, giving an effective cost of capital of 16–18%.

3. The opportunity cost of money is related to the return that the person might get with a complete spectrum of investment alternatives. Where is the most effective place to invest money? The opportunity cost of money is typically 25–35% or higher, depending on the investment goals of the individual. Commercial enterprises evaluate energy investments based on the opportunity cost and consequently often reject energy investments that have a payback period of greater than two or three years.

8.2.1 Return on Investment Goals

The return on investment (ROI) goals are used to calculate the twenty-year present value of energy savings of the solar energy systems. To be cost-effective, the incremental solar system cost must be equal to or less than the present value of the energy savings. Assuming a fuel cost escalation rate of 3%, if a maximum allowable incremental cost is equal to the twenty-year present value of fuel savings at a desired rate of return, the discounted-cash-flow payback period is related to the ROI as shown in figure 8.4.

By using the dependence of market penetration on payback (figure 8.1), a market penetration goal for any year can be converted into a payback goal. The payback goal is converted into a required ROI using figure 8.4. The required payback period decreases and the real ROI increases as the desired market pentration increases as shown in table 8.1. The ROI goal and payback goals are plotted in figure 8.5 and establish the economic performance goals for the period between the first significant market penetration in 1986 and the year 2000. Achieving a 20% market penetration would require a system with a payback of about nine years or a real ROI of 11% or greater.

Two methods can be used to compare different systems: (1) incremental system cost versus first-year energy cost saved (in dollars), which is given by the modified uniform present worth factor, and (2) the incremental system cost for unit of annual energy saved.

The ratio of maximum allowable incremental system cost to first-year energy cost saving for the case of a single fuel being used is given by the modified uniform present worth factor. The present value is based on the modified uniform present worth factor, which is a function of the fuel escalation rates and the expected real ROI. Figure 8.6 shows the uniform present worth factor for gas and for electricity for a system life of twenty years, assuming CONAES A escalation rates (Brooks and Ginzton, 1980). It is assumed that a solar cooling or cooling and heating system is cost-effective when the incremental solar system cost is equal to (or less than) the present value of the energy savings.

To establish cost goals, we used local 1980 energy prices and fuel escalation rates as given by Brooks and Ginzton (1980). These values are 5.7% for natural gas and 3.3% for electricity; fuel prices for 1980 of $3.16/10^6 Btu ($0.316/therm) for natural gas and $19.29/10^6 Btu ($0.066/kWh) for electricity and a general inflation rate of 10% were also used. Figure 8.7 shows the value of saving 10^6 Btu (1.05 GJ) of electricity or gas for twenty years as a function of ROI and starting year assuming CONAES A fuel escalation rates. Present values at some future date are expressed in constant 1985 dollars. The shift in the curves toward higher present value

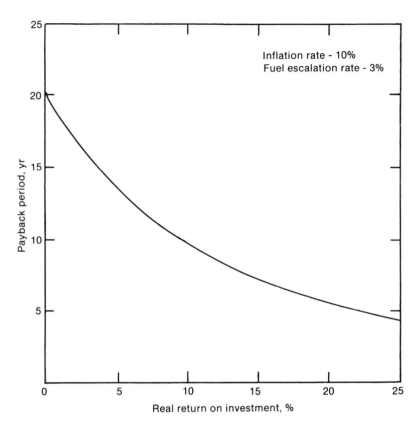

Figure 8.4
Payback period as a function of real return on investment assuming 10% inflation rate and 3% fuel escalation rate (Warren and Wahlig, 1983).

Table 8.1
Economic performance goals for a representative system

Goal	Year				
	1986	1988	1990	1995	2000
Market penetration (%)	0	2.8	5.9	12.8	20.0
Payback period (yr)	20	16.3	13.4	10.5	8.9
Real ROI goal (%)	0	2.7	5.2	8.8	11.4

Source: Warren and Wahlig (1985).

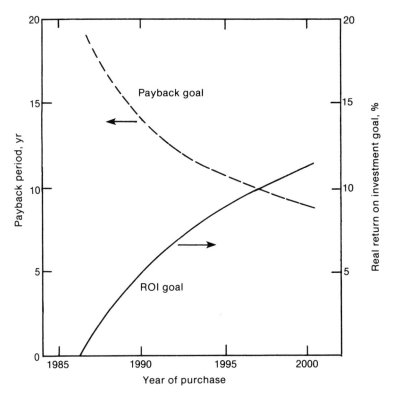

Figure 8.5
Payback goal and real ROI goal as a function of year to reach a 20% market penetration in
the year 2000 (Warren and Wahlig, 1983).

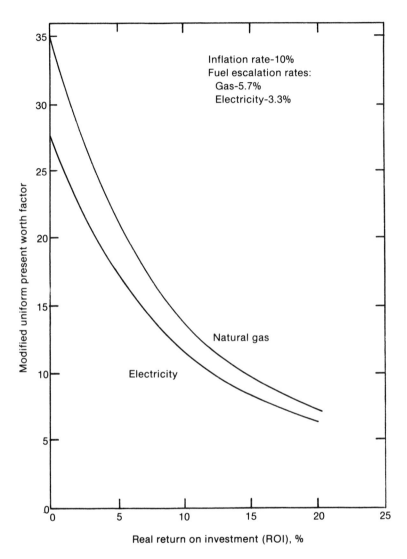

Figure 8.6
Twenty-year modified uniform present worth factors for electricity and natural gas (Warren and Wahlig, 1983).

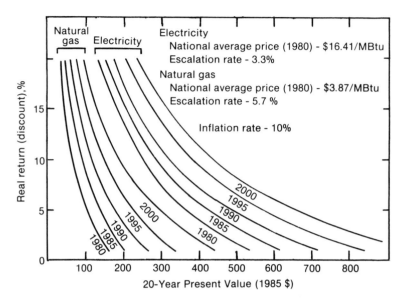

Figure 8.7
Real return on investment (discount rate) as a function of the twenty-year present value of
10⁶ Btu (1.05 GJ) fuel savings and starting year using CONAES A fuel escalation rates and
10% inflation rate (Warren and Wahlig, 1983).

reflects the assumptions made concerning escalation of fuel prices from
1980 to 2000.

By using either of the methods depicted in figures 8.6 and 8.7, the allowable incremental solar system costs can be calculated as a function of real
rate of return and year of purchase. These calculations have been carried
out for solar cooling and heating systems on representative residential
and commercial buildings in Atlanta, Fort Worth, Miami, Phoenix, and
Washington, D.C. Figure 8.8 displays the allowable incremental costs as
a function of ROI for different initial years of purchase for a residential
solar cooling and heating system in Fort Worth compared with a conventional heat pump system.

8.2.2 Net Present Value versus Simple Payback

The analysis of the economics of active solar systems can be turned
around to ask the question, based on the value of energy savings, How
much can be spent on the system? or What is the cost goal for the system?

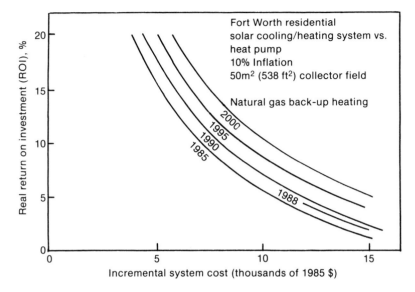

Figure 8.8
Real return on investment as a function of incremental residential solar cooling and heating system cost (1985 dollars) and year of purchase (Warren and Wahlig, 1983).

Results show that for a range of economic assumptions a simple payback analysis assuming a five-, seven-, or nine-year payback is an adequate figure of merit in establishing cost goals for active solar heating and cooling technologies. The value of the energy saved in the first year of system operation multiplied by the payback period establishes the cost goal (Scholten and Morehouse, 1983).

I evaluated the range of economic assumptions and scenarios by determining the cost goal multiplier obtained from a net present value of future energy savings over a five-, seven-, or nine-year period under a range of economic assumptions. With a range in escalating rates for electricity from 0.3% to 3.3% and for natural gas from 1.0% to 5.7%, the cost goal multiplier for a five-year payback ranges from 4.7 to 5.4 and, for a nine-year payback, from 8.2 to 10.3.

With the relatively short paybacks required for market acceptance of a new technology, the fuel escalation rates do not greatly change the payback period. Certainly for estimating the approximate costs of future technologies, the uncertainties generated by assuming a simple payback are smaller than the uncertainties of future costs of collectors and other components. Future fuel escalation rates, however, have a significant impact on the first-year energy cost savings at the time the future systems are built. One solar investment strategy is to defer purchase of a solar system until the first year that energy cost savings are greater than the annualized cost of owning and operating the system. A less conservative strategy is to wait until levelized annual energy cost savings over the life of the project are less than the annualized operating cost.

A wide range of incentives are available that can make investing in an active solar energy system more attractive. For the commercial owner there have been energy and investment tax credits, depreciation and loan interest expense tax deductions, and disincentives of income tax to be paid on energy cost savings (Hirshberg, 1977; Moden, 1981). For the residential owner there have been energy tax credits and tax deductions on loan interest expense.

The effects of tax rate, depreciation, operation and maintenance, and leveraging the investment with borrowed money can be shown to be a multiplier of the simple cost goal (Scholten and Morehouse, 1983). For residential systems this multiplier is typically 1.03–1.10 without energy or investment tax credits. For commercial systems the negative impact of income tax paid on energy savings makes the factor 0.5–0.7 without energy and investment tax credits. Large tax credits reduce the capital outlay and the effective cost of the system by as much as 40–50%.

8.3   System Cost Goals

The ROI of a solar heating and/or cooling system depends on many variables. The ratio of the allowable incremental cost of the solar system to the cost of a conventional cooling system depends on the thermal performance of the solar system (which determines the amount of conventional fuel displaced in meeting the building load), the efficiency of the conventional cooling system being replaced, the cost of conventional fuel (fossil fuel or electricity) being displaced, the value of money (discount, interest, and inflation rates), maintenance expenses, the system life, and the expected rate of return on investment. Most of these parameters change with time and geographic location. System life for this analysis is assumed to be twenty years.

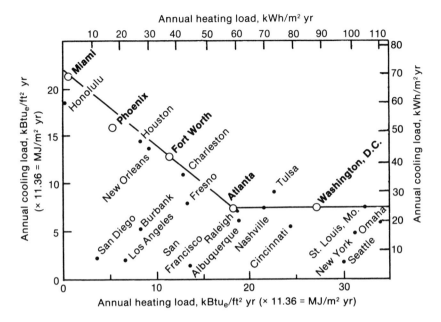

Figure 8.9
Annual residential electric cooling and gas heating energy consumption for selected cities based on DOE 2.1 simulation analysis (Warren and Wahlig, 1983).

8.3.1 Test Locations

The geographical distribution of the space cooling market was analyzed by examining the current cooling market and making projections (Warren and Wahlig, 1982). Figure 8.9 shows the geographical distribution of the space cooling load as represented by annual residential heating and cooling loads for thirty-two cities (based on DOE 2.1 simulation runs). From this analysis five cities—Miami, Phoenix, Fort Worth, Atlanta, and Washington, D.C.—were identified as representing a wide range of cooling and cooling and heating climates.

8.3.2 Thermal Performance Analysis

Simulations of the annual thermal performance of active solar Rankine and absorption cooling and heating systems (Versteegen and Morehouse, 1979; Hughes et al., 1981a) have been conducted using TRNSYS (University of Wisconsin, 1983). These calculations have been carried out for residential solar cooling and heating systems in four cities (Fort Worth, Phoenix, Miami, and Washington, D.C.) and for commercial solar cooling-only systems in three cities (Fort Worth, Phoenix, and Miami), which are representative of the cooling market.

Three types of systems were evaluated: residential 3-ton absorption (ARKLA Corp.), commercial 25-ton absorption (ARKLA Corp.), and commercial 25-ton Rankine (AiResearch Corp.). One ton of refrigeration is equivalent to 12,000 Btu/h (12.7 MJ/h).

The residential buildings used in the analysis are taken from LeBoeuf (1980). Typical single-family residences were chosen for southern cities, represented by Fort Worth, and for more northerly locations, represented by Washington, D.C. Hourly building load calculations were based on a TRNSYS-compatible, standardized residential load model. Hourly residential load calculations and system performance calculations proceed simultaneously in the hourly TRNSYS simulation.

The small, well-constructed, seven-zone office building used in the analysis has a nominal design cooling load of 25 tons and meets or exceeds ASHRAE 90–75 standards (ASHRAE, 1975). Additional energy conservation features, such as low total lighting levels and minimum ventilation rate, are incorporated. The building was originally described for Washington, D.C.; however, the description is adequate in other geographic locations if the gross air circulation value is changed for each location. Building loads were generated using the computer code DOE 2.1 (Lawrence Berkeley Laboratory, 1980) and were used as inputs to the TRNSYS simulations.

8.3.3   Establishing Cost Goals

For a given solar system there is an annual fuel savings. For a given first year of operation there is a minimum ROI to meet the market penetration goal, shown in figure 8.5. At that ROI, with assumptions about annual fuel savings and fuel costs, the twenty-year present value of fuel savings can be calculated and set equal to the incremental system cost goal. Incremental system cost goals as a function of year have been generated for residential solar cooling systems for cooling and heating, as shown in figure 8.10. The performance of systems in Washington, D.C., is typical of these residential systems. Figure 8.11 shows the incremental system cost goals for commercial systems for cooling only.

Considerable care should be taken in extrapolating these representative system costs to other systems. These incremental cost goals have

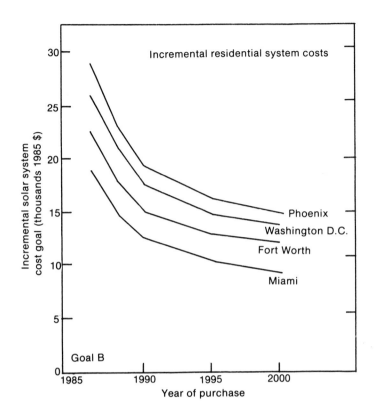

Figure 8.10
Residential incremental solar system cost goals for representative 3-ton cooling systems in four cities to achieve 20% market penetration in the year 2000 (Warren and Wahlig, 1983).

been developed for specific systems and buildings in representative geographic locations. For example, two commercial systems (absorption and Rankine) with a 25-ton chiller serving a building with a 25-ton peak cooling load have been modeled in Phoenix, Arizona. The collector array has been sized to 1,600 ft² (150 m²) in each case. Using as a rule of thumb for driving a chiller the figure of 130 ft² (12 m²) of collector per ton of cooling, the collector array might appear to have been sized for an approximately 12.5-ton solar chiller. The array certainly will not be able to run the chiller at peak capacity. However, 80% of the time the chiller load will be 50% of the peak capacity or less, and the lower collector area is adequate to produce the annual solar cooling achieved. The peak building load of 25 tons could be met in various ways, for example, by driving

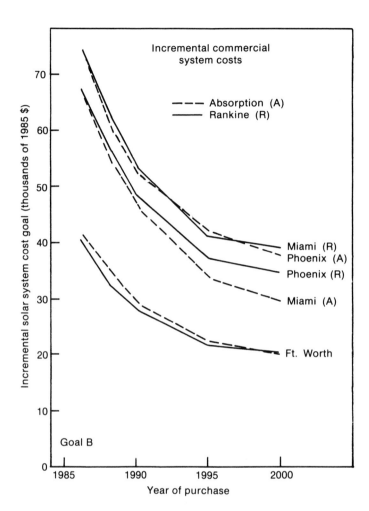

Figure 8.11
Rankine and absorption commercial incremental solar system cost goals for representative
25-ton cooling systems in three cities to achieve 20% market penetration in the year 2000
(Warren and Wahlig, 1983).

Table 8.2
Incremental system cost goals for commercial and residential absorption systems in Phoenix

	Year	
System	1986	2000
Commercial absorption	$75,000	$42,500
(150 m², 12.5 ton)	$500/m²	$283/m²
	$6,000/ton	$3,400/ton
Residential absorption	$28,800	$15,200
(50 m², 3 ton)	$576/m²	$305/m²
	$9,600/ton	$5,080/ton

Source: Warren and Wahlig (1983).

the solar chiller to 25 tons capacity with auxiliary (a gas boiler for the absorption system and purchased electricity for the Rankine system) or by using a 12.5-ton solar chiller in parallel with a 25-ton or 12.5-ton conventional electric or gas-fired chiller, with or without cold storage.

For the 3-ton residential system analyzed here, which includes cooling, heating, and hot water loads, a collector area of 538 ft² (50 m²) or 179 ft²/ton (17 m²/ton) was selected. Thus this collector array should be capable of operating the chiller at full capacity.

I did a commercial analysis for cooling only, although I did not include the additional economic benefit derived from using the same solar collectors to provide heating. Inasmuch as the collector area determines the amount of solar energy collected and the collectors constitute the single largest cost item, the incremental solar system cost per unit collector area is a commonly used measure of the cost goal. Similarly, the incremental solar system cost per ton of cooling is also sometimes used as a cost goal measure. Accordingly table 8.2 presents the incremental solar system cost goals, the cost goals per unit collector area, and the cost goals per ton of cooling based on simulation analysis in Phoenix, Arizona.

8.3.4 Comparing Costs and Goals

Figure 8.12 shows preliminary subsystem cost estimates for a residential system. These values are based on estimates of subsystem cost and performance improvements. Figure 8.13 plots the projected solar system costs for a given year and the projected cost goal for a typical residential system. The difference between the projected costs and the cost goals is

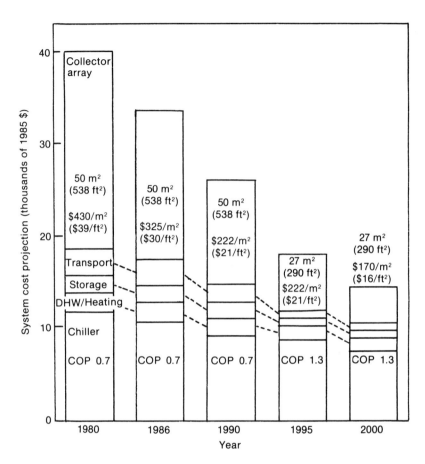

Figure 8.12
System cost projections for a 3-ton residential absorption cooling, heating, and hot water
system showing subsystem costs (Warren and Wahlig, 1983).

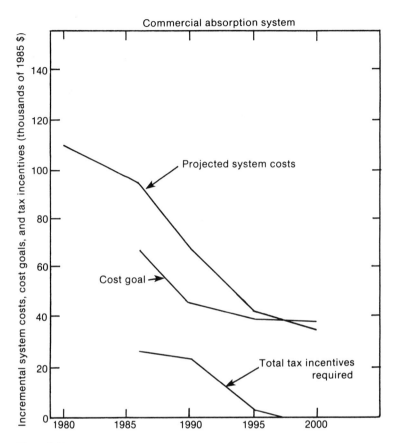

Figure 8.13
Typical incremental cost goals, projected incremental system costs, and required incentives
for a commercial absorption cooling system as a function of year of purchase to achieve a
20% market penetration in the year 2000 (Warren and Wahlig, 1983).

the amount of incentives or subsidies required for active solar cooling to achieve the desired cost to the residential consumer and thus to achieve the desired market penetration. These required incentives are also plotted in figure 8.13 and are projected to decrease steadily from 1985 to 2000.

8.4   Subsystem Cost and Performance Goals

The solar cooling system cost goals for different locations can, in turn, be subdivided into subsystem cost and performance goals. Such a breakdown is not unique, in that the subsystem cost allocations can be individually varied so long as the overall system cost goal is achieved. Subsystem costs probably will be reduced substantially by technical improvements in subsystem performance (for example, increased chiller efficiency resulting in reduced collector subsystem array size), by volume production economies, and by improved packaging that will reduce system engineering and installation costs.

Estimating costs of active solar systems requires a detailed design, including the size of the collector array and storage, the system configuration and control sequences, and the piping layouts. As part of the Active Program Research Requirements (APRR) assessment, Science Applications, Inc. (Scholten and Morehouse, 1983) developed a unified cost catalogue showing present and projected future costs of different solar components (Science Applications Inc., 1983). This catalogue should not be used to estimate costs of actual construction projects, but it does establish a uniform basis for comparing the costs of different technologies. With the experience developed in the many active solar installations, the costs of collector arrays, mounting, piping, storage, etc., are becoming better established, and standard engineering cost estimation can be used.

Principal subsystems include collector, energy transport, storage, heating, domestic hot water, and chiller. The collector subsystem includes the collectors with mounting structure and installation, piping, and insulation. The energy transport subsystem includes loop piping, fittings, insulation, pumps, valves, heat exchangers, fluids, electrical work, and controls. The storage subsystem includes hot or cold sensible or latent heat storage. The chiller subsystem includes either a Rankine chiller with auxiliary motor and cooling tower, an absorption chiller with auxiliary boiler and cooling tower, or a desiccant chiller.

The cost, performance, and reliability of all active solar technologies must be improved if solar energy is to play a major role in the marketplace. The cost of collecting solar energy must be reduced if domestic space and water heating are to remain economically attractive as the tax credits are phased out. Advanced absorption, Rankine, and desiccant cooling technologies must be developed for solar energy to contribute to space cooling.

8.4.1 Collector Array Costs

The collector array is usually one of the major costs of any active solar cooling system. A key to cost-effective cooling systems is reducing the collector array costs. For residential systems with less than about 500 ft^2 (50 m^2) of collector area, the collectors can usually be mounted directly on the roof, with no support structure necessary. For commercial systems, however, a support structure is often needed to tie the building structure to the collectors themselves. For the high-performance chillers operating at higher temperatures, evacuated tubes with reflectors or parabolic trough collectors will likely be used.

A key to low-cost collectors is the use of lightweight, inexpensive materials that reduce the costs of the collectors, the supporting structure, and the installation labor. The cost to a contractor of a flat-plate collector can conceivably be reduced from a present cost of typically $15/ft^2 ($160/m^2) in 1985 dollars to about $3.2/ft^2 ($34/m^2) (Kutscher et al., 1984). The Low-Cost Collector Program has projected the manufacturing cost of a linear trough collector with a lightweight reflector and iron pipe absorber at $6.50–$8.50/ft^2 ($70–$90/m^2). One vendor currently has developed such a collector with an installed cost of about $21/ft^2 ($225/m^2) (1985 dollars).

Jacobsen and Ackerman (1981) summarized the current costs of manufacturing, distributing, and installing residential water heating and space conditioning systems using flat-plate and evacuated tube collectors. They concluded that developing packaged and standardized subsystems will reduce system engineering and installation costs. The historical price of the residential heat pump shown in figure 8.3 indicates that price has been significantly reduced as the volume of production increases and learning curve experience accumulates. They examined the cost reduction opportunities and developed a range of system costs that may be anticipated by a mature solar industry. They projected that costs could be dramatically reduced for both flat-plate and evacuated tube collectors as the

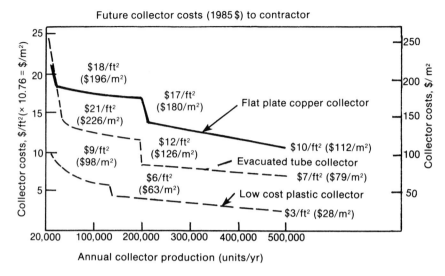

Figure 8.14
Future solar system collector costs as a function of annual collector production (Jacobsen and Ackerman, 1981).

volume of collectors produced increased, as shown in figure 8.14 (Jacobsen and Ackerman, 1981). The potential for reducing costs is significant. With automation and a production volume of greater than 200,000 panels per year in a single facility, the cost of evacuated tube collectors can be reduced to $8.50/ft^2 ($90/m^2) (1985 dollars). The sudden drops in price shown in figure 8.14 result from economies of scale when annual production reaches sufficient levels.

An alternative future development path for active solar cooling uses low-temperature, lower-performance technology. Current absorption chillers with a coefficient of performance (COP) of 0.7 or desiccant coolers have the distinct advantage of operating at temperatures below 203°F (95°C). Performance of these systems could be adequate if low-cost, high-performance collectors could be developed. Referring to figure 8.14, low cost can be defined as collector costs of about $7.80/ft^2 ($85/m^2) (1985 dollars). High performance in this context means an efficiency of about 40–50% at a typical absorption chiller driving temperature of about 185°F (85°C). Work at Brookhaven National Laboratory has developed solar collectors using polymer film technology (Wilhelm and Andrews, 1982) that could have sufficiently high performance and low cost, but serious material bonding problems must be solved before the technology is acceptable to the market.

8.4.2 Thermal Performance Improvement

The thermal performance of active solar cooling systems can be improved by increasing the thermal COP of the absorption chillers, thereby reducing the solar heat input, collector area, and heat rejection requirements for meeting a given fraction of the cooling load. The efficiency of the collector systems also can be improved.

Technical improvements in subsystem performance are anticipated. For the residential 3-ton absorption chiller and for the commercial 25-ton absorption chiller, the current COP of about 0.7 should stay relatively constant until double-effect absorption chillers are developed and introduced in these sizes, at which time the COP should increase to about 1.3. Today's lithium bromide, single-effect absorption chillers have a maximum COP of 0.72, whereas the COP of double-effect chillers can approach 1.15. Advanced ammonia-water absorption chillers presently under development, if successful, will operate at higher temperatures with improved COP of 1.25–1.55 at 280°F (138°C) (Dao, 1978).

The size of the collector array depends on the COP of the chiller. At a COP of 0.7, a peak heat input of 125 kW (430 kBtu/h) is needed to operate a chiller at 25 tons. With a solar input of 300 Btu/h ft^2 (946 W/m^2) and a collector efficiency of 45%, the collector area required to drive the chiller at full capacity is about 130 ft^2/ton (12 m^2/ton). As the COP of the chiller is increased, the required energy collected and collection area will be reduced correspondingly. For example, if the COP is increased from 0.7 to 1.3 and if the collector efficiency is maintained at 45%, then the required collector area would drop from 129 ft^2/ton (12 m^2/ton) to 70 ft^2/ton (6.5 m^2/ton).

8.4.3 Electrical Performance Improvement

Crucial to the development of cost-effective solar cooling systems is reducing parasitic mechanical and electrical energy consumption. The electric energy consumption of the chiller system consists of energy to run the collector loop and to run the generator, solution, and chilled-water pumps on the absorption chiller or the boiler and chilled water pumps on the Rankine system; energy is also needed to reject heat. The electrical energy consumption can be reduced by proper design and sizing of pumps, piping, and heat exchangers in the collector loop and the chiller. The energy required to reject heat to the environment depends on the type of sink used (wet cooling tower, air-cooled dry coil, or wet coil evaporative cooler), the fan power to provide air movement, and the pumping power required. The evaporatively cooled direct condenser coil can reject heat near the outdoor wet bulb temperature and has the lowest combined fan and pumping power requirement; however, it has additional maintenance requirements. Because today's active solar cooling system must reject almost twice the heat as a conventional system, it requires twice the fan energy for the same efficiency of heat rejection. Clearly the need for heat rejection limits the possible energy savings for the solar cooling systems.

Computer simulations by Warren and Wahlig (1985) using TRNSYS (University of Wisconsin, 1983) predict smaller than expected energy savings for active cooling and heating systems primarily because of increased estimates of parasitic power consumption and the use of more efficient conventional air conditioning for comparison to the solar system performance. Figure 8.15 shows the predicted annual total electrical energy use for a 25-ton system as a function of collector area. Also shown are the solar auxiliary use, the conventional energy use to serve the same cooling load, and the F-CHART (University of Wisconsin, 1981) predictions for the auxiliary energy use. It is important to validate the TRNSYS simulation against real system data to establish that the projected energy savings are correct. In developing new chillers, parasitic power consumption must be reduced and COP of the absorption chillers improved for the system to be cost competitive.

8.5   Commercial Absorption Cooling System Example

The Active Program Research Requirements (APRR) assessment (Scholten and Morehouse, 1983) established a methodology to compare the performance and economic viability of different active solar space conditioning systems. Today, with an established solar industry, the costs and performance of active solar heating and hot water systems are better defined so that the economic decisions are straightforward. The technologies of solar absorption heat pumps, solar desiccant cooling systems,

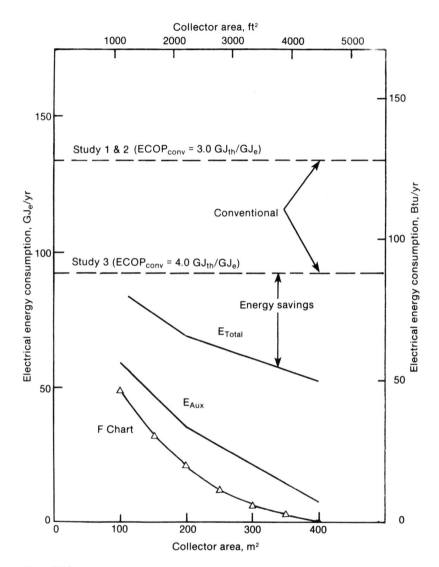

Figure 8.15
Electrical energy consumption (GJ_e/yr) for a 25-ton solar absorption cooling system with electric vapor compression backup in Phoenix as a function of collector area (Warren and Wahlig, 1985).

and heat-engine-driven heat pumps, which are still being developed, can be evaluated as better performance and cost data become available. Establishing cost goals for future active solar systems is useful for evaluating different solar cooling technologies. The system incremental cost goals for the year 2000 have been calculated based on the present value of future energy savings discounted at a rate of ROI sufficient to stimulate significant market penetration. The present value of future energy savings is calculated assuming a twenty-year system life and a nine- or five-year simple payback period. Because my analysis focuses on determining rough cost goals for future technologies and because tax considerations change with legislation, I have not included the effect of tax credits, tax rates, depreciation, etc. in this analysis.

In the commercial market the cost of a solar cooling system should be paid back from energy cost savings in no more than five years. This payback is equivalent to a 23% real ROI over the twenty-year life of the system. In the residential market the cost of the solar cooling system should be paid back out of savings in no more than nine years.

The energy savings potential of both residential and commercial active solar cooling systems needs to be assessed to establish future cost goals and to define research needs. The energy savings potential of current absorption and Rankine cooling systems was estimated by analysis of detailed computer simulations using TRNSYS in four cities (Phoenix, Miami, Fort Worth, and Washington, D.C.) and by projecting electrical and thermal performance improvements that can and must be made for economically viable systems (Warren and Liers, 1983).

The energy savings produced by an active solar heating and cooling system are determined in part by the energy requirements for space conditioning by conventional means. Considerable advances in fossil-fuel fired heating and cooling equipment expected between now and the year 2000 will reduce the predicted energy savings. The performance of conventional heat pumps and vapor compression air conditioners will be improved by more efficient motors and optimized heat exchanger areas. The development of pulse combustion gas-fired furnaces will increase the heating efficiency to 90% or greater. For my analysis the seasonal COP of commercial cooling systems was assumed to be 4.0.

Figure 8.16 shows a representative solar-fired commercial absorption system. The baseline 25-ton absorption chiller operates with a COP of 0.7

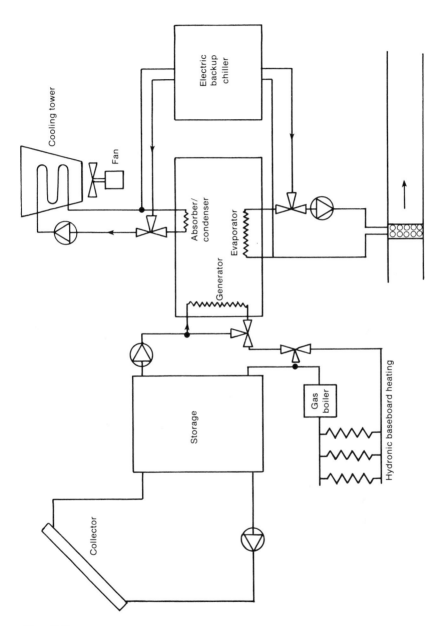

Figure 8.16
Water-cooled commercial absorption cooling system (Warren and Liers, 1983).

Table 8.3
Energy savings in Phoenix for a commercial solar-fired absorption cooling system
(1985 dollars)

Savings	Backup energy source			
	Electric baseline	Electric improved	Electric advanced	Gas advanced
Conventional electrical, GJ_e	91.4	91.4	91.4	91.4
Solar electrical energy, GJ_e	52.3	40.0	34.4	23.7
Net electrical energy savings, GJ_e	39.1	51.4	57.0	67.8
(10^6 Btu_e)	37.2	48.7	54.1	64.3
Solar natural gas required, GJ_g				30.6
Natural gas energy savings, GJ_g				30.6
(10^6 Btu_g)				29.0
First year 200 fuel cost savings	$1,685	$2,210	$2,455	$2,575
Cost goal 20-year payback	$32,800	$42,390	$47,585	$48,590
Cost goal 9-year payback	$16,645	$21,780	$24,240	$24,910
Cost goal 5-year payback	$8,715	$13,395	$12,735	$13,180

Source: Warren and Liers (1983).

at a firing temperature of 195°F (90°C) and has an electrically operated backup chiller. Advanced chillers proposed or under development are projected to operate with a COP as high as 1.55 at an operating temperature of 280°F (140°C). For the baseline systems the collector loop consists of about 2500 ft² (232 m²) of evacuated tube collectors. For the advanced systems providing the same solar fraction, the collector loop consists of about 915 ft² (85 m²) of advanced, integrated CPC evacuated tube collectors or equivalent high-performance, linear trough collectors. Backup can be provided with a conventional electric chiller or with a gas boiler firing the absorption chiller. For advanced systems the gas backup boiler can be integrated into the absorption chiller.

Performance estimates of the baseline system have been obtained from the detailed simulation by Choi and Morehouse (1982). The performance of the improved systems have been estimated assuming a reduction in the parasitic power consumption for the chiller by about 20% and for heat rejection by about a factor of 2. The energy savings of the baseline and improved solar-fired absorption cooling system are shown in table 8.3. The energy conventional cooling system is a vapor compression chiller with a seasonal electrical COP of 4.0 (thermal/electrical).

System performance with advanced collectors and chillers has been estimated assuming that (1) the thermal COP of the absorption chiller is increased from 0.7 to about 1.55 and the operating temperature is increased from 195°F (90°C) to 280°F (140°C), based on projections of advanced chiller performance to be achieved by research in progress; (2) the collector area is reduced to give the same solar fraction for cooling, considering both the increased chiller COP and the use of the integrated compound, parabolic reflector, evacuated tube collector that has been proposed and tested (Winston and O'Gallagher, 1981); and (3) the backup cooling is provided by an integrated gas boiler that operates at a COP of 1.55.

Figure 8.17 shows the projected improvement in electrical energy savings for a 25-ton solar-fired absorption cooling system as the electrical efficiency for heat rejection, chiller operation, and collector loop operation are improved. The effect of advanced system performance on electrical energy savings is also shown. The main benefit of the advanced systems arises from reducing in the collector area and cost and reducing in the heat rejection required by the chiller. Integrating the gas backup into the chiller also decreases the total system cost because separate backup equipment is not needed.

The cost of today's active solar cooling systems has been estimated by engineering analysis. Clearly this cost must be reduced. Assumptions must be made about the cost reductions for future systems. The precise costs that can be reached are a matter of speculation. Table 8.4 shows the projected cost of present, improved, and advanced solar-fired absorption cooling systems in the year 2000, as expressed in constant 1985 dollars.

Today's installed collector and support cost of $38/ft^2 ($415/m^2) (1985 dollars) must be reduced to about $10/ft^2 ($110/m^2). Currently, 25-ton gas-fired water-cooled absorption chillers have an installed cost of about $530/ton. If the cost of advanced solar-fired absorption chillers with integral gas backup can be reduced to that of today's chillers, the ultimate cost of the advanced absorption chillers would be reduced from about $980/ton to $530/ton. Based on this analysis, an active solar-fired, high-performance absorption heat pump in a cooling-only mode with an integrated gas-fired heat pump backup should achieve costs that give approximately a nine-year payback. This is seen by comparing the final estimated incremental system cost of $25,000 with the final-column cost goals in table 8.3.

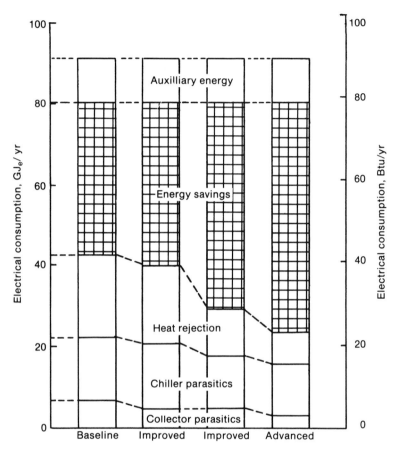

Figure 8.17
Projected electrical energy savings with improvements in heat rejection, chiller and collector parasitic energy usage, and chiller thermal improvement (Warren and Liers, 1983).

Table 8.4
Component and systems costs for commercial solar-fired absorption cooling systems (1985 dollars)

	Backup energy source			
Costs	Electric baseline	Electric improved	Gas advanced	Advanced
Collector area (m^2)	230	232	93	93
Collector area (ft^2)	2500	2500	1000	1000
Chiller COP	0.7	0.7	1.55	1.55
Collectors (installed)	$95,500	$25,000	$10,000	$10,000
Storage	9,200	9,200	9,200	9,200
Solar-fired chiller	27,400	13,700	13,700	13,700
Cooling tower	6,500	6,500	4,800	4,800
Backup chiller	14,700	14,700	14,700	
Controls	11,800	3,400	3,400	3,400
Miscellaneous pumps, piping, etc.	10,500	10,500	10.500	10,500
Total system cost	$175,100	$83,000	$66,000	$51,400
Conventional cost	$26,400	$26,400	$26,400	$26,400
Estimated incremental cost	$149,200	$56,600	$39,700	$25,000

Source: Waren and Liers (1983).

8.6   Conclusions

At present, solar heating and cooling technology is not cost-effective in the marketplace. Even though the application of solar energy is known to save renewables, the solar industry is driven primarily by market forces that are currently unfavorable to it. Not only has the cost of fossil fuels reached lows undreamed of in the recent past, but solar tax incentives are gone as well.

Market forces will continue to shape the future of solar cooling and heating technology. Adoption of solar space cooling will depend on the development of reliable and effective energy conversion devices, such as thermally fired heat pumps and desiccant dehumidification with evaporative cooling. These devices will be more efficient than conventional cooling systems, will make thermally fired systems competitive with electric vapor compression cooling, and will reduce our consumption of nonrenewable energy presently used to provide cooling. In terms of heating, solar will have to compete with fossil fuels, requiring the development of highly efficient and low-cost solar collection and transportation systems. Likewise, solar technology for space and domestic hot water heating will depend on the development of cost effective packaged systems to deliver the solar energy to the application. Ultimately widespread use of solar technology for solar heating and cooling depends on achieving cost and performance goals and on developing reliable systems that can be easily maintained. The cost requirements for a successful solar technology will depend on the factors discussed in this chapter. These include the thermal and electrical performance of the solar technology and the competing conventional alternative to meet the specific end-uses, the return on investment or payback expectation of the potential user, and the component and installation costs. These cost requirements apply equally to open- and closed-cycle absorption systems, Rankine systems, and desiccant systems. Work continues to define new technology options to displace nonrenewable energy with solar energy for solar cooling and heating systems (Bankston and Breger, 1986).

References

ARI. 1980. *Industry Shipments of Unitary Heating and Cooling Equipment by Product (1946–1979)*. New York: American Refrigeration Institute.

ASHRAE. 1975. *Energy Conservation in New Building Design*. Standard 90–75. New York: American Society of Heating, Refrigeration and Air Conditioning Engineers.

Bankston, C. A., and D. S. Breger. 1986. *Comparative Systems Analysis of Solar Building Heating, Cooling and Hot Water R&D Alternatives*. Washington, DC: CBY Associates, Inc.

Brooks, H., and E. L. Ginzton, eds. 1980. *Energy in Transition 1985–2010*. San Francisco, CA: W. H. Freeman and Co.

Choi, M., and J. Morehouse. 1982. *Analysis of Commercial and Residential Solar Absorption and Rankine Cooling Systems*. McLean, VA: Science Applications, Inc.

Dao, K. 1978. *A New Absorption Cycle: The Single-Effect Regenerative Absorption Cycle*. LBL-6879. Berkeley, CA: Lawrence Berkeley Laboratory.

Hirshberg, A. S. 1978. "Overview of the impact of federal incentives on solar heating and cooling economics," in *Proceedings of the Conference on Role of Utility Companies in Solar Energy*. Chicago, IL: Institute of Gas Technology, 115–121.

Hughes, P. J., M. K. Choi, W. B. Scholten, J. R. Stokley, and J. H. Morehouse. 1981a. *Status of Several Solar Systems with Respect to Cost and Performance Goals*. McLean, VA: Science Applications, Inc.

Hughes, P. J., J. H. Morehouse, M. K. Choi, N. M. White, and W. B. Scholten. 1981b. *Evaluation of Thermal Storage Concepts for Solar Cooling Applications*. Subcontract report SERI/TR-09083-1. Golden, CO: Solar Energy Research Institute.

Jacobsen, A. S., and P. D. Ackerman. 1981. "Cost reduction projections for active solar systems," in *Proceedings of the American Section, International Solar Energy Society Conference*. Newark, DE: International Solar Energy Society, American Section, 129–1295.

Kutscher, C., R. Davenport, R. Farrington, G. Jorgensen, A. Lewandowski, and C. Vineyard. 1984. *Low-Cost Collectors/Systems Development Progress Report.* SERI/RR-253-1750. Golden, CO: Solar Energy Research Institute.

Lawrence Berkeley Laboratory. 1980. *DOE-2.1: The Users Guide.* LBL-8689, Rev. 1. Berkeley, CA: Lawrence Berkeley Laboratory.

LeBoeuf, C. 1980. *Standard Assumptions and Methods for Solar Heating and Cooling System Analyses.* SERI/TR-351-402. Golden, CO: Solar Energy Research Institute.

Lilien, G. L., and P. E. Johnston. 1980. *A Market Assessment for Active Solar Heating and Cooling Products, Category B: A Survey of Decision Makers in the HVAC Market Place.* Final Report DOE/CS/30209-T2. Washington, DC: U.S. Department of Energy.

Moden, R. 1981. "Effects of the provisions of the corporate and personal income tax codes on solar investment decisions," in *Proceedings of ASME Solar Engineering Conference.* New York: American Society of Mechanical Engineers, 475–484.

Planco, Inc. 1981. *Assessment of Active Solar Air Conditioning, 1980–2000.* Washington, DC: U.S. Department of Energy (unpublished).

Scholten, W. B., and J. H. Morehouse. 1983. *Active Program Research Requirements.* McLean, VA: Science Applications, Inc.

Science Applications, Inc. 1983. *Active Solar Component Cost Catalog, Version 3.* McLean, VA: Science Applications, Inc.

University of Wisconsin. 1981. *F-CHART 4.0: The University of Wisconsin Solar Energy Design Program.* Madison, WI: University of Wisconsin.

University of Wisconsin. 1983. *TRNSYS, a Transient System Simulation Program, Version 12.1.* Madison, WI: University of Wisconsin.

U.S. Department of Energy. October 1983. *Energy Projections to the Year 2010; a Technical Report in Support of the National Energy Policy Plan.* DOE/PE-0029/2. Washington, DC: U.S. Department of Energy.

Versteegen, P. L., and J. H. Morehouse. 1979. *A Thermal and Economic Comparative Analysis of Absorption and Rankine Solar Cooling Systems for Commercial Buildings.* McLean, VA: Science Applications, Inc.

Warren, M. L. 1982a. *Active Program Research Requirements for Commercial Solar Rankine Cooling Systems.* LBID-650. Berkeley, CA: Lawrence Berkeley Laboratory.

Warren, M. L. 1982b. *Active Program Research Requirements for Residential Solar Absorption and Rankine Cooling Systems.* LBID-652. Berkeley, CA: Lawrence Berkeley Laboratory.

Warren, M. L. 1982c. *Active Program Research Requirements for Water and Air Cooled Closed Cycle Commercial Absorption Systems.* LBID-651. Berkeley, CA: Lawrence Berkeley Laboratory.

Warren, M. L., and H. S. Liers. 1983. "Research goals for active solar cooling systems," in *Proceedings of the 1983 American Solar Energy Society Annual Meeting.* Vol. 6. Boulder, CO: American Solar Energy Society, 169–174.

Warren, M. L., and M. A. Wahlig. 1982. *Cost and Performance Goal Methodology for Active Solar Cooling Systems.* LBL-12753. Berkeley, CA: Lawrence Berkeley Laboratory.

Warren, M. L., and M. A. Wahlig. 1983. "Methodology to determine cost and performance goals for active solar cooling systems." *Journal of Solar Energy Engineering* 105:217–223.

Warren, M. L., and M. Wahlig. 1985. "Cost and performance goals for active solar absorption cooling systems." *Journal of Solar Energy Engineering* 107:136–142.

Wilhelm, W. G. 1981. *Low Cost Solar Energy Collection for Cooling Applications*. BNL-51408. Upton, NY: Brookhaven National Laboratory.

Wilhelm, W. G., and J. W. Andrews. 1982. *Development of Polymer Film Solar Collectors: A Status Report*. BNL-51582. Upton, NY: Brookhaven National Laboratory.

Winston, R., and J. O'Gallagher. 1981. "Engineering development studies for integrated evacuated CPC arrays." in *Proceedings of Active System Heating and Cooling Contractor's Review Meeting*. Washington, DC: U.S. Department of Energy, 8-8.

9 Cost Requirements for Passive Solar Heating and Cooling

Charles R. Hauer

Passive solar technology involves the systems and components integrated into a design of a building for solar heating, cooling, and lighting. It also encompasses such diverse factors as building orientation, site preparation, building design, and landscaping. Because passive concepts reflect integrated design rather than just solar collector area and thermodynamic efficiency, economic or performance analyses of passive solar systems are difficult and must be based on total building performance; that is, performance must be related to building comfort and the effective use of solar energy. Investigators have addressed this need by studying the analytical and economic issues of passive solar systems using such correlations as solar savings fraction, unutilizability, and cost/performance objectives. These correlations vary from one technology to another in their applicability and represent an attempt to evaluate passive design performance in some standard format. In addition to the technical and economic aspects of passive solar, one must also consider market penetration, builder and market considerations, utility interface, and economic modeling.

Here the discussion of passive solar economics summarizes a broad range of technical, analytical, and economic research. First, I describe briefly passive design methods and terms to establish a frame of reference for the reader. I include major references so that readers interested in particular facets of passive solar applications may pursue their interests further. I do not discuss detailed concepts of passive heating, cooling, and daylighting technologies because volumes 4, 6, 7, and 9 of this series describe these concepts fully. In particular, the economic evaluation of a passive design is illustrated in section 2.7.4 in chapter 2.

Passive solar technologies are used to heat or cool building space and to provide daylight to the interior of structures. The concepts listed in what follows are often used in discussing passive systems; their definitions will prove helpful in the discussions in other sections.

Direct gain occurs when sunlight enters a living space and heats the absorbing surfaces, which in turn heat the space and the occupants by re-radiating the energy and convectively heating the air.

Indirect gain occurs when sunlight is absorbed by thermal mass interposed between the glazing and the conditioned space. The conditioned space is partially enclosed and bounded by the mass, creating a strong thermal coupling between the thermal mass and the conditioned space. A storage wall system (for example, Trombe wall) is one type of indirect gain system.

Attached sunspace essentially combines the direct gain and storage wall (indirect gain) approaches, although most often the thermal mass is in the slate or concrete floor of the sunspace. Often the sunspace also serves as a greenhouse.

Daylighting uses sunlight to reduce the electric lighting loads of a structure. Direct gain is part of this concept if an infrared reflective coating is not used on the glazing, permitting both heat and light to enter the living space. In those regions where cooling loads predominate, infrared-coated glass is used to exclude heat and the cooling load is reduced by the heating that would be attributed to the electric lighting required if daylighting were not used.

These passive solar concepts generally apply to both residential and commercial buildings. Other terms needing definition are passive cooling, which involves coupling the building to the ground, and using the ground as a heat sink for the building, radiative heat transfer to the night sky, and thermosiphon systems, which are generally used for water heating. Thermosiphon systems are similar to active solar water heaters except they do not use a pump and the fluid circulates through the solar collector by convection, driven only by the temperature gradient of the system. A thermosiphon air system heats spaces using an absorber surface, which heats the surrounding air, causing it to circulate through the building. Finally, a "hybrid system" also is often used and denotes a system, such as an attached greenhouse or a thermal storage wall, that uses a fan to transfer heat to the living space.

Unlike active solar energy systems, in which the heat absorbed by the solar collectors can be fairly accurately measured by the working fluid pumped into the building, passive systems constitute an integrated design in which various building elements perform various functions related to heating, cooling, or energy storage. Passive solar energy systems are as much an architectural concept as they are a technology. Therefore it is difficult to evaluate the economics of passive systems in classic terms. Questions on passive system component costs and energy saved by the system are difficult to answer. At best we can hope to say how much more a passive building costs (marginal cost) when compared with a conventional building and how much less energy is consumed as a result of the passive components. Considerable variation exists in estimating both of these factors.

To address this issue, Mueller Associates (1981) proposed the concept of cost/performance goals, which assigns costs and performance standards to various passive components. These cost/performance goals are different for residential and commercial buildings applications. One principal reason for this difference is the "use cycle" of residential or commercial buildings. Commercial buildings are devoted to office, light manufacturing, or warehouse use and are most often vacant at night. The objective of the cost/performance goal concept is to establish limits for the cost of passive system elements and thereby define economically acceptable design.

9.1 Discussion of Technical and Economic Factors

In passive building systems heat is gained and retained by using building elements in novel ways. Heat is distributed through the building using temperature differences for conduction, convection, and radiation. A well-designed passive structure integrates five functions for heating collection, absorption, storage, distribution, and regulation.

Generally the collector consists of transparent apertures on the south-facing side of the building. These apertures may range from vertical windows to skylights in the roof. The absorber is a dark surface that converts the incident sunlight into heat, which is then available for transfer to the interior of the building or into storage. Thermal energy may be stored in two ways: by increasing the temperature of the storage medium, which may be the building mass (referred to as sensible heat storage), or by melting the storage material (latent heat storage or phase-change storage). Next the heat must be distributed to the air in the interior of the building. This may be accomplished by radiation or by natural convection. The air flow may be assisted by small fans. In this case the temperature in the space can be controlled by the fan. Other means of regulation involve the gains or losses of heat through the aperture. Night insulation reduces losses in the winter, and overhangs and shading limit gains in the summer.

Several nontechnical factors strongly affect economic analyses for passive solar residential and commercial buildings. These factors can be summarized under the general class of problems associated with utility costs and characteristics in a region and builder perceptions of the market.

Electric utility rates and the availability and price of natural gas as a fuel significantly affect the economics of passive solar. Moreover, the insolation characteristics of the region determine how large a solar savings fraction may be conveniently attained; this in turn sets the requirements for backup utility service. Builders must consider utility rates and rate structures, as well as the demand profile the utility experiences when preparing passive solar designs. The utility interface issues are discussed in several references. In a study of energy price subsidies Battelle Pacific Northwest Laboratory (1980) discusses the impact of federal fuel subsidies on electric utilities. Bezdek, Hirshberg, and Babcock (1979) and Coleman and Ford (1982) relate local electric utility rates to the economic viability of solar water heating; Coleman and Ford (1982) discuss the impact of local utility costs on the economics of roof ponds; and the MITRE Corporation (1978) links utility rates and capacity to the prospects for solar energy adoption through the year 2000.

Builders and developers perceive their local market based on many years of experience in their region. These perceptions, which are generally quite accurate, relate to what will sell in their region and how important fuel savings can be in giving them a market edge. Two corporations have researched this subject (Real Estate Research Corp., 1980; Market Facts, Inc., 1980). One of these studies for the Department of Housing and Urban Development reports on solar home sales in the 1980s (Real Estate Research Corp., 1980). Market surveys were also carried out to poll consumer attitudes toward solar housing features to assist builders in formulating market strategies (Market Facts, Inc., 1980; AIA Research Corp., 1980).

9.2   Performance Modeling Methods

As noted previously, to evaluate the economics of passive solar technologies, we must be able to quantify their benefits. The Solar Energy Research Institute (SERI) and several national laboratories, notably Los Alamos National Laboratory (LANL), developed several levels of building design and analysis tools. These tools ranged in complexity from computer simulations of building performance, requiring mainframe computers, to manual calculation methods, yielding rules of thumb for the builder. These design tools are described in many references (Treichler, 1976; Kusuda, 1974; Henninger, 1975; Lokmanhekim, 1978; Horak et al., 1979; Diamond et al., 1979; Hittle, 1977; ASHRAE, 1975; SERI, 1980; Arens, Nall, and Carroll, 1979; Hall et al., 1979; and Balcomb et al., 1980) and in volume 7 of this series. I focus here on the manual methods, largely developed by Balcomb et al. (1980) at LANL because they are simple and furnish the degree of resolution required to make the economic cost/benefit decisions that determine the choices in using passive solar. Moreover, these methods yield such general design parameters as glazing area and thermal storage capacity, making it possible to calculate the cost impact of the solar component. The main methods are the solar load ratio method and the load collector ratio method.

Although the methods are described in detail in the preceding references, it is still useful to define some of the terms used. First, the solar savings fraction (SSF) is the percentage of energy saved by a passive solar building as compared with a conventional building of similar characteristics in a particular region. Similar characteristics mean that the buildings are identical except for the passive solar features. Thus, if a conventional building required 50×10^6 Btu/yr (53 GJ/yr), and its solar counterpart required only 20×10^6 Btu/yr (21 GJ/yr), the solar savings fraction would be $30 \times 10^6/50 \times 10^6$ or 60%.

Second is the building load coefficient (BLC), which is similar to a standard steady-state heat load calculation except that it incorporates the degree-day characteristics of the site, with the solar collection area losses not being counted. Thus the BLC measures heating demand and also indicates how well insulated a building is. Third, the load collector ratio (LCR) is simply the BLC divided by the number of square feet of solar collection aperture provided in a building. A significant part of the work carried out in the computer modeling of building heating and cooling (including BLAST and DOE-2) is discussed in chapter 3 of this volume and given by the references listed at the beginning of this section.

9.3   Differentiation between Residential and Commercial Buildings

Economic analyses require an understanding of the economic perspective of the individual making the calculations. The perspectives may be those of the builders, owners, tenants, or utility. The boundary conditions fac-

Table 9.1
Primary energy use in buildings

Energy use	Residential (%)	Commercial (%)
Space heating	48	44
Space cooling	7	21
Lighting	6	23
Hot water	14	2
Other	25	10
Total	100	100

Source: Neeper and McFarland (1982).

ing each group must be defined to make economic analyses useful. Moreover, the significant differences between residential and commercial buildings must be addressed that relate to the use cycle, codes, utility rates, taxes and subsidies, return on investment, and distribution of energy use. Table 9.1 illustrates primary energy use in buildings in 1977.

Data indicate that for both residential and commercial buildings more than 55% of the energy is used for heating and cooling and that in commercial buildings about 25% of the energy is used for lighting. Furthermore, energy-use patterns are significantly different for residential and commercial buildings, particularly in terms of cooling and lighting. These patterns and the fact that commercial buildings are often vacant at night dictate that different economic conditions exist between commercial buildings and residences. Further complications stem from tax laws, codes, and utility rate differences between residential and commercial property and the requirement for a return on investment for commercial property. Residential energy use comprises the same elements as commercial energy use but in different proportions, with the entire cost generally carried by the user or owner of the building.

Builders must be able to "benefit" from the application of solar technology. For residential builders, where the housing is built on speculation, the builder hopes to benefit because the building now has greater "salability" because of its lower heating costs. Commercial buildings are more often built to an owner's specifications, where the owner may or may not be the occupant. When passive solar measures are considered for commercial buildings, daylighting is the primary reason to opt for passive solar, because lighting represents about a quarter of the building's energy consumption. Heating and lighting together account for almost 70% of the energy consumed. However, daylighting benefits depend significantly on the geographic location, building use, and utility rates. Selkowitz, Villecco, and Griffith (1981) have studied both the technical and economic aspects of daylighting. Other sources (Rosenfeld and Selkowitz, 1976; Duguay and Edgar, 1979; Commission Internationale de l'Eclairage, 1970; Daylighting Committee of the IES, 1962; Libbey-Owens-Ford Company, 1976; Kusuda and Collins, 1978; Arumi, 1977; Architectural Aluminum Manufacturers Association, 1977) discuss this technology and its economic implications in great detail. Daylighting is also discussed in volume 9 of this series. Utility rate structures cannot be overemphasized in discussing the economics of daylighting in commercial buildings, particularly when demand charges are incorporated in the rate structure or when seasonal rates resulting from air conditioning peak loads are used by a utility.

The concept of the "payback" time is often used as an economic indicator for solar energy systems. Boer (1978) developed a payback methodology that defines payback time as the length of time until initial cost and operating expenses with compound interest are equal to the cumulative fuel cost savings with anticipated fuel cost escalations. If daylighting is the main passive solar feature, then the marginal cost of the daylighting serves as the capital investment to be amortized, and the reduced lighting costs adjusted for heating and air conditioning load changes serve as the benefit. Payback time is a good economic indicator in the planning and design stages of any passive solar project. For owner-occupied residences the payback time measures the value of the passive investment. Application of passive solar to rental apartment buildings is generally not made because the renter pays the energy bill and there is no income to offset the additional investment.

9.4 Economic Models of Passive Solar Applications

The terminology of passive solar methods were outlined in the preceding sections. Analytical techniques that measure both the anticipated performance and the marginal costs of implementing passive solar options were referenced. It should be noted that these marginal implementation costs can vary from almost no additional investment to a significant fraction of the building's cost. This variation in cost is a function of the fraction of the building's energy load that the solar component is expected to carry, the location of the building, and the building use, namely residential or commercial.

The costs for a passive design may even approach zero, for example, in the case where only building orientation and/or fenestration is required to obtain substantial energy savings. However, these marginal costs may also be significant if a substantial fraction of a building energy load is to be carried by the solar component. When marginal costs are low, a complex economic analysis is clearly not needed. But, as the marginal costs approach 5% of the cost of the building, detailed analyses on life-cycle costs and energy savings become necessary. To apply energy cost analysis methods, we must examine the relationship between passive and conservation measures and consider the differences in marginal energy usage between residential and commercial buildings. In general, conservation relates to "energy saved" by reduced consumption; passive methods relate to "solar energy added" to the building. Both approaches result in reduced demand.

Passive solar energy systems can supplement as well as compete with energy conservation methods in decisions on building investments. Each option reduces the operating costs of a building associated with comfort conditioning; used together, they are more effective. Indeed, without carefully applying energy-conserving design strategies, passive solar designs are less effective. Thus a trade-off exists between how much of the money available for improving the energy efficiency of a building should be assigned to preventing energy loss or producing energy gain. This decision depends on whether the energy costs for a building are dominated by heating or cooling. Regardless of this trade-off, the owner or user of the building must perceive an increase in performance by this "marginal expenditure."

For every marginal expenditure of $1,000 for the first cost of conservation or passive solar, the building owner encounters a monthly expense of $10–$20 for amortization and interest. This expense must be compensated by energy cost savings if an economic advantage is to be realized by undertaking the passive measures and the associated marginal cost. This savings is particularly significant in the residential sector of the building industry, where every incremental cost reduces the market for the housing because the mortgage lenders use standard debt-to-income guidelines in approving mortgages. Thus a marginal cost increase of $1,000 to accommodate energy saving designs affects the market size available to the builder, despite the fact that the investment pays for itself in energy savings and the net result is zero cash flow. This adverse impact on the potential marketability of the house results mainly from the failure of lenders to credit housing cost with energy savings. If the marginal costs of passive solar designs are significant, then this "market shrinkage factor" may negatively influence home builders against using these technologies.

In the commercial buildings sector the effect of marginal cost increases is felt in a different way. For commercial buildings interest and fuel costs represent operating expenses and as such reduce tax liability for the landlord or the business that owns and/or occupies the building. Only the return on investment is affected by additional capital costs for the building. If the full amount of the marginal cost can be mortgaged, then commercial investors are at little risk in evaluating their returns. Indeed, a benefit may be realized in terms of cash flow. However, if the marginal costs of solar or conservation components cannot be fully mortgaged, then the return on investment declines and commercial builders are less likely to employ the technology. In any event total outlay may be increased, resulting in a balance between operating cost and the profitability of the investment.

Although chapter 2 of this volume covers economic analysis methods in detail, it is useful to consider interest rates, cash flow, and inflation factors in a general way. As the solar investment becomes a larger fraction of the building cost, it becomes more difficult to meet the zero cash flow criterion, that is, the condition where the interest and amortization cost of the investment is balanced by the energy savings. The payback period can be used to illustrate the economic effect on the investment from changes in interest rates, inflation rates, and fuel escalation rates. These economic calculations relate to the amount of energy the system can be expected to save over the year and over its lifetime. The calculations must account for the value of future savings in present terms to reach conclusions based on economic worth. This results in some form of present value analysis. The equation used to compute the constant annual payment Y (in current dollars) necessary to repay a capital loan IC in N years at a fixed annual interest rate i is

$$Y = IC \frac{i}{1 - [1/(1 + i)]^N}.$$

This calculation assumes no inflation. The greater the inflation rate, the more worthwhile it becomes to make a capital investment now to reduce recurrent costs over the years. Fuel costs, which may either rise or decline, are good examples of such recurrent costs. One way to address this issue is to deal in terms of real interest rates, that is, interest rates adjusted for anticipated inflation. Adjusting costs for inflation is more complex, although the "true" interest rate may be approximated by using the difference between the interest rate i and the inflation rate in the given equation. This approximation becomes better as the inflation rate approaches the interest rate. Moreover, should fuel costs escalate faster than the inflation rate, additional benefits accrue to anyone having invested in solar technologies to reduce fuel costs. Should these fuel costs decline, the savings decline. Another approach to adjusting solar investment costs for inflation is to deflate the anticipated annual payments by the anticipated rate of inflation, yielding costs in some measure of constant dollars for comparing with fuel cost savings.

The economic viability of solar energy heating and cooling systems strongly depends on regional or local fuel costs and fuel type as well as on the local climate and insolation. Researchers at LANL have analyzed these factors for passive solar space conditioning systems and have documented much of this work (Balcomb and McFarland, 1978; Noll and Robson, 1980; Kirschner, 1979; Noll and Wray, 1978; and Roach, Noll, and Ben-David, 1979). Many analyses emphasized life-cycle costs and developed extensive methodologies in this area; however, these models depend heavily on future economic conditions. The uncertainties regarding futures of these methods tend to make these analyses less attractive to developers, builders, and owners who are interested in the near-term economic viability of a technology. Builders and developers in particular want to be able to tell potential customers how long it will take for energy savings to result in simple payback or how the passive system marginal costs are effectively balanced by the annual fuel savings. The issues of importance for builders and buyers relate to the solar savings fraction, maximum marginal cost to attain this savings fraction, and payback time. Noll and Kirschner (1980) developed a simplified economic analysis that addresses these issues. Their economic model is no substitute for the more extensive life-cycle cost analysis described in previous chapters. Rather, their work considers the perspective of the near-term user of the technology first, permitting a near-term decision to be made in the marketplace.

Detailed economic analysis is also used in assessing market potential.

The economic research in support of passive solar technologies addressed market potential by using market penetration modeling, market potential indexing, and market surveys. The MITRE Corporation (Bennington et al., 1978), under a DOE contract, developed the most prominent market penetration model, SPURR, for solar energy technologies. Chapter 5 of this volume also discusses market penetration studies.

Market penetration analysis generally postulates substituting an existing technology with a new technology using cost, availability, or improved performance as the driving force. For solar technologies operating cost and nondependence on fossil fuels are most often used. Costs are based on capital expenditures. Although not necessarily true, one of the most significant shortcomings of most market penetration models is their failure to consider local variable conditions, such as fuel costs, building and land cost, housing demand, and the availability of the solar resource. Market penetration models are useful on a national scale for planning purposes, but builders' needs can be met only by using methods most sensitive to local conditions and variables.

In response to these local needs, Noll and Robson (1980) developed an analytic method that gives a market potential index. This method, based on multiattribute decision analysis, localizes market potential for passive residential application to 223 sectors throughout the country, covering new and retrofit construction. For example, researchers considered the following attributes for new construction in each sector: solar performance, building heat loss characteristics, solar system costs, fuel costs, housing starts, financial parameters, government incentives, income characteristics, population growth, and consumer liquidity. Because each attribute has different dimensions, the method must transform each into a dimensionless ratio, which indicates a positive or negative value in adopting passive solar for that sector. This method is a complex analytic framework that can indicate where the market potential for adoption is likely or unlikely.

Addressing the builder's problem of "what sells" requires an analysis or survey of the market to ascertain the degree to which passive solar technologies meet the perceived needs of the residential buyer in the various parts of the country. Market Facts, Inc. (1980) undertook the task of addressing this question for SERI. Their analysis yielded general planning data rather than hard economic data that the builder or developer could use in making site- or area-specific economic or marketing decisions. The Regional Solar Energy Centers (RSECs) made significant progress on the local economic issues of interest to the builder or developer. In addition, these centers assisted local architects and developed passive residential designs incorporating those passive solar features that were economically viable in a particular region. Probably the most significant contribution of these centers was their consideration of regional issues.

Although Bezdek, Hirshberg, and Babcock (1979) analyzed the economic feasibility of solar water and space heating, they did not specifically address passive solar technologies. Their economic analysis, however, is useful to the home builder or developer because it deals directly with those economic parameters most relevant to the industry, such as local insolation and local utility rates. Their calculations emphasize the balance between the solar energy collected and used and the energy costs saved for various levels of solar self-sufficiency. Their analysis is useful for anyone interested in basic economic evaluations of solar energy systems, either active or passive.

9.4.1 Life-Cycle Cost Analysis

Life-cycle cost analysis is covered in detail in chapter 2 of this volume; only several limited issues relating to commercial buildings and limited partnerships are discussed here. Life-cycle costing (LCC) is a methodology used for the economic evaluation of a project wherein all relevant cost factors over the anticipated lifetime of the project are included and accounted for in assessing its economic efficiency. This work, which outlines the methodology as applied to solar energy, is given in several references (Marshall and Ruegg, 1980; Powell, 1980; Ruegg and McConnaughey, 1980; Ruegg et al., 1978; Ruegg, Petersen, and Marshall, 1980; and Ruegg and Sav, 1980).

For commercial property, whether an office building or multifamily residences, several other factors must be considered, namely, whether the builder becomes the owner of the property (or the operator) or whether the tenant or the landlord is responsible for the utility costs. Often, small office buildings are occupied by owners, who are in a limited partnership. Multifamily residential buildings are also used for offices with limited partner ownership. In these cases economic considerations are governed to a large extent by the tax shelter objectives of the limited partners. Laquatra (1982) analyzed this situation with regard to passive solar on multifamily dwellings in upstate New York.

9.4.2   Cost Performance Characteristics

The objectives of developing cost/performance characteristics for passive solar components range from providing guidance on the economic value of a particular component in a given application to providing scientific and engineering guidance on the potential of a "conceptual component." The major concern is the economic value or the cost/performance characteristics of a given component of a passive system. Note that the cost/performance ratios for a component are dependent on the climatic region in which the component is used and on the application.

The incremental energy savings resulting from the use of a given component depends on the building type and the region. Mueller Associates (1981) developed an incremental projected performance method using the passive system simulation tools CALPAS 3 and TEASOL. For each component, a typical base-case passive system was simulated and compared to a second run that incorporated the component with the improved performance. This method yielded marginal performance data and ascribed a net improved productivity to the new component. Mueller Associates gave cost/performance ratios on a "relative" basis, using only the benefit produced above the base case. The cost/performance ratios were calculated for three regional types: cold, hot and dry, and hot and humid. Although this method is not as useful for economic analysis as a builder might wish, the work of Mueller Associates shows what has been accomplished in the area of cost/performance characteristics for passive components. In summary, the work on cost/performance evaluations has most value in a programmatic context rather than in commercial applications, although industry may wish to judge the marketability of their components based on these cost/performance guidelines.

9.4.3   Solar Energy-Utility Interface

For the builder considering the use of passive technologies, the technical and economic issues of solar energy that utilities must consider are of little importance. However, some discussion of utility issues is nevertheless important to understand the economics of passive systems. Historically utilities sought to subsidize or, at a minimum, influence builders in their selection of fuels for space conditioning. In some regions and for some electric utilities the increasing use of electricity for heating or cooling presented a severe burden because it exacerbated peak load problems.

Thus conservation and the use of passive solar energy in new construction could help utilities meet peak demand in some regions.

A commonly asked question concerns the use of electric power as backup for solar water and space heating systems. Utilities maintain that the marginal costs of serving solar energy users with backup electrical energy are greater than those associated with other utility customers. However, Bright and Davitian (1982) studied three utilities and showed that the marginal cost of providing the backup for solar (both new and retrofit) was not greater than the cost of providing energy for electric water or space heating in general. The authors studied the long-run marginal costs for providing electric service to both solar- and non-solar-assisted water and space heating systems.

The relationship between utilities and solar energy development raises many questions. These questions are investigated in a special issue of *Energy* (Cambel, Bezdek, and Hauer, 1982). Although many of the questions discussed in this journal are significant, their economic impact on passive solar applications is for the most part quite small. However, one of the important problems is the large variability of electric utility rates in the United States. On the low end of the rate scale are electric utilities of the Pacific Northwest or the regions served by the Tennessee Valley Authority; on the high end are the utilities of the Northeast. The variability of electric power costs is great enough to affect economic viability of passive solar installations. Moreover, some utilities that experience peak demand during the summer months structure their rates, using demand charges and other methods, to discourage consumption during peak demand period. Over the years there has been discussion on the question of time-of-day pricing, where the price of electricity varies with the time of day to reflect the demand that customers place on the utility in relation to the utility's generating capacity. This pricing strategy is designed to shift customer demand to a time when the utility has excess generating capacity available at lower kilowatt hour cost. Although this strategy has been used in some experimental programs in the United States, its use here is not as common as in Europe.

9.5   Builder/Designer, Market, and Cost Issues

The builder or designer faces both technical and economic design questions that must be answered before making a decision on whether to incorporate passive solar features and, if so, to what extent. In this section I summarize the research findings and reference some publications that the builder or designer can use to answer these questions. First, a solar component in a building must enhance the marketability of the property through its ability to effect savings in operating costs (heating, cooling, and lighting). Second, the design innovations needed to accomplish these cost savings must be aesthetically acceptable to local buyers. Third, the design methodology must exist for calculating the energy savings resulting from these design innovations and connecting these to annual cost benefits so that payback or life-cycle cost calculations can be carried out. The three major research tasks discussed and referenced to guide builders and designers in this regard are SERI's Denver Metro Passive Homes project (SERI, 1982), the LANL *Passive Solar Design Handbook* (Balcomb et al., 1980); and the Total Environmental Action (TEA) handbook (Total Environmental Action, Inc., 1980).

9.5.1   Denver Metro Market Experience

Under the SERI Denver metro program a number of local builders and designers received design support to develop a number of passive solar, detached, single-family homes in Denver and its suburbs (SERI, 1982). In all, more than ten designs were developed and built as one-of-a-kind units. The project was a controlled experiment that considered the marginal costs for the passive solar components and calculated the anticipated fuel savings for passive solar and for an enhanced conservation package. An economic analysis was carried out and the completed buildings were monitored during the 1981 heating season. The analyses and presentations were designed to assist the homebuyer in understanding the energy saving process for passive solar and conservation and the annual and long-term potential benefits that the owner accrues. Indeed, the reports for each design are consistent in format from one design to another and represent a step-by-step documentation of the anticipated and actual building performance. Although the study may be flawed for reasons of various subsidies, both the designs and the thermal and economic analyses may serve builders interested in pursuing passive solar applications.

Figures 9.1 through 9.5 and tables 9.2 through 9.7, describing the Heritage house, delineate the methodology used in the economic analysis. Note that the costs of the conventional home are identical with those for the passive solar home, except that the equivalent of the "passive invest-

Figure 9.1
Denver metro Heritage home.

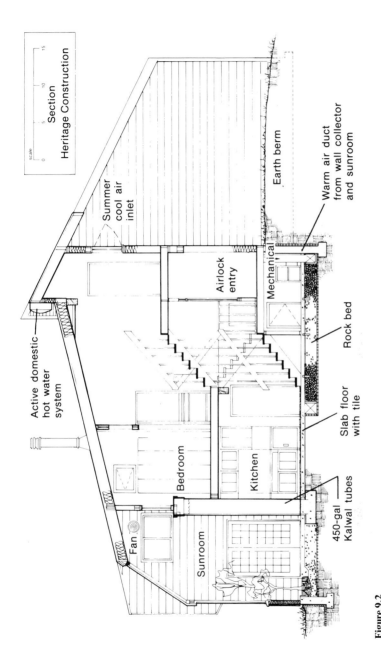

Figure 9.2
Heritage house section illustrating passive features.

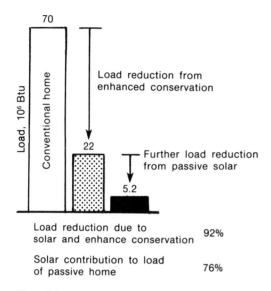

Figure 9.3
Heritage house, annual load reduction.

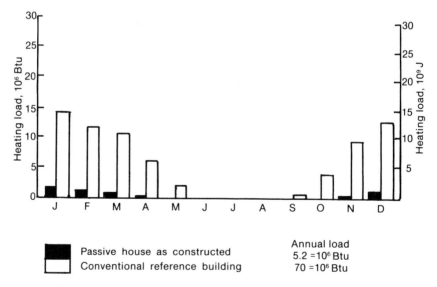

Figure 9.4
Heritage house, monthly heating load.

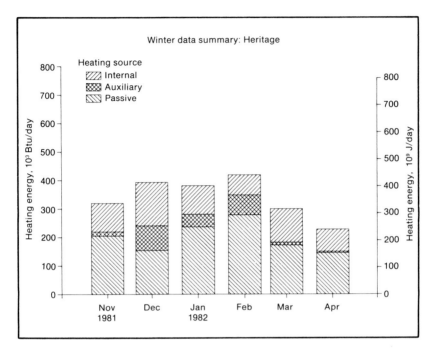

Figure 9.5
Measured thermal performance.

ment" was used in the conventional home for the installation of amenities not related to energy. Note also that SERI did all of the analyses shown in these figures.

9.5.2 *LANL Solar Passive Design Handbook*

Economic or cost analyses of passive solar energy designs mean nothing without a technological design base. Passive solar designs involve a range of technological disciplines ranging from heat transfer to architecture. Architects are rarely engineers, nor are engineers generally facile in architecture. Moreover, the users of passive solar technology are generally neither; they are builders. Thus Balcomb et al. (1980) took on a difficult task when they embarked on writing a passive design handbook. Two of the volumes are significant for the builder, namely, volumes 2 and 3. Volume 3 largely updates volume 2 and therefore deserves the most discussion. This work has been republished in summary form (Balcomb et al., 1984).

Table 9.2
Heritage House, Denver, Colorado: standard analytical assumption[a]

Variable	Value
Finished area	1900 ft² (176 m²)
Unfinished area	726 ft² (67 m²)
Heating set point	65°F (18°C)
Applicable shading of solar apertures (except overhangs)	None
Infiltration rate	0.6 air changes/h
Ground reflectance	0.3
South double-glazing properties	$U = 0.55$ Btu/ft² h °F = 3.1 W/m²K
	$R = 1.82$ ft² h °F/Btu = 0.32 m² K/W
Transmittance	0.747 at normal incidence
Extinction coefficient	0.419
Internal heat generation	53,000 Btu/day (56 MJ/day)
Masonry, concrete properties	
Conductivity	1.0 Btu/ft h °F (1.73 W/m K)
Density	150 lb/ft³ (2400 kg/m³)
Specific heat	0.2 Btu/lb °F (0.84 kJ/kg K)
Nonmass absorptance	0.2
Mass absorptance	
Trombe walls	1.0
Direct gain	0.8
Reflective backlosses	0.0
Night insulation	
Trombe walls	5 pm–8 am
Direct gain	5 pm–7 am

Source: SERI (1982).
a. Southern exposure faces due south.

Table 9.3
Heritage House[a]

Variable	Value
Heat capacitance	45 Btu/°F ft^2 glass (0.92 MJ/m^2 K)
Insulation	
Ceiling	R-40
Walls	R-20
Subgrade	R-8
Glazing area	
South	151 ft^2 (14 m^2)
North	10 ft^2 (0.9 m^2)
East	22 ft^2 (2.0 m^2)
West	26 ft^2 (2.4 m^2)
Solar wall area	179 ft^2 (16.6 m^2)

Source: SERI (1982).
a. Assumptions: 400 ft^3 (11 m^3) rock bed charged at 550 cfm (0.26 m^3/s). Rock bed has 42% void ratio (capacitance of 8,127 Btu/°F (15.4 MJ/K). Solar wall heat loss coefficient assumed to be 6.8/m^2 K.

Table 9.4
Conventional home: comparative assumption

Variable	Value
Ratio of glass area to gross floor area (glass equally distributed on all walls)	10%
Air infiltration rate	1.0 air changes/h
Insulation	
Ceiling	R-19
Walls	R-11
Passive interior mass	Masonry thermal storage removed
Basement insulation	None

Source: SERI (1982).

Table 9.5
Heritage House: incremental costs for passive solar[a]

Feature	Incremental cost of passive features[b] ($)	Total construction cost of passive features ($)	Federal tax credit (40% on nonstructural components) ($)	State tax credit (30%) ($)
Site-built solar collector wall	2,727	3,825	1,314	986
Greenhouse glazing	1,203	1,203	0	360
Concrete mass wall	94	425	0	64[c]
Water tubes for thermal mass	100	445	178	134
Rock storage dampers	1,070	1,070	428	321
Mechanical heat distribution and ducts	2,585	2,585	0	775
Total incremental cost	7,779			
Total cost after credits			3,219	

Source: SERI (1982).
a. Costs are given in 1982 dollars; to convert to 1985 dollars multiply by 1.117.
b. Includes only incremental cost over that for conventional home with standard building materials, such as drywall and carpet floors.
c. 50% of cost eligible for state tax credit.

Table 9.6
Heritage House: incremental costs for conservation upgrade[a]

Insulation	Builder's cost[b] ($)	State tax credit (20%) ($)
Walls	169	34
Ceiling/attic	169	34
Foundation/slab	456	91
Airlock entry	625	—
Total incremental cost of conservation upgrade	1,419	
Total cost after credits		1,260

Source: SERI (1982).
a. Cost are given in 1982 dollars; to convert to 1985 dollars multiply by 1.117.
b. Includes only incremental cost over that for insulation levels required by code.

Table 9.7
Heritage House: comparative fuel bills[a]

Type of home	Annual heating bill (1982 rates)
Passive solar home	$\dfrac{5.2 \times 10^6 \text{ Btu}}{3{,}414 \text{ Btu/kWh}} \times \dfrac{\$0.055}{\text{kWh}} = \$84$
Conventional home	$\dfrac{70 \times 10^6 \text{ Btu}}{3{,}414 \text{ Btu/kWh}} \times \dfrac{\$0.055}{\text{kWh}} = \$1{,}128$
	Annual fuel savings $= \$1{,}044$

Source: SERI (1982).

a. Annual heating bill (electric) $= \dfrac{\text{Heating load}}{\text{Delivered heat/kWh}} \times$ Electricity price.

Volume 3 of the *Passive Solar Design Handbook* (Balcomb et al., 1983) is divided into five chapters covering introduction, scope, and terminology; balancing conservation and solar; direct gain passive systems; sunspaces and their design; and monthly calculations (the SLR method).

In following the calculations, builders or architects would consider local insolation, climate, and cost-related factors. Instead of discussing each chapter, I will convey here a sense of what Balcomb and his coworkers were trying to achieve. Their objective was to take the reader through a stepwise process to promote an understanding of the design principles involved in minimizing auxiliary heat within a level of cost constraint using passive solar and conservation technologies. In addition to the five chapters, there are eight appendixes (mostly tables). The data in the appendixes cover insolation, conservation factors, solar load ratio correlations, weather data, and much more. By following the solar load ratio (SLR) calculation process for a particular design, one develops a good idea of what that design can be expected to achieve. The calculations can be done by hand and, once mastered, can be repeated innumerable times for design variations to improve performance.

The SLR method averages the pertinent factors affecting heating requirements for a building over a one-month period. This method uses tabular data provided in the appendixes of volume 3 of the Passive Solar Design Handbook. In general, the SLR method requires the following steps:

1. Obtain building information

   a. Building load coefficient (BLC)

   b. Projected solar collector area (Ap)

   c. Load/collector ratio (LCR = BLC/Ap)

2. Obtain site and climate information (from vol. 3, appendix D)

   a. Latitude

   b. Latitude minus mid-month declination

   c. Clearness ratio, monthly, K

   d. Incident solar radiation, horizontal surface, monthly

   e. Heating degree days (DD), monthly

3. Obtain absorbed solar radiation, monthly, S

4. Obtain monthly solar savings fraction (SSF)

   a. Calculate S/DD monthly

   b. Determine SSF, monthly, two options (appendix C)—graphic or analytic

5. Calculate auxiliary heat requirement Q

   a. Monthly, $Q = (1 - SSF) \times BLC \times DD$

   b. Annual, $Q = $ sum of monthly Q

6. Calculate annual solar savings fraction, $SSF = 1 - Q/(BLC \times DD)$

By using the SLR method, one can assess how much energy a passive solar heating design could save in a particular region of the country, given that the assumptions on climate are correct. This method gives both energy costs for that region and estimates of the economic benefits.

In summary, the basic work carried out by Balcomb and his coauthors in volumes 2 and 3 of the *Passive Solar Design Handbook* or the ASHRAE summary provides the guidance necessary for making an acceptable economic and technical evaluation of a solar design. These documents should be highly valued by builders, developers, and architects.

9.5.3 Total Environmental Action (TEA) Passive System Costs

In addition to discussing the LANL methodology for deriving passive solar system performance, it is also necessary in assessing economic value to consider the other side of the equation, namely the cost (in the marginal sense) of a particular passive system. The TEA handbook final report (Total Environmental Action, Inc., 1980) is oriented toward the homebuilder and residential construction. A passive system's cost depends on many factors, such as the design of the system and its level of detail, the materials and the trade skills employed, the type of building, the wall materials used or replaced, the fenestration layout of the south wall and the relationship of these windows to the interior spaces, and the extent of conservation methods employed. TEA analyzed each of these cost factors to determine the extra cost of the passive case, rather than the general construction case. This is the "extra" cost because credit has been taken for the cost of the standard element for which the passive element has been substituted.

The TEA methodology, which is based largely on the standard building industry costing practices, uses uniform construction industry (UCI) divisions. The cost of each division is given as a percentage of the final cost, after accounting for the credit of the component or system replaced. Final costs are given by building type based on the gross square feet of aperture area for a direct gain system. In the case of a sunspace the costs are shown in terms of the sunspace floor area. The TEA report is quite detailed and gets into the specifics of such cost elements as glazing type and installation labor. This report serves as a good guideline for builders who need to assess passive system costs.

9.6 Summary

Cost requirements for passive solar heating and cooling are derived from the physical realities of the particular design and its energy "productivity" coupled with the economic realities of the location and the market. Although builders may not be fully familiar with the design features, they are quite sensitive to the economic realities of the marketplace. On the technical side, cost analyses of passive systems depend most critically on how much solar energy is collected and put to use. Considering the economics side of the cost analysis, interest rates, inflation rates, discount rates, and fuel cost escalation rates all represent significant levels of uncertainty. These uncertainties permeate the economic and analytic approach to use of passive solar energy. Given these uncertainties, the reader will find much useful information in the reference material cited.

References

AIA Research Corp. 1980. A *Survey of Passive Solar Homes*. Washington, DC: U.S. Department of Housing and Urban Development.

Architectural Aluminum Manufacturers Association. 1977. *Voluntary Standard Procedure for Calculating Skylite Annual Energy Balance*. AAMA 1602.1. Chicago, IL: Architectural Aluminum Manufacturers Association.

Arens, E. A., D. H. Nall, and W. L. Carroll. 1979. *The Representativeness of TRY Data in Predicting Mean Annual Heating and Cooling Requirements*. ASHRAE Symposium Paper PH-79-8, no. 1. Atlanta, GA: American Society of Heating, Refrigerating, and Air-Conditioning Engineers.

Arumi, F. 1977. "Daylight as a factor in optimizing the energy performance of buildings." *Energy and Buildings* 1(2):175–182.

ASHRAE. 1975. *Bibliography on Available Computer Programs in the General Area of Heating, Refrigeration, Air Conditioning and Ventilation*. GRP 153. New York: American Society of Heating, Refrigeration, and Air Conditioning Engineers.

Balcomb, J. D., and R. D. McFarland. 1978. "A simple technique of estimating the performance of passive solar heating systems." *Proceedings of the 1978 Annual Meeting*. Newark, DE: International Solar Energy Society, American Section, Vol. 2.2:89–96.

Balcomb, J. D., R. W. Jones, R. D. McFarland, and W. O. Wray. 1984. *Passive Solar Heating Analysis: A Design Manual*. Atlanta, GA: American Society of Heating, Refrigeration, and Air Conditioning Engineers.

Balcomb, J. D., C. D. Barley, R. D. McFarland, J. E. Perry, Jr., W. O. Wray, and S. Noll. 1980. *Passive Solar Design Handbook*, vol. 2. DOE/CS-0127/2. Los Alamos, NM: Los Alamos National Laboratory.

Balcomb, J. D., R. W. Jones, C. E. Kosiewicz, G. S. Lazarus, R. D. McFarland, and W. O. Wray. 1983. *Passive Solar Handbook*, vol. 3. Boulder, CO: American Solar Energy Society.

Battelle Pacific Northwest Laboratory. 1980. *Analysis of the Results of Federal Incentives Used to Stimulate Energy Production*. Richland, WA: Battelle Pacific Northwest Laboratory.

Bennington, G. E., P. A. Kurto, C. G. Miller, K. Rebibo, and P. Spewak. 1978. *Solar Energy: A Comparative Analysis to the Year 2020*. MTR 7579. McLean, VA: MITRE Corporation, METREK Division.

Bezdek, R. H., A. S. Hirshberg, and W. H. Babcock. 1979. "Economic feasibility of solar water and space heating." *Science* 203:1214–1220.

Boer, K. W. 1978. "Payback of solar systems." *Solar Energy* 20:225–232.

Bright, R., and H. Davitian. 1982. "The marginal cost of solar backup." *Energy: The International Journal* 7(1):75–84.

Cambel, A. B., R. H. Bezdek, and C. R. Hauer, eds. 1982. "The solar energy-utility interface." Special issue of *Energy: The International Journal* 7(1).

Commission Internationale de l'Eclairage. 1970. *International Recommendations for the Calculation of Natural Daylight*. Publication 16 (E-3.2). Paris: Commision Internationale de l'Eclairage.

Daylighting Committee of the IES. 1962. "Recommended practice of daylighting." *Illuminating Engineering* 57:517–557.

Diamond, S. C., H. L. Horak, B. D. Hunn, J. L. Peterson, M. A. Roschke, and E. F. Tucker. 1979. *DOE-2 Program Manual.* LANL Report LA-7688-M. Los Alamos, NM: Los Alamos National Laboratory.

Duguay, M. A., and R. M. Edgar. 1979. "Lighting with sunlight using sun tracking concentrators." *Applied Optics* 16(5): 1444–1446.

Hall, I. J., R. R. Prairie, H. E. Anderson, and E. C. Boes. 1979. *Generation of Typical Meteorological Years for 26 SOLMET Stations.* ASHRAE symposium paper DE-79-2, no. 3. Atlanta, GA: American Society of Heating, Refrigerating, and Air-Conditioning Engineers.

Henniger, R. H., ed. 1975. *NECAP: NASA's Energy-Cost Analysis Program.* NASA-CR-2590. (Parts 1 and 2). Niles, IL: General American Transportation Corp.

Hittle, D. C. 1977. *BLAST: The Building Loads Analysis and Systems Thermodynamics Program. Vol. I, Users Manual.* CERL-TR-E-119. Champaign, IL: Army Construction Engineering Research Laboratory.

Horak, H. L., B. D. Hunn, J. L. Peterson, M. A. Roschke, E. F. Tucker, and D. A. York. 1979. *DOE-2 Reference Manual.* LANL Report LA-7689-M. Los Alamos, NM: Los Alamos National Laboratory.

Kirschner, C. 1979. *Passive Solar Economics in 15 Northwest Locations.* LANL Preprint LA-UR-79-2025. Los Alamos, NM: Los Alamos National Laboratory.

Kusuda, T. 1974. *NSBLD: Computer Program for Heating and Cooling Loads in Buildings.* NBSIR 74-574. Gaithersburg, MD: National Bureau of Standards.

Kusuda, T., and B. Collins. 1978. *Simplified Analysis of Thermal and Lighting Characteristics of Windows: Two Case Studies.* BSS 109. Gaithersburg, MD: National Bureau of Standards.

Laquatra, J., Jr. 1982. *An Economic Analysis of a Passive Solar Multifamily Dwelling for Upstate New York.* LA-9212-T. Los Alamos, NM: Los Alamos National Laboratory.

Libbey-Owens-Ford Company. 1976. *How to Predict Interior Daylight Illumination.* Toledo, OH: Libbey-Owens-Ford Co.

Lokmanhekim, M. 1978. *CAL-ERDA, a New State-of-the-Art Computer Program for Building Energy Utilization Analysis.* Paper presented at the Third International Symposium on the Use of Computers for Environmental Engineering Related to Buildings, 10–12 May 1978, Banff, Alberta, Canada. Availability: K. Charbonneau, 3rd International Symposium on the Use of Computers for Environmental Engineering Related to Buildings, National Research Council of Canada, Ottawa, Ontario, K1A OR6, Canada.

Market Facts, Inc. 1980. *Preliminary Situation Analysis for Passive Solar Market Surveys.* Washington, DC: Market Facts, Inc.

Marshall, H. E., and R. T. Ruegg. 1980. *Simplified Design Economics: Principles of Economics Applied to Energy Conservation and Solar Energy Investments in Buildings.* Special Publication 544. Gaithersburg, MD: National Bureau of Standards.

Mueller Associates. 1981. *Cost/Performance Goals Report.*

Neeper, D. A., and R. D. McFarland. 1982. *Some Potential Benefits of Fundamental Research for Passive Solar Heating and Cooling of Buildings.* LA-9625-MS. Los Alamos, NM: Los Alamos National Laboratory.

Noll, S. A., and C. Kirschner. 1980. *Thermal and Cost Goal Analysis for Passive Solar Heating Designs.* LA-UR-80-248. Los Alamos, NM: Los Alamos National Laboratory.

Noll, S. A., and W. M. Robson. 1980. *Market Potential Indexing for Passive Solar Technologies.* LA-UR-80-1025. Los Alamos, NM: Los Alamos National Laboratory.

Noll, S. A., and W. O. Wray. 1978. *A Microeconomic Approach to Passive Solar Design: Performance, Cost Optimal Sizing, and Comfort Analysis.* LANL Preprint LA-UR-78-1722. Los Alamos, NM: Los Alamos National Laboratory.

Powell, J. W. 1980. *An Economic Model for Passive Solar Designs in Commercial Environments: NSB Building Science Series 125.* NBS-BSS-125. Gaithersburg, MD: National Bureau of Standards.

Real Estate Research Corp. 1980. *Selling the Solar Home '80.* (Market Findings for the Housing Industry). Chicago, IL: Department of Housing and Urban Development.

Roach, F., S. A. Noll, and S. Ben-David. 1979. *Passive and Active Residential Solar Heating: A Comparative Economic Analysis of Select Designs.* LANL Preprint LA-UR-79-57. Los Alamos, NM: Los Alamos National Laboratory.

Rosenfeld, A. H., and S. E. Selkowitz. 1976. "Beam daylighting: Direct use of solar energy for interior illumination," in *Proceedings of the 1976 International Solar Energy Society Annual Conference.* Cape Canaveral, FL: American Section of the International Solar Energy Society, 375–391.

Ruegg, R. T. 1980. *Life-Cycle Costing Manual for the Federal Energy Management Program: A Guide for Evaluating the Cost Effectiveness of Energy Conservation and Renewable Energy Projects for New and Existing Federally Owned and Leased Buildings and Facilities.* NBS-Handbook-135. Gaithersburg, MD: National Bureau of Standards.

Ruegg, R. T., and G. T. Sav. 1980. "Microeconomics of solar energy." *Solar Energy Handbook.* J. E. Kreider and F. Kreith, eds. New York: McGraw-Hill, Ch 28.

Ruegg, R. T., S. R. Petersen, and H. E. Marshall. 1980. *Recommended Practice for Measuring Life-Cycle Costs of Buildings and Building Systems.* NBSIR80-2040. Gaithersburg, MD: National Bureau of Standards.

Ruegg, R. T., J. S. McConnaughey, G. T. Sav, and K. A. Hockenbery. 1978. *Life-Cycle Costing: A Guide for Selecting Energy Conservation Projects for Public Buildings.* Building Science Series 113. Gaithersburg, MD: National Bureau of Standards.

Selkowitz, S., M. Villecco, and J. W. Griffith, 1981. "Daylighting," in *Solar Design Workbook.* SERI/SP-62-308. Golden, CO: Solar Energy Research Institute, Ch. 6.

SERI. 1980. *Analysis Methods for Solar Heating and Cooling Applications.* SP-35-232. Golden, CO: Solar Energy Research Institute.

SERI. 1982. *Solar Design Briefs,* briefs 1–11. Golden, CO: Solar Energy Research Institute.

Total Environmental Action, Inc. 1980. *Determination of Passive System Costs and Cost Goals.* DOE/CS/35233-T1. Washington, DC: U.S. Department of Energy.

Treichler, W. W. 1976. "A non-computerized rational energy analysis procedure," in *Proceedings of Conference on Improving Efficiency and Performance of HVAC Equipment and Systems for Commercial and Industrial Buildings.* West Lafayette, IN: Purdue Research Foundation, Vol. 1, 214–224.

10 Cost Requirements for Solar Thermal Electric and Industrial Process Heat

Ronald Edelstein

In this chapter I present cost requirements for solar thermal systems and subsystems, delineate the methodologies used to establish these cost requirements, and outline the barriers and pitfalls to successful development of cost/performance goals for energy R&D projects. The cost requirements presented are from U.S. Department of Energy (DOE) (1984). The discussion of methodologies is based mainly on the Pacific Northwest Laboratories analysis (Williams, Dirks, and Brown, 1985) that led to the derivation of these cost requirements and on the procedures used by the interlaboratory Solar Thermal Cost Goals Committee (STCGC). The discussion of problems is intended to help other organizations improve their goals development process.

This chapter does not provide a cookbook formula for developing goals for energy-related research. There are too many different markets to be satisfied, too many different types of research required (from basic research to applied research to engineering development to field testing to technology transfer), and too many varied technological opportunities to oversimplify the R&D goal-setting procedure. Instead, this chapter is designed to relate to researchers, program managers, and planners some of the experiences encountered in the Solar Thermal Program in its goal selection process. For related background in economic evaluation, see sections 2.7.5 and 2.7.6 in chapter 2.

10.1 Definition and Use of Cost Goals

What exactly is a goal and of what use are goals in a research program? In discussing program planning and strategy, researchers often use the words "objectives," "goals," "requirements," "aims," and "purpose" interchangeably (Blanchard, 1976; Massie, 1971; Newman, Summer, and Warren, 1972; Twiss, 1974). However, usually a distinction is made between broad, general, long-term targets and focused, specific, short-term, time-related targets. So, for clarity, I use Vancil's (1979) distinction between goals and objectives. He describes objectives as timeless and enduring and as being stated in general terms. Goals, on the other hand, are "temporal, time-phased ... more specific, and stated in terms of a particular result that will be accomplished by a specified date." I use the terms "goals" and "requirements" interchangeably in this chapter.

The Gas Research Institute (GRI) (1982) describes goals as "measurable indicators of expected performance and cost of the technology under development ... characterized by such factors as efficiency, production rates, product costs, levels of environmental emissions, and the like." Furthermore, GRI distinguishes final from intermediate goals in that "final performance goals define the target at which the R&D is aimed, while intermediate performance goals provide information at key decision points that aid in deciding whether and how to proceed to the next phase of R&D."

For a research program, then, goals should provide performance and cost targets or requirements for R&D activities. Furthermore, the time frame for achieving these goals should be specified in advance of the research activities to provide a target for the scientists and engineers engaged in the research and to serve as a measure of their progress.

The degree of specificity of the goals strongly depends on the stage of R&D. A basic research program, for instance, might have general targets more closely aligned with the previous definition of objectives. On the other hand, an applied R&D program with specific technological approaches would need specific quantitative goals. Although solar thermal R&D involves some basic research, its system-level technology options are well enough delineated to require these quantitative targets.

Finally, these research targets must be compatible with and competitive in the markets that the technologies and systems are expected to serve. It is of no use to complete R&D activities successfully only to find that no manufacturer is interested in adopting the technology because it has no market.

Thus, as described by Schimmel [in Edelstein (1981)], solar thermal goals should be viewed as

1. A set of capital cost and performance figures that, if achieved, will permit solar thermal systems to be accepted by some portion of the target market

2. A yardstick by which progress in a specified direction and time frame can be measured

3. An implicit appraisal of future conditions and what will be required to compete with other energy sources

4. Values in both cost and performance that the researchers and technology developers will be challenged to attain

10.2   Description of Solar Thermal Technology

In this section I describe the hardware involved in a solar thermal system to provide a context for the solar thermal goal-setting process. Much more detail on solar thermal hardware can be found in volume 1 of this series.

A solar thermal system concentrates and converts the sun's radiant energy into thermal energy or, by subsequent conversion, to electrical or mechanical energy. Potential end uses include process heat or steam, electricity generation, cogeneration (combined thermal and electrical energy production), and mechanical power. Advanced solar thermal systems, under development, convert photon energy into photochemical or electrochemical energy for fuels and chemicals production.

Solar thermal systems are composed of a number of subsystems, which differ either slightly or significantly depending on the end-use energy (heat, electricity, or mechanical power) they produce. The basic process heat system is composed of (1) a concentrator subsystem to collect the sun's energy and concentrate, reflect, and focus it on the next subsystem, the receiver; (2) a receiver subsystem to absorb the reflected solar radiation as thermal energy; (3) a thermal storage subsystem (optional) to store the thermal energy from the receiver, which provides for either a buffering capability to protect the receiver or to allow for load matching with the user's end-use system; and (4) a transport subsystem to deliver the thermal energy to the user. If the user requires electrical or mechanical energy, then a power conversion subsystem is required to convert the thermal energy into electrical or mechanical power. The use of certain subsystems, such as concentrators and receivers, for both process heat and electricity generation can complicate the goal-setting process and is discussed in section 10.4.8.

Several primary systems options evolved through the R&D process and manufacturer interest. The first of these is the line-focus system, typified by the one-axis tracking parabolic trough, which is a medium-temperature [up to 750°F (400°C)] system using a linear receiver tube mounted at the focal point of the parabolic trough to absorb the incoming insolation. Thermal energy transport occurs through the receiver tubes and through connecting piping to a field location for thermal storage or power conversion. This system is typically applicable in sizes up to tens of megawatts electric.

A second system is a point-focus technology, the heliostat-central receiver system, which uses a field of two-axis tracking and relatively flat mirrors (called heliostats) to reflect the sun's energy to a central receiver mounted atop a tower. The power conversion subsystem is located in or near the central tower. This system is capable of temperatures up to 2,000°F (1,100°C) and large system size ranges (up to hundreds of megawatts electric).

A third system, also a point-focus system, is the parabolic dish, which uses a a two-axis tracking paraboloid to focus the solar radiation on the focal point of the concentrator. For electricity generation the thermal energy is then converted at each dish into electrical energy by a heat engine. An alternative parabolic dish electrical generation configuration uses thermal transport to carry the heat energy to a single heat engine (instead of having a heat engine located on each dish). For process heat applications no heat engine is needed and the thermal energy is transported to the process. This system is also capable of temperatures over 2,000°F (1,100°C) and is available in modular sizes of about 25–40 kW_e.

10.3   Results

This section is based on information from the solar thermal five-year plan (DOE, 1984).

10.3.1   System Goals

Solar thermal goals were derived from two primary end-use markets: bulk electricity generation and industrial process heat (IPH). Electricity generation system goals were based on a utility-owned, grid-connected solar thermal power plant; IPH goals were based on a privately owned, industrial solar thermal system producing process heat in the range of 400°–1,100°F (200°–600°C). Therefore two system-level goals were required instead of just one because utility electric and industrial IPH represent substantially different market segments. These markets differ in terms of competing energy sources, economic, financial, and regulatory considerations, required backup systems, and other factors. Other applications, such as remote solar thermal power plants, cogeneration systems, electricity generating systems owned by third parties, and fuels and chemicals production, represent potentially large market segments for solar thermal systems but were not considered in developing the long-term system goals.

System goals were based on the projected levelized energy cost of competing energy sources in the middle-to-late 1990s. These goals were not meant to represent threshold energy costs that solar thermal must achieve initially to penetrate the market but were even more ambitious targets that would allow solar thermal to compete with many energy sources and thus achieve significant market penetration.

The system goal for electricity was based on solar thermal competing with a coal-fired power plant. Because coal power plants are expected to be one of the most economical electrical energy sources in the coming decades, the energy cost from an intermediate-load, coal-fired power plant was used as the competing conventional technology target for establishing system goals for solar thermal electricity generation. Base-load coal plants were not selected because solar thermal technology is more cost-effective as an intermediate-load displacement option. By setting system goals based on one of the lowest energy cost technologies, planners expected solar thermal to be competitive with several other alternatives (for example, oil and gas) if cost goals were met.

Several studies investigated the characteristics of coal-fired power plants (Delene et al., 1984; EPRI, 1982; Reynolds, 1982; Ringer, 1984) and reported that for intermediate-load coal plants (with capacity factor ranges of 0.4–0.5), levelized energy costs ranged from slightly under $0.05/kWh to nearly $0.08/kWh. The capacity factor is the fraction of a year that a power plant will be run at its nameplate rating. Therefore the solar thermal system cost goal selected for electric applications was $0.05/kWh. This goal, in addition to being much lower than projected energy costs from oil- or gas-fired power plants, would allow solar thermal to compete with intermediate-load coal plants to be built in the late 1990s.

The system energy cost target for solar thermal IPH was based on solar thermal competing with such premium fossil fuels as natural gas, residual oil, and distillate oil. However, solar thermal was expected to supplement fossil fuel systems for process heat (rather than serve as a stand-alone source of heat), so the system goal for this application was determined from the value of the competing fuel displaced by solar thermal.

The primary source of fuel price projections and escalation rates used to develop system cost goals for IPH was a report on the National Energy

Table 10.1
Solar thermal system energy cost goals

Application	Levelized energy cost (1984 dollars)
Electricity	$0.05/kWh
Industrial process heat	$9/10^6$ Btu (10^9 J)

Source: DOE (1984).

Policy Plan (DOE, 1983), covering the period from 1990 to 2010 (which is close to the planning period for solar thermal technology goals). This study gave a range of fuel prices based on low, base-case, and high world oil price scenarios. The levelized energy cost projections were $7–$14/10^6 Btu (10^9 J) for residual oil, $9–$17/10^6 Btu (10^9 J) for distillate, and $7–$12/10^6 Btu (10^9 J) for natural gas. Note that more recent fossil fuel price projections [for instance, Holtberg et al. (1985)] and the current (1988) low price for world oil would place the levelized energy cost for these systems at considerably lower levels than those indicated here. Using the energy price projections indicated (DOE, 1983), planners developed a levelized energy cost target of $9/$10^6$ Btu (10^9 J) for delivered solar thermal energy (accounting for efficiency).

Table 10.1 summarizes energy cost goals for solar thermal systems.

10.3.2 Component Goals

Although market-based energy or system goals focused on and were determined by the price of competing energy sources, component or subsystem goals focused on the efficiency and cost of the candidate solar thermal technology options. These component goals follow a strategy for solar thermal technology that will eventually result (if component development is successful) in achieving the energy goals. Thus component goals were developed based on improvements in the expected component performance and cost through R&D, within the constraints of the overall system goals.

Component goals were developed for those portions of the solar thermal plant that were major cost drivers or whose performance was critical to the system's overall efficiency. Based on these criteria eight areas were selected for specification of component goals: concentrators, receivers, energy transport, energy storage, energy conversion, balance of plant, total installed system cost, and operations and maintenance (O&M)

costs. The first six of these areas represent components or subsystems in the classical sense of hardware, the total installed system cost represents the sum of component costs plus installation, and the O&M portion represents the annual cost for routine O&M.

Table 10.2 gives component goals for electric and IPH applications for the three major system configurations (central receiver, dish, and trough). For the majority of components the goals are the same for IPH and electric applications, reflecting the fact that many of the components required for the two applications are substantially the same. Any difference in a component goal between the two applications is a result of significantly different design specifications for the component. For example, the goal for the receiver component of the dish electric system has both a lower efficiency and a higher cost than the dish process heat receiver. This difference is caused by the requirement for much higher receiver operating temperatures for the electric application than for the process heat use.

Many degrees of freedom exist in developing component goals. First, goals can be traded off among components to achieve the same system goal. Second, capital cost and performance can be traded off within the same component to achieve an identical effect on levelized energy costs. As a result, no unique set of component goals alone will satisfy the system goal; a number of solutions are feasible. This is discussed further in section 10.4.8.

All component goals were developed with the objective of being attainable through future R&D efforts. Predicting the amount of improvement possible is subject to uncertainty; therefore the best allocation of component goals may change in the future as solar thermal R&D proceeds or as improvements in some solar components occur faster than in others.

10.4 Nature of the Problem: Pitfalls in Setting R&D Goals

In this section I discuss the major issues involved in setting cost/performance goals for solar thermal systems. One set of issues raised such philosophical questions as, What is the real basis for setting cost goals? What targets of opportunity should R&D shoot at? In what time frames will the products of this R&D be applied? Additional issues arose when the strategy or approach to be used was developed. Another set of issues included the selection of baseline financial assumptions for fossil fuel prices, return on investment (ROI), and technical assumptions on product perfor-

Table 10.2
Solar thermal long-term component goals

Component	Electric		Industrial process heat	
	Annual efficiency (%)	Cost (1984 dollars)[a]	Annual efficiency (%)	Cost (1984 dollars)
Concentrators				
Central receiver	64	$100/m^2	64	$100/m^2
Dish	78	$140/m^2	78	$140/m^2
Trough	—	—	65	$110/m^2
Receivers				
Central receiver	90	$45/m^2	90	$45/m^2
Dish	90	$70/m^2	95	$30/m^2
Trough	—	—	90	$30/m^2
Transport				
Central receiver	99	$25/m^2	99	$25/m^2
Dish	99	$7/m^2	94	$65/m^2
Trough	—	—	98	$30/m^2
Storage				
Central receiver	98	$20/kWh$_t$	98	$20/kWh$_t$
Dish	—	—	98	$20/kWh$_t$
Trough	—	—	98	$20/kWh$_t$
Conversion				
Central receiver	39	$350/kW$_e$	N/A	N/A
Dish	41	$350/kW$_e$	N/A	N/A
Trough	—	—	N/A	N/A
Balance of plant	N/A[b]	$50/m^2	N/A	$50/m^2
System				
Central receiver	22	$1,600/kW$_e$	56	$460/kW$_t$
Dish	28	$1,300/kW$_e$	68	$470/kW$_t$
Trough	—	—	56	$390/kW$_t$
O&M				
Central receiver	N/A	$6/m^2/yr	N/A	$5/m^2/yr
Dish	N/A	$10/m^2/yr	N/A	$5/m^2/yr
Trough	—	—	N/A	$5/m^2/yr

Source: DOE (1984).
a. System costs in $/kW are based on theoretical power available from receiver. $10.76/m^2 = $1.00/ft^2.
b. N/A, not applicable.

mance. The next set came about during the selection of goals based on the data gathered. Finally, R&D management issues arose concerning responsibility for goal setting and tracking, measurability (How do you know when you're there?), and goals dissemination.

10.4.1   Attainability-Based versus Value-Based Goals

The issue of the fundamental basis for setting cost goals is perhaps the most controversial one faced during the solar thermal goal-setting process. In setting targets for a research activity, a program manager can examine the state-of-the-art hardware in the field and set goals to compete with this hardware. Alternatively a program manager can use an engineering cost study of current R&D hardware to set a cost target, taking into account mass production and other cost reduction techniques. Goals derived from these (or similar) processes are known as attainability-based goals (Edelstein, 1981). Attainability-based goals are confined by the current state of knowledge.

Program managers can also assess the current or projected future market for their hardware and base their goals on the value of the technology in the marketplace. Similarly, general market needs in a specific target market may suggest new research initiatives for developing competitive hardware. These market-derived goals are known as value-based goals (Edelstein, 1981).

A sharp difference of opinion exists between scientists and economists over whether to base goals on what current technology could attain, given engineering development and mass production, or on the market value of the technology. Technical personnel generally favor attainability-based goals, whereas economists tend to favor value-based goals. The advantages and disadvantages of each type of goal are described briefly in the following paragraphs.

The advantages of the attainability-based approach are that it represents a realistic appraisal of solar thermal system costs based on the best engineering estimates and highlights those high-cost components that might need additional development. Because this approach uses projections from current R&D hardware, it can predict a technology's probability of success better than a straight value-based approach, which has little technology basis. Furthermore, being hardware and not market based, attainability-based goals do not fluctuate each time world oil prices change. A disadvantage of this approach is that it tends to lock in one design relatively early in the development cycle, restricting future innovations that may lead to a totally new product not related to the current hardware. More important, because these goals are not market based, there is no guarantee that the hardware based on them would be competitive with fossil-fuel-fired systems, even if the targets were met.

The distinct advantage of the value-based approach, which is market oriented, is that goals can be chosen based on market requirements; if the goals are reached, there is a good chance of a market existing for the hardware developed. Establishing a defined market also allows for a reasonably quantitative cost/benefit calculation to assess whether or not the research is worthwhile. This type of analysis is not easily performed using the attainability-based approach (because the total size of the market or even total number of units sold is not known). In addition, because value-oriented goals are not based on incremental improvements in current hardware, they allow for more innovative, possibly higher-risk research. Of course, choosing a market sector and completing a market potential analysis is not a trivial exercise. A disadvantage of this approach is that goals need to be reevaluated periodically to take into account changing market conditions, and markets can change dramatically, as evidenced by the tenfold increase in the price of oil between 1970 and 1980. The uncertainty inherent in projecting future market conditions makes extrapolating fuel prices and fuel-use patterns difficult. Although the success of government-sponsored R&D is not necessarily measured by market penetration, it is important that the hardware thus developed be competitive.

The value-based approach was chosen to select solar thermal energy and system-level cost goals because planners felt that the ability of the technology to compete in the future energy marketplace was critical to its ultimate success. System goals (DOE, 1984) were based on energy price targets that solar thermal had to meet to achieve a significant economic impact in a given market (for example, utility electric). These goals were determined by solar thermal's primary competitor-fossil fuels.

The attainability-based approach was chosen to select component goals because these goals were directed at providing a target for solar thermal technology development and were developed based on projected improvements in component efficiency and cost through R&D. However, the component goals had to be not only technically attainable through

R&D but also compatible with the value-based system and energy goals. Researchers in government laboratories and industry judged the attainability of component goals through assessment and analysis. Program managers and economists judged the competitiveness of the value-based system and energy goals through levelized cost and market potential analyses.

Some trade-offs in the development of goals were made because of the different approaches taken in developing system and component goals. For example, the desire for the lowest possible system goal (that is, being economically superior to all other energy sources to maximize the return for technology development) was balanced by component goals, which were expected to be technically achievable through R&D. The set of goals that emerged represented a compromise between maximizing the potential return, represented by the system goal, and making acceptable the risk inherent in attaining the component goal.

10.4.2 Target Markets, Fuel Types, and Regions

To select value-based goals, one must analyze and select markets to compete in, fossil (or other alternative) fuels to compete against, and regions of the country in which to compete. It is possible to set up a hypothetical fossil-fuel-fired plant in a good insolation region of the United States, pick a premium fuel (such as distillate oil) to compete against, and set a cost target based on displacing this ideal competitor. However, the technology developed toward such a goal might not be cost competitive in lower insolation regions or against lower-priced, more heavily used fuels. It is quite important, then, to look at the integrated national picture to set value-based goals. The major energy-use sectors can later be disaggregated to an appropriate level of detail to ensure that the technology will in fact be competitive across a broad enough spectrum of markets, regions, and fuel types.

Historically, as far back as 1973 in preliminary planning for solar R&D, only a few solar thermal applications were targeted. These markets included central station electricity generation, irrigation pumping, and cogeneration (or as it was called then, total energy systems). By 1977, as the Solar Thermal Program was divided into central and dispersed systems, the target markets had evolved to include central station electricity generation with the heliostat-central receiver system and irrigation pumping, dispersed electricity generation, and total energy applications for dispersed systems (including the trough and parabolic dish). Many of these applications had been under development for decades. For instance, a solar-powered steam engine was exhibited by Mouchot at the Paris Exposition in 1878 and an irrigation pumping system had been demonstrated in Egypt by Schuman and Boys in 1913 (Meinel and Meinel, 1977).

The DOE Industrial Process Heat Program pursued IPH applications of solar thermal technologies until its integration with the Solar Thermal Program in 1980. The Solar Thermal Energy Systems Program then pursued R&D targeted toward IPH, electricity generation, cogeneration, and irrigation pumping. In addition, the research and advanced development subprogram was targeted for the long term toward the production of gaseous and liquid fuels from solar thermal to enable penetration of the transportation and residential and commercial market sectors.

Market selection was based principally (Braun and Edelstein, 1980) on solar thermal's technical compatibility with electricity generation (at Coolidge, Arizona, with a parabolic trough 150-kW$_e$ system and at Barstow, California, with the 10-MW$_e$ central receiver system), numerous IPH experiments using parabolic troughs, and conceptual design studies of repowered electric plants and IPH boilers (to retrofit the fossil-fuel-fired systems with solar thermal central receiver systems). The major market sectors selected were those in which energy generation was already demonstrated by field tests, where user interest was indicated, and where potential, based on preliminary studies, existed, that is, the electricity generation and IPH markets.

Holtberg et al. (1985) projected primary energy consumption in the United States for the year 2000 with use by sector (in descending order) as follows: industrial, electricity generation, transportation, and residential and commercial. The industrial and electricity markets together represent 31.5 quads (10^{18} J) for industrial primary fuel use and 23.8 quads (10^{18} J) for net (primary energy use minus generated energy) electrical generation. These two sectors, which represent 58% of projected energy consumption in the United States for the year 2000, were chosen to establish value-based system goals for solar thermal.

A key issue in developing goals for electricity generation applications was that of identifying the competing conventional power plants and fuel types for solar thermal (Williams, Dirks, and Brown, 1985). Premium-fueled (oil and natural gas) plants were one potential target because they used relatively expensive fuels; however, they did not form a good basis

Table 10.3
Energy consumption to generate electricity (in quads)[a]

Fuel type	1984	1990	2000
Petroleum	1.3	1.0	0.9
Gas	3.2	2.5	2.1
Coal	14.1	15.9	19.8
Nuclear	3.6	6.4	6.5
Hydro	3.7	4.0	4.0
Renewables	0.2	0.4	0.8
Total	26.1	30.2	34.1

Source: Holtberg et al. (1985).
a. 1 quad = 10^{18} J.

for setting system goals. This is because premium fuel use for electricity generation represents only about 9% of the total primary fuel used for electricity generation, as shown in table 10.3 for the year 2000. In addition, premium fuels were expected to be phased down over time, so that the amount of plant capacity that solar thermal could displace would not be large. As table 10.3 shows, the type of conventional plant that is expected to provide the largest fraction of primary energy (over 58% in 2000) and to continue to grow over time is the coal-fired power plant.

System goals for electricity generation were thus developed using the energy costs from a conventional coal-fired power plant as the competing cost target. An intermediate-load plant was chosen as being most closely aligned to the capabilities of a solar thermal system. Because planners chose one of the lowest energy cost plants as the basis for setting solar thermal goals, they expected solar thermal to be competitive with several other plant types (for example, premium-fuel-fired) as well if the goals were met. The energy cost from a coal-fired plant was used as a target for system goals to indicate that the cost represented a reasonable target for widespread adoption of solar thermal, not to indicate that solar thermal could displace a large amount of coal-fired capacity. Utility-specific, detailed grid-mix analyses considering a range of power plants and fuel types were required to predict the types of plants and fuels solar thermal would actually displace.

Industrial process heat represents a large potential market for solar thermal technology. If we view that portion of the total industrial end-use market for fuel and power consumption only (ignoring feedstocks), we can see from table 10.4 that consumption of all fuel types is expected to

Table 10.4
Industrial sector energy consumption for fuel and power (in quads)[a]

Fuel type	1984	1990	2000
Petroleum	4.6	5.5	6.1
Gas	6.8	7.7	7.9
Coal	1.7	2.1	3.5
Electricity	2.9	3.5	4.0
Renewables	3.1	3.6	4.1
Total	19.1	22.4	25.6

Source: Holtberg et al. (1985).
a. 1 quad $= 10^{18}$ J.

increase by the year 2000. The largest primary fuel sources in the industrial market are natural gas and petroleum, which account for almost 55% of the fuel used for IPH and power.

Natural gas and oil were the competing conventional fuels used as the basis for IPH system goals (Williams, Dirks, and Brown, 1985). Unlike electric applications, coal was not targeted as a competing fuel for the IPH market. Because the required return on investment was higher in the industrial sector than in the electric utility sector, solar thermal would most likely be unable to compete with coal in IPH applications.

The insolation level selected for setting solar thermal system goals was equivalent to the Barstow, California, value (Williams, Dirks, and Brown, 1985). This was an optimistic assumption, as this level of incoming direct solar radiation is not typical for most of the United States.

Setting value-based goals for other possible applications, such as cogeneration or liquid fuels production, can be even more difficult. In setting goals for cogeneration, one has to add thermal and electrical energy, decide whether energy at high temperatures [for example, 932°F (500°C)], is worth more than at medium temperatures [say, 482°F (250°C)], and determine whether or not to count energy units twice if the energy is recycled at lower temperature to another use. For liquid fuel production, does one set the value based on energy into the process, plant cost savings, or on the value of energy out of the process?

10.4.3   Strategic Time Frames

The temporal dimension is a key element in quantifying R&D goals. Two important time-related parameters are the projected date(s) for goal achievement and the period for technology transfer, that is, commercialization. The target date is under the control of the organization sponsoring the research; the transfer date is only loosely under such control. Both are discussed next.

In setting a target date for achieving research goals, program managers must first set a definite date for completing the R&D activity. Of course, given that a research program has several different technology options at different stages of development, the parts of its entire program are not expected to end at the same time. Therefore different time frames might be selected for each technology option, depending on its stage of development, market need, and technical difficulty in reaching the goal. Program managers must also track research progress periodically. For example, if the only program goal they set is one of achieving a $46/ft² ($500/m²) capital cost target and 25% efficiency by the year 2000, they will find it difficult to determine if their research is on schedule or not before that time. Thus intermediate goals are needed to determine if the research is on schedule and if it is achieving its desired aims. Intermediate goal setting should consider both the current R&D status and long-term goals.

The Gas Research Institute (Gottlieb, 1984) developed performance and cost goals to be achieved by its hardware-oriented research at different stages of R&D. The final goal was called the commercial goal, which represented the performance and cost of a mass-produced system. An intermediate goal, to be achieved at a specified point in time at the so-called proof-of-concept (POC) test, was called the POC goal, which represented the actual performance of the prototype device and the projected cost of that device if produced on a limited, continuous basis.

A word of caution must be expressed about setting intermediate goals. The manager or planner should be careful to set intermediate goals that will not cause the R&D to veer toward an intermediate product that can divert funds and effort from the truly commercial target at the end of the path. The intermediate goal should represent a checkpoint on the path toward the long-term goal and not an alternative, more expensive concept. For example, if the commercial cost goal of $46/ft² ($500/m²) for a solar technology requires developing second-generation systems needing composite materials and an intermediate goal of $186/ft² ($2,000/m²) is set, then R&D toward the intermediate goal should not proceed based on first-generation hardware using metals. In such a case the intermediate goal would become an end in itself, instead of acting as a progress monitor on the way to the final target, and could result in the development of unmarketable hardware.

An appropriate time frame, a "window," for technology transfer or commercialization should be considered in setting a market-oriented target. Considerations should include the requirements of laws, such as the Power Plant and Industrial Fuel Use Act of 1978 (now repealed), and the projected introduction of competing technologies, such as photovoltaics, into the marketplace. The market for solar thermal repowering (solar thermal use in a fuel-saver mode only) represented just such a window for solar thermal. Introducing a technology into the marketplace is beyond the control of a research organization or federal agency. Private industry must carry the technology forward to commercialization. Still, setting a time frame based on a market window aids an R&D effort by identifying a specific potential market at the end of the development stage.

In setting solar thermal goals, it was decided (DOE, 1984) to focus on the middle-to-late 1990s to set value-based goals. Between now and the mid-1990s, solar thermal systems were expected to be sold in many specific applications where an energy cost higher than the system goals was economically feasible. Near-term sales of solar thermal systems in these high-value applications are vital in achieving the technology goals because reaching a significant production level in solar collectors is an important element in achieving the component cost goals.

10.4.4 Methodologies

In this section I summarize the electric and IPH market analyses and assumptions used to set the value-based system goals. Many differences existed between these two market sectors and the assumptions used as a basis for analyzing them. For both market sectors system goals were based on the projected levelized energy cost of competing conventional energy sources. However, different ownership assumptions were used for each sector: electric utility ownership for the electric sector and private industry ownership for the IPH market.

The levelized energy cost (LEC) approach (DOE, 1984) considered all relevant costs, including initial capital costs, annual O&M costs, and return on investment (ROI). The method is a standard approach often used in the utility industry and is described in more detail elsewhere (Doane et al., 1976; EPRI, 1982). After it assesses the net present value of all costs, the method calculates an equivalent energy cost, which is level (that is, constant from year to year) over the system's lifetime. Williams, Dirks, and Brown (1985) used the LEC approach to develop solar thermal goals because it provided an economically valid approach to evaluating revenues and expenses, had significant use for past evaluations of solar thermal technology, and was a technique familiar to the electric utility industry.

To determine the value of solar thermal in both the electric and IPH market sectors, it was important to reflect on how solar thermal would actually be incorporated into the market in question. This required using assumptions that reflected the market's capital investment decision criteria. Private industry tends to be more conservative than the electric utilities in their decision making. As a result, the assumptions selected for the IPH analysis were more conservative than those selected for the utility study. This did not represent a difference in philosophies between the two studies but rather a difference between the conditions actually encountered in those markets. The primary effect of these differing market conditions was reflected in the real discount rate chosen for the calculation of life-cycle costs: 3.2% for the electric utility market and 10% for the IPH market.

The influence of different market conditions was also felt in selecting "economic" lifetimes. Whereas the IPH analysis used a twenty-year stream of fossil fuel costs in calculating the value of solar thermal systems, the utility studies used a thirty-year stream of fuel costs. The rationale was that an industrial plant manager would want payback sooner than a utility company would. (Utilities were accustomed to thirty-year life-cycle costing in their present economic calculations.) Some even suggested that a five- to seven-year payback assumption for the industrial sector would be more appropriate.

Note that lifetime requirements raise another dilemma, because value-based goals are supposed to be independent of the specific technology used; yet here we have a case in which the solar thermal systems and thus the major components (such as heliostats) must have two different lifetimes to meet the value-based targets of the IPH and electric market sectors. What would the performance goals (in terms of lifetime) be for heliostats under these circumstances? Should program managers choose a twenty-year lifetime initially to penetrate the IPH market (which has a higher required ROI) with a less expensive product? Should they design to thirty years to capture both markets? Should they aim for the higher-value market in the near term and the market requiring a less expensive heliostat in the long term? Or perhaps two different heliostats would be required in the same time frame, a less expensive, maybe less durable and less accurate model, possibly with replaceable components, for the IPH market and a higher quality, more durable, and hence more expensive model for the electric utility market. Thus it can be seen that value-based goal selection cannot really be totally divorced from hardware (attainability) considerations.

The component goals that were set show the impact of different market sectors in the design of the receiver component. That is, the dish system in the IPH market required a lower-cost, lower-temperature (and thus higher-efficiency because of less radiation losses) receiver than the dish system in the electric sector. On the other hand, heliostat goals by system type were identical for the electric and IPH markets. The heliostat lifetime selected was thirty years to achieve penetration in both markets.

In addition, the IPH analysis considered solar to operate in a "fuel-saver" mode only. That is, it used only the value of fossil fuel and not the capital cost of the fossil boiler to compute the value of solar. This is legitimate in the IPH market, where a fossil-fuel-fired backup is almost always required (depending on load matching and solar system storage) to ensure that the plant runs at all times. The electric utility sector analysis, however, did consider the value of the displaced capital costs of the fossil-fuel-fired equipment.

Finally, in the IPH sector analysis natural gas and oil were targeted as the competing fuels; coal was selected in the electric sector analysis. These selections were made based on solar thermal having the potential for making a major impact in the different market sectors.

In summary, many of the differences cited were simply the result of program managers choosing two market sectors as targets for solar thermal systems that had differing investment criteria and energy-use patterns.

10.4.5 Financial Assumptions

When we set financial and economic parameters, we widen the uncertainty in future projections to a national and world scale. What will the inflation and discount rates be in the year 2000? What will the world oil price be in the year 2000? What about institutional constraints and opportunities, such as fuel price decontrol legislation, environmental regulations, and energy tax credits? What about the future surplus or shortage

Table 10.5
Levelized energy cost assumptions (fossil fuel system)[a]

Parameter	Electricity	IPH
Plant construction time	3 yr	N/A
Economic life	30 yr	20 yr
Depreciation time	15 yr	N/A
Depreciation schedule	ACRS[b]	N/A
Investment tax credit	0.1	N/A
Discount rate	0.0315	0.10
(real, after tax)		

Source: DOE (1984).
a. Financial assumptions for solar thermal systems were the same as the fossil system except a 10% investment tax credit was assumed for the solar thermal system for IPH.
b. ACRS = accelerated cost recovery system.

of premium fuels such as natural gas? Almost all these factors are beyond the control of government- or private-sector-sponsored R&D, yet they must be considered in selecting assumptions for goal-setting analyses.

Table 10.5 summarizes the levelized energy cost assumptions for the electricity and IPH sectors. The discount rate for the electric utility sector was taken from *Technical Assessment Guide* (EPRI, 1982) and is assumed to be the real after-tax cost of capital for that sector. The discount rate for the IPH sector, also the real after-tax cost of capital, included a risk premium for the variability in the price of premium fossil fuels. The investment tax credit for solar IPH was presumed to be in force during the period of investment, an optimistic assumption. The reasons for the differences in the financial assumptions between sectors was discussed in section 10.4.4.

10.4.6 Technical Assumptions

Several technical issues that affected the value to be placed on the solar thermal technology had to be resolved. The two principal issues were storage and capacity credit.

The first issue was whether or not to consider the solar thermal system as containing an inherent amount of thermal storage. A solar thermal system's value could change significantly, based on its storage capacity, because of such factors as load matching. The IPH system goals were developed using a fossil fuel backup system. As a result, thermal storage requirements for the IPH market were reduced considerably from an alternative scenario where no fossil-fuel-fired backup was provided.

Thermal storage needs for the IPH sector were set by such considerations as thermal buffering to protect the solar thermal system from transients. In actually designing a solar thermal system for a specific IPH application, engineers might economically justify much more storage than just buffering. For instance, if a plant operates five days a week, thermal storage of the weekend solar energy would probably be cost-effective. Considerably more storage was needed for the electric utility sector than for the IPH case because the solar thermal systems were designed to compete with an intermediate-load coal-fired power plant. Storage requirements were set by the intermediate-load specifications, that is, the ability to operate the plant at capacity for 40–50% of the time.

Capacity credit, that is, the value of that portion of the fossil-fuel-fired power plant capital costs that a solar system gets credit for displacing, is an important parameter in setting goals in the electric utility market sector. The capacity credit is usually determined by utility-specific grid-mix analysis; the alternative technology is given credit for displacing portions of capital costs for new conventional plants that are not built because of the alternative technology. The solar thermal system goals were based on full capacity credit for displacing an intermediate-load coal-fired power plant. This was an optimistic assumption used in setting system goals, as it was not determined by a grid-mix analysis. The capacity credit assumption accounted for approximately one-third of the energy costs of a fossil-fuel-fired plant.

No capacity credit was considered for the IPH value because it was based on a fuel-saver mode. However, if for instance a solar thermal-natural gas hybrid system would have been compared against a new coal-fired plant, then a capacity credit could have been considered.

10.4.7   Development of System Goals

In selecting system goals, ideally one would want to consider trade-offs among R&D costs, system costs, probability of success, and benefits. Benefits to the nation are usually determined by such factors as cost savings to the public compared to conventional fuel options. This requires additional information on levels of market penetration, future fuel costs, and consumer and industry choices of fuel types. Most of these data are generally unavailable or highly speculative. As a result, the actual system goals selection process must be considerably simplified from the ideal case.

System goals selected (DOE, 1984) for solar thermal were energy price targets that had to be met for solar thermal to be economically viable in the electric utility and IPH markets. These goals were therefore determined primarily by the characteristics of competing energy sources in these markets.

A more detailed approach using more than two market sectors to establish system goals could have been chosen, and other applications, such as cogeneration and fuels and chemical production, were potential candidates. However, the decision was made to base goal selection on two primary market sectors, electricity generation and IPH, based on user interest, design studies, field tests, and R&D progress made thus far. If systems and components could be successfully developed for the goals selected, then other markets could also be penetrated.

Similarly, more detailed analyses could have been conducted in the utility and IPH sectors on such factors as regional differences for utilities and for solar insolation and differences by IPH application temperature. However, such an approach would have greatly compounded the problems and cost of goals development without necessarily increasing the usefulness of the final set of goals. The approach of having one system goal for each sector was a compromise between the desire to consider all important market differences among applications and the need to set applicable and easily used goals for research managers.

In developing goals, the essential driving force behind each type of goal had to be recognized. System goals represent the value-based payoff for developing solar thermal technology because they define the final delivered energy price of the solar thermal system. To maximize the market potential for the technology, the system goals should ideally be set as low as possible to allow solar thermal to be competitive with all other energy sources. In developing component goals (discussed in the next section), one would ideally set them high (for example, high cost, low efficiency) to increase their probability of success. However, in the compromise between setting value-based system goals and attainability-based hardware goals, some iteration is required. The desire for the lowest possible system goals must be balanced by component goals that are technically achievable through R&D. The set of goals selected is a compromise between reducing the risks inherent in attaining component goals and maximizing the potential payoff represented by the system goals.

Table 10.6
Range of coal plant characteristics used in estimating the levelized energy cost

Plant characteristic	Range
Plant size	500–1,200 MW$_e$
Net annual heat rate[a]	9,650–10,200 Btu/kWh
Capital cost	$1,000–1,900/kW$_e$
Annual O&M cost	$0.0039–0.0065/kWh
First-year coal cost (1990–1995)	$1.60–$2.80/10^6 Btu
Real price escalation for coal	1.5%–2.9% annual

Source: Williams et al. (1985).
a. 1 Btu = 1,055 J.

Table 10.7
Levelized energy cost projections for IPH from fossil fuels ($/10^6 Btu)[a]

Fuel type	Low oil price scenario	Base-case scenario	High oil price scenario
Residual oil	7	11	14
Distillate oil	9	13	17
Natural gas	7	9	12

Source: DOE (1984).
a. 10^6 Btu = 10^9 J.

As previously indicated, system goals for electricity were developed using an intermediate-load coal-fired power plant as the competing technology. Table 10.6 summarizes the assumptions used for the coal-fired plant. For intermediate-load plants, the calculated levelized energy costs (LEC) ranged from slightly under $0.05/kWh to nearly $0.08/kWh. A system goal of $0.05/kWh was selected. As indicated earlier, this goal would be considerably lower than the LEC from oil- or gas-fired power plants.

Oil and natural gas were the competing fuels selected for IPH. Table 10.7 summarizes the LEC projections used for the IPH analysis. The primary source for fuel price projections and escalation rates used in the IPH analysis was a report supporting the National Energy Policy Plan (DOE, 1983). The plan's estimates for three scenarios (base-case, high world oil prices, and low world oil prices) for residual fuel, distillate fuel, and natural gas were analyzed using the LEC methodology. Based on results of this analysis, a system goal of $9.00/10^6 Btu (10^9 J) was selected, which would allow solar thermal to compete with all three fuel types in the base-case and high world oil price scenarios. However, based on the 1988 low world oil prices and more recent projections of oil and natural gas prices than the NEPP (Holtberg et al., 1985), a reevaluation of the system goal for IPH will probably be necessary.

In summary, the energy and system-level goal selection process was not an automatic or easy one, despite or because of the large quantity of data and sensitivity studies on which the process was based. Research organization(s) using the goal selection process should realize this difficulty from the outset. It takes a long time to gather and interpret market data (which still may not be complete or perfectly clear); the selection process can be indirect and may have to satisfy diverse parties and interests (and will still not achieve a consensus on all points). Even then the goals selected will not be unassailable (either from after-the-fact experts or unforeseen changes in the real-world market situation, such as a reduction in oil prices). This is not meant to discourage the setting of R&D goals, because they are a necessary precursor to performing research, but rather to put the process in perspective.

10.4.8   Development of Component Goals

As described earlier, Williams, Dirks, and Brown (1985) used two important criteria to develop component goals. These criteria were that the goals must result in achieving the system-level energy cost goal and that the component goals must be attainable through R&D. It is relatively easy to determine whether the component goals satisfy the system goal by using LEC analysis. However, it is more difficult to determine the probability that a given component goal is achievable, and that the degree of risk has been evenly distributed among the components (if desired). The achievability of the component goals was judged on several key assumptions regarding the future of solar thermal R&D efforts from the present through the middle 1990s. Included in these key assumptions (Williams, Dirks, and Brown 1985) were that government funding for the Solar Thermal Program would continue roughly at its present levels, the solar thermal industry would gradually expand sales into the 1990s, research programs directed at increasing the performance and reducing the cost of components would continue to be successful, and industries would achieve a significant annual sale of solar-thermal-specific components.

For the last assumption, a solar-thermal-specific component was one that would not be produced except for use in a solar thermal plant. Obviously not all components of the solar thermal plant would fall into this category, although some of the more critical ones, for example, concentrators and receivers, would. The last assumption, then, is most important for those distributed items, such as concentrators, where sales on the order of 10,000–50,000 units/yr may be needed to achieve the component cost goals.

The probability of achieving the component goals will not be the same for all technologies except under ideal conditions. Technologies that are at an earlier stage of R&D will inherently have more risk than those that are more developed. For this reason, no effort was made to equalize the probabilities of achieving component goals among the three solar thermal technologies.

DOE (1984) established component goals in an allocation process, in which it examined performance and cost trade-offs of components. This approach led to some iteration in the process of developing goals. The general approach used (Williams, Dirks, and Brown 1985) was to first provide the lead R&D centers (Sandia Laboratories in Livermore, California, and Albuquerque, New Mexico) with technology data sheets that define the centers' best estimate of attainable efficiency and cost for each component. Initial LEC calculations were made based on these data and compared to the system goals. This initial set of goals was then modified in a lengthy review process that asked industry, R&D centers, and DOE to reflect on the following criteria: standardization among technologies, comments from reviewers, improved data, and the need to meet system goals. Table 10.2 shows the final component goals chosen.

Many degrees of freedom exist in the selection of component goals. First, trade-offs are possible between different components, such as receivers and concentrators, to achieve identical system goals. Second, for the same component, trade-offs between such parameters as cost, efficiency, lifetime, and O&M costs are also possible. As a result, no unique set of component goals alone satisfies the system goals; many solutions are feasible. A creative feature of the analysis (Williams, Dirks, and Brown 1985) used to set solar thermal goals was a detailed set of equations that allows researchers to choose alternative cost/performance goals for a given component or to trade off between components within the framework of the system goal. Thus researchers had the degree of freedom necessary to perform innovative R&D, still within the boundary of achieving approved system and component goals. All component goals were developed with the objective of being attainable through future R&D and yet making the system goal achievable. Predicting the amount of improvement in a component's performance or cost is subject to uncertainty. The allocation of component goals may thus change in the future as solar thermal technology proceeds through R&D stages and as certain components improve faster than others.

10.4.9   Measurability: How Do You Know When You Have Succeeded?

How does one determine when goals are reached? This is not as simple a question to answer as it seems at first. Research organizations by nature do not often commercialize the devices they develop. At best, they might transfer the technology to the private sector. As a result, the research organizations have little control over the commercial price of the device. How then does one verify when cost goals are achieved?

A difference of opinion often arose among those who favored a soft approach and those who favored a hard approach to resolve this issue. The proponents of the soft method wished to use learning curves or other manufacturing study projections to extrapolate from prototype hardware costs to mass production costs and thus determine if their technology had reached its goals. The advantage of this technique is that goals can be assessed using actual laboratory or field test hardware without waiting for mass production to be achieved. The disadvantage is that this methodology will produce a soft number (difficult to verify) that is highly dependent on such noncausative approaches as learning curve theory. More important, there would be no verification of manufacturer interest to produce the hardware at that price.

The hard approach would evaluate the goals based on the actual laboratory or manufactured cost of the hardware as developed or produced. Although such cost data might be difficult to obtain, it would bring the responsibility of price to where it ultimately belongs, the marketplace.

For performance goals (for example, system efficiency) actual field testing of a small-scale system for a reasonable period (not measurement of peak efficiency) can be used to verify attainment. This is the approach often taken for measurable parameters. Other alternative approaches, such as computer simulation of system efficiencies, are not usually considered reliable enough (without experimental verification of the computer model) to be the ultimate arbiter of whether or not goals are reached. Obviously performance goals, at least for solar thermal systems, are easier to verify using experimental hardware than are cost goals.

The measurability issue must be resolved early in the goals selection process, so that those who are setting and/or those who will be held to the goals will know how the judgment process will take place. Measurability could affect the level of goals chosen.

10.5   Historical Perspective

In this section I discuss the development of goals for solar thermal technology and systems before the selection of goals in the 1985 Five-Year Plan (DOE, 1984).

10.5.1   Pre-1980 Goals

Solar thermal program goals set before 1980 tended to be for energy and system levels and for the concentrator component. For instance, Braun and Edelstein (1980), in an early discussion of program goals, indicated that the market would impose the necessary conditions for the selection of the "right price" for solar thermal. Cost goals were set for 1990 for electric applications at $1,250–2,500/kW$_e$ (this and all following costs are given in 1984 dollars), for IPH heat applications at $6.25–12.50/10^6$ Btu (10^9 J), and for concentrators at $9.30–18.60/ft^2 ($100–200/m^2). They also indicated that concentrator production rates on the order of 10,000/yr were required to achieve these levels of cost reduction. Goals for other components were the responsibility of the R&D centers, within the framework of overall system and energy goals.

In the backup for the Sunset Review document (Jet Propulsion Laboratory, 1981), cost goals were indicated as $1,500/kW$_e$ for utility electric power applications and $10–12.50/10^6$ Btu (10^9 J) for IPH applications. These numbers were based on production rates of 20,000 concentrators per year.

Program managers set these early goals with a value-based perspective in mind. However, translating these goals into individual technologies and R&D centers was difficult. Furthermore, differences in interpreting what the goals actually represented, diverse assumptions, and varied analytical techniques often led to an inconsistent delineation of goals to the component level. Thus an interlaboratory Solar Thermal Cost Goals Committee (STCGC) was established in 1980 to resolve these difficulties.

10.5.2 Results of the Solar Thermal Cost Goals Committee

The STCGC chose target energy-use sectors, selected a consistent set of parameters, assessed the competition, completed market potential studies, and derived energy cost goals, system level goals, and then subsystem or component goals. The STCGC chose the electric utility and IPH markets as the target market sectors. The committee decided that market potential analyses should be conducted on a regional basis to ensure that solar thermal systems could compete across the nation and with a variety of fossil fuels. It is worthwhile to compare the results of these studies with follow-up analyses (for example, Williams, Dirks, and Brown 1985).

The Jet Propulsion Laboratory analyzed the electricity sector using a utility grid mix (Gates and Terasawa, 1982). Their model compared an electric utility grid with and without solar thermal power systems over the thirty-year life of a solar plant. The value of the solar plants was the difference in net present values between the conventional grid system and the grid with solar thermal. The analysis was run over a range of solar penetration of the grid from 1% to 20%. Based on this analysis, the STCGC selected a goal of about $3,200/$kW_e$ (1984 dollars) for the electricity sector.

The Solar Energy Research Institute (SERI) conducted the IPH analysis based on a regional market potential analysis developed by L. Flowers (unpublished). Based on this study, the STCGC selected a system goal of $220/$kW_t$ for the IPH sector.

Note that current (DOE, 1984) targets as compared to the STCGC goals are much more ambitious for the electric utility sector than for the IPH sector.

10.6   Implementation of Goals

In this section I detail an approach to implement the goals development process described previously.

10.6.1   Subcomponent Cost Targets

Upper-level market and systems goals must be broken down into levels that can be used for R&D program management. A breakdown to the component level was described in sections 10.3.2 and 10.4.8. In this section I discuss finer differentiation of the goals to the subcomponent level.

Williams, Dirks, and Brown (1985) took a major step toward disaggregating goals by breaking them down into their component levels. However, more disaggregation was needed so that researchers could use these goals to set down requirements for hardware development projects. For example, a concentrator goal of $9/ft^2 ($100/m^2) might be an appropriate target for a concentrator development project, but it would not be detailed enough for research on drive or mirror module components. Furthermore, it would certainly not be at a fine enough level for materials scientists in providing a criterion for their selection of new or modified materials for solar thermal concentrators. However, a mirror module goal of $3/ft^2 ($32/m^2) or 540 lb/ft^2 (50 lb/m^2) could be detailed enough to target materials research activity.

Thus more effort is required to further disaggregate these data to the major subcomponent level. A cost goal for every nut and bolt is not advocated because this would most certainly stifle research and innovation. Rather, a reasonable level of depth is needed in setting technical goals so that research across disciplines (materials, thermal research, concentrator development) can be consistent and so that materials research can mesh with subcomponent work. This in turn must lead to cost-effective components so that, when a whole package is put together, it can still meet the overall targets.

Individual research laboratories should bear the responsibility for further differentiation of goals so long as their goals are at least as stringent as those recommended by the program managers. Of course, the research laboratories must consider trade-offs between component goals and R&D costs.

SERI took a significant step in differentiating goals when it developed a heliostat component cost breakdown (Murphy, 1982) to set research requirements for their Advanced Concentrator Research Program. Cost goals for third-generation subcomponents used an installed heliostat component cost of about $5/ft^2 ($54/m^2). These subcomponents included the reflector assembly, a support structure, the drive assembly, the controls, and the foundation or pedestal. The reflector assembly goal targeted subcomponent research and was at a level of detail sufficient for the SERI Materials Research Branch to use in developing cost targets for their polymer materials research.

10.6.2   R&D Resource Requirements

Effort is needed to determine the budget allocations required to reach the goals. The achievement of goals in their target years cannot be separated from the dollars required to perform the research. If budget levels are halved, the chances that the goals will be reached in those years (or at all) will have been considerably reduced.

10.6.3   Probabilities of Success

The third remaining task is to develop data on probabilities of success tied to both the targets and the resource requirements. The probabilities could be divided into two categories: technical and commercial. As R&D dollars allocated to the program increase and as technical progress is achieved, the technical probability of success should increase accordingly. The commercial probability, on the other hand, is related to such factors as manufacturer involvement, customer interest, and removal of potential institutional barriers and may not be much influenced by federal funding. The total probability of success would be the product of the technical and commercial probabilities. The Gas Research Institute takes a similar approach (Gas Research Institute, 1982) in its planning and appraisal activities.

A research program that completes these three activities can fully integrate the cost goals into the mainstream of its research activities and can quantifiably justify incremental budget changes in specific directions. In fact, it could lay the basis for a marginal benefits analysis of the program by relating R&D costs (allocations) and R&D benefits, including success probabilities.

10.6.4   Periodic Adjustment of Goals

Because energy and system goals are value based, they should be reevaluated once every few years based on the latest projections of conventional fuel prices and other relevant parameters beyond the control of the Solar Thermal Program. When solar thermal R&D is completed, use-

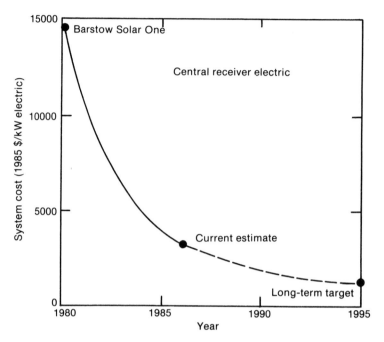

Figure 10.1
Tracking of solar thermal system costs toward goals: central receiver electric.

ful, efficient, cost-effective subcomponent, component, or system results should emerge. This can occur only if researchers remain aware of the competition. Researchers must also consider factors internal to the program, such as budgeting and scientific progress, in reevaluating goals.

10.6.5   Tracking

Finally, to ensure that R&D is kept on schedule, researchers must periodically check the progress of the research against the targets. It would be extremely beneficial to develop incremental goals, earlier than 1990 or 2000, that would project to the market-based targets. Depending on earlier decisions on measurability, the actual goals themselves could be used for performance goals or "projected" cost goals or for some mean value between present costs and ultimate goals if the measurability criterion uses actual costs.

Figures 10.1 through 10.3 show the tracking of system costs for the central receiver electric system, the parabolic dish electric, and the para-

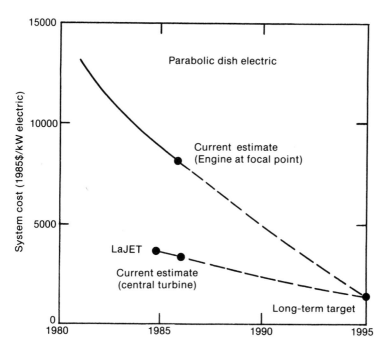

Figure 10.2
Tracking of solar thermal systems costs toward goals: parabolic dish electric.

bolic trough IPH systems (DOE, 1984). The Barstow Solar One point shown in figure 10.1 represents the actual cost of the 10-MW$_e$ solar thermal system in Barstow, California. The two lines in figure 10.2 are for two different technology options. The upper curve is a track of costs for a heat engine mounted at the focal point of each dish; the lower curve for a central turbine. In figure 10.3, the IPH Projects point is for pre-1980 IPH experiments; the MISR point represents the Modular Industrial Solar Retrofit IPH project. As indicated, the Solar Thermal Technology Program made considerable progress toward the achievement of its long-term goals for all three of its major system options.

10.7   Conclusions

One might draw from the discussion that the development of goals for an R&D program is a cut-and-dried process with a fixed set of protocols to

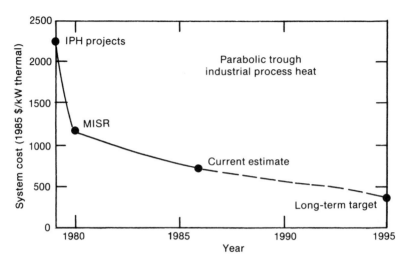

Figure 10.3
Tracking of solar thermal system costs toward goals: parabolic trough IPH.

follow and a static set of assumptions to delineate, or an "ivory tower" task suitable for economists or planners (but certainly not researchers or engineers). Neither of these apparent conclusions is true.

The process of developing goals for a research program is fraught with pitfalls, some of them already discussed. From the initial moment of establishing goals through the final conclusion of the process, someone (or many people) will always be there to point the way down other paths. These paths will undoubtedly include new target markets or applications, more appropriate user decision criteria (payback, return on investment, life-cycle costing, cost/benefit ratio), different fuel price tracks or different alternative fuels, a nearer-term or longer-term window of opportunity (which might require a shift in R&D toward development or basic research, respectively), innovative system or technological approaches, and more appropriate component cost breakdowns.

Those charged with developing the R&D goals will have to choose a methodology and a set of assumptions from among these suggestions. They will also have to reconsider their selections based on changing market conditions (for example, world oil prices), new projections, new senior management, new regulations and repeal of regulations (like the Powerplant and Industrial Fuel Use Act of 1978), and research breakthroughs. This is not to say that the development of goals is an impossible task but rather that it is a dynamic process. In addition, the technology under development will have to compete with not only state-of-the-art equipment but also advanced alternative (and conventional) technologies. A number of approaches to handle alternative futures are available, including sensitivity analysis, scenarios development, and technology forecasting. The key to these approaches is that the research goals (and the research program) should be resilient under a variety of probable market and technological conditions.

Eventually, a point will be reached when goals will have to be selected. This should not be done in isolation by a select staff group or a single individual, either the researcher or senior manager. Both a top-down approach to developing value-based (or market-based) targets and a bottom-up approach of projecting technology development should be used, with the energy and system-level goals being driven by market forces and the component goals and allocation of system goals among the components being forged from a technology basis. Obviously a compromise between these approaches will have to be reached. However, it will do no good to develop technology that will not compete in the future marketplace or, conversely, to set goals so low that neither the R&D dollars nor the technology will be available within the strategic time frame.

The researchers and engineers involved in developing the hardware, as well as staff-level personnel, must take part in the goals development process from the beginning to set up a consistent framework that can be applied across research organizations and technologies. Consistency of the goals development process and resiliency of the goals in the face of changing market conditions are critical to both the selling of the research program (not a negligible consideration) to governing boards and funding agencies and to the potential competitiveness of the technology in the market. Bringing in a group of prospective manufacturers or end users to provide some real-world advice on present and future market conditions would also be helpful. Finally, the researchers themselves must be part of the activity. If they do not accept the goals and internalize them into the R&D process, the goals cannot be reached, regardless of what the reports or plans say. Moreover, the appropriate level of goal must be communicated to research contractors; these targets must be consistent with system-level goals.

In summary, although a research program can be conducted with no goals, with goals developed by each researcher independent of the other players along the R&D path or by senior management who do not communicate these goals to the research organization, staff, and contractors, such a program will have questionable success in developing products that will be able to compete in the marketplace. A research program with clearly defined goals can develop a cogent strategy, communicate these goals to the outside world, and receive credit for achieving them when the research is completed.

References

Blanchard, B. S. 1976. *Engineering Organization and Management.* Englewood Cliffs, NJ: Prentice-Hall.

Braun, G. W., and R. B. Edelstein. 1980. "Review of the U.S. Department of Energy's Solar Thermal Program." *Revue de Physique Appliqueé* 15:1219–1228.

Delene, J. G., G. R. Smolen, H. I. Bowers, and M. L. Myers. 1984. *Nuclear Energy Cost Data Base. A Reference Data Base for Nuclear and Coal-Fired Power Plant Generation Cost Analysis.* DOE/NE-0044/2. Oak Ridge, TN: Oak Ridge National Laboratory.

Doane, J. W., R. P. O'Toole, R. G. Chamberlain, T. B. Bos, and P. D. Maycock. 1976. *The Cost of Energy from Utility-Owned Solar Electric Systems—A Required Revenue Methodology for ERDA/EPRI Evaluations.* ERDA/JPL-1012–76/3. Pasadena, CA: Jet Propulsion Laboratory.

Edelstein, R. B. 1981. *Solar Thermal Cost Goals.* SERI/TP-633–1063. Golden, CO: Solar Energy Research Institute.

EPRI. 1982. *Technical Assessment Guide.* EPRI P-2410-SR. Palo Alto, CA: Electric Power Research Institute.

Gas Research Institute. 1982. *R&D Planning Manual.* Chicago, IL: Gas Research Institute.

Gates, W. R., and K. L. Terasawa. 1982. *Size of the Electric Utility Market for Solar Thermal Technologies.* JPL-5105–116. Pasadena, CA: Jet Propulsion Laboratory.

Gottlieb, M. 1984. "Proposed agenda for 1984 senior research council meetings." Gas Research Institute Interoffice Memo, Myron Gottlieb to R&D Directors. Chicago, IL: Gas Research Institute.

Holtberg, P. D., T. J. Woods, and N. B. Ashley. 1985. "1985 GRI baseline projection of U.S. energy supply and demand to 2010," *Gas Research Insights* (Nov.): 1–64.

Jet Propulsion Laboratories. 1981. *A Review of the Solar Thermal Energy Systems Program: Accomplishments to Date and Future Plans.* Pasadena, CA: Jet Propulsion Laboratories.

Massie, J. L. 1971. *Essentials of Management,* second edition. Englewood Cliffs, NJ: Prentice-Hall.

Meinel, A. B., and M. P. Meinel. 1977. *Applied Solar Energy: An Introduction.* Reading, MA: Addison-Wesley.

Murphy, L. M. 1982. *Recommended Subelement Cost Goals for Advanced Generation Heliostats.* SERI/SP-253-1790. Golden, CO: Solar Energy Research Institute.

Newman, W. H., C. E. Summer, and E. K. Warren. 1972. *The Process of Management: Concepts, Behavior, and Practice*, third edition. Englewood Cliffs, NJ: Prentice-Hall.

Reynolds, A. 1982. *Projected Costs of Electricity from Nuclear and Coal-Fired Power Plants*. DOE/EIA-0356/1. Washington, DC: U.S. Department of Energy.

Ringer, M. 1984. *Relative Cost of Electricity Production*. P300-84-014. Sacramento, CA: California Energy Commission.

Twiss, B. C. 1974. *Managing Technological Innovation*. London: Longman Group.

U.S. Department of Energy. 1983. *Energy Projections to the Year 2010*. DOE/PE-0029/2. Washington, DC: U.S. Department of Energy.

U.S. Department of Energy. 1984. *National Solar Thermal Technology Program Five-Year Research and Development Plan 1985–1989*. Washington, DC: U.S. Department of Energy.

Vancil, R. F. 1979. "Strategy formulation in complex organizations," in Sloan Management Review 17(2):1–18.

Williams, T. A., J. A. Dirks, and D. R. Brown. 1985. *Long-Term Goals for Solar Thermal Technology*. PNL-5463. Richland, WA: Pacific Northwest Laboratory.

11 Historical Cost Review

Charles E. Hansen and Wesley L. Tennant

We present the historical cost trends of solar thermal systems from 1972 to 1982 and discuss possible reasons behind these trends. We also present a summary of selected projections from federal government sources to compare anticipated and actual solar systems costs over this period. Our intent is to compile historical cost data from the numerous studies, reports, and surveys and to give a general overview of total cost for installed systems by generic system type and major market sector for the entire United States. Such data obviously must ignore the many variations in system cost caused by location and many other factors. By using average cost data, these all are considered in the aggregate. The result, then, is not a technically usable cost history for any particular system, application, or regional market sector but a cost history that provides a general overview of solar systems costs since their widespread market introduction in the early 1970s. More detailed cost data for particular applications are found in other volumes of this series. For example, Shingleton discusses costs of hot water systems and King discusses costs of space heating systems in volume 6.

Although the economic feasibility of solar heating, cooling, and power generation depends on several factors, the cost of the solar collectors and related system components is significant. Historically solar system costs have hovered above the point of widespread economic viability. In the early 1970s, primarily under government sponsorship, a concentrated effort was launched to decrease these costs. This effort involved the private sector and the government in systems modeling, performance testing, full system demonstrations, component and materials research, development and testing, marketing information dissemination, and tax credits and other subsidies. These activities and the normal competitive market forces have resulted in cost changes of either a direct or an indirect nature, such as efficiency improvements, increased durability, and easier maintainability. Other factors, such as learning and experience in manufacturing, design, and installation and greater production volumes, have had impacts, some of which have been offset by changes in marketing and distribution costs. By 1980 the costs of delivering solar energy had become competitive enough with conventional energy sources (primarily electricity) to sustain an industry. However, the cost trends of installed solar systems have not been the sole economic driver; the escalating costs

of oil, gas, and nuclear energy—and especially delivered electricity—
have contributed to the economic viability of solar energy. Although real
cost trends of solar energy do not always display the hoped-for long-term
reductions, the increasing use of solar systems over the time studied
strongly implies that real economic improvements in solar thermal technologies have occurred, allowing these systems to compete with traditional energy technologies in the marketplace.

11.1 Historical Cost Perspective

The many definitions of costs of solar systems used in research and applications from 1972 to 1982 indicate the need for a stable platform of reference. The variable definition of cost for solar systems and the rapidly
changing products and technology for solar require use of estimating techniques and generic groupings that can assist in an objective comparison.

11.1.1 Various Definitions of Cost

Solar thermal systems have traditionally been costed by several standards.
First, the cost is usually considered to be the total economic cost including profit and is therefore normally termed the "price." Cost is used in
this context in this chapter. Second, the cost usually includes all costs
associated with installation to the point of actually delivering energy on
site. This also is the definition used in this chapter unless stated otherwise. Third, the cost is frequently expressed as the price of delivered energy (in dollars per 10^6 Btu) over the life cycle of the system, considering
all economic variables, for example, tax credits, alternative energy prices,
interest rates (if the system is financed), expected life, and geographic
location. This life-cycle costing method also considers the efficiency, durability, and maintainability of the total system. It is the cost indicator
used at the individual decision level. Life-cycle costing is, however, a
poor indicator for comparing the longer-term trends in costs of solar
hardware. It is too susceptible to influence from geographic factors such
as insolation and economic factors such as tax laws and interest rates that
have nothing to do with the cost of the technology itself.

The fourth common standard of costing solar systems, and a more useful cost criterion for comparative purposes, is the cost (of total systems
installed) per square foot of collector area. However, because of large in-

creases in system efficiencies and design over the preceding ten years, this criterion is not perfect as a comparative measure. A 1982 square foot of collector area is likely to perform much more efficiently than a 1972 square foot, and it is likely to be much more durable and easier to maintain. The reader should be aware that the bulk of the data and research literature discussed in this chapter use the total installed system cost per square foot without consistent adjustments for efficiency, durability, maintainability, or consistent conversion to the actual cost of delivered energy. Such adjustments are impossible to make given the type of data available.

11.1.2   Rapidly Changing Technologies

The historical perspective must include the review of the costs of many rapidly changing solar technologies. Hundreds of different solar products (or different versions of products) exist, each with its own state-of-the-art technology and its own cost/performance profile. It is therefore not possible to track the cost history of "identical" products; the real world requires that we address a selected technology at one point in time and proceed to a *similar* technology at a different point in time for which "all else" besides cost difference is *not* necessarily equal. The selections must be those from the research literature that address a broad range of such changing technologies; a statistically more valid method is not achievable. For example, low-temperature systems are used to describe a wide range of technical configurations with considerable variations over time in efficiency, durability, maintainability, and other characteristics.

11.2   Working Definition of Cost

Given the variation in cost terminology and the changes in technologies just described, the reader should be aware of the rationale behind and the need for the working definition of cost presented in this chapter.

11.2.1   Variables Affecting Solar Technology Costs

For any individual installation the cost of a solar energy system ideally should be measured in terms of cost per unit of energy delivered, that is, dollars per 10^6 Btu (10^6 Btu equals 1.055 GJ). However, this measure is subject to many variables, including installation costs, down payment, loan terms and rates, purchaser's income tax rate, maintenance costs, system heat load, solar radiation, collector area and orientation, alternative fuel cost and expected future costs, collector system efficiency, heat storage capacity, tax credits, and property tax rate. In a typical estimate of cost per unit of energy delivered, these factors must be defined or calculated over the life of the solar system—perhaps twenty-five or thirty years. Many of these factors are considered in the purchase of *any* energy system; however, because solar normally has high equipment costs (three to ten times higher) compared with conventional fueled systems, these factors must be specifically considered to compare or "size" a collector and storage system so it pays for itself over a period commensurate with payback periods for investments at going market rates of return.

The life-cycle cost approach allows one to cost solar systems for a particular installation and compare these costs with those for conventionally fueled systems. This method of analysis ensures that the factors discussed previously are considered over the life of the system. The life-cycle cost method essentially calculates the net present value of the cash investment in the solar system, whether it is just a down payment (perhaps part of a down payment on a home) or the entire hardware and installation purchase price. This method is a decision model for a particular installation, not a method of establishing the precise cost per delivered unit of energy for comparison among various sites and installations. In fact, the life-cycle cost approach stresses that the delivered cost of energy from any particular solar system varies substantially over the system's lifetime—ranging from "very costly" in the first years to "very cheap" once the initial investment payback point is reached.

11.2.2 The Working Definition to Be Used: Installed Cost per Unit or Capital Cost

We chose the immediate capital cost per unit of energy-producing capacity as our working definition of solar system cost. This is the best solution for the purposes of comparison. The cost includes materials, design, installation, and (normally) testing or balancing, including all associated marketing and overhead costs. The cost may be expressed as cost per square foot or square meter of collector area for an active system or, as in the case of passive, cost per annual energy-saving capability. The latter ratio, frequently used for passive systems, is called the capital productivity.

However, using the total installed system cost per unit of collector area is not without its problems. A system's collector area is not directly convertible to cost per energy unit produced or delivered because of the great variation in insolation among locations and other factors, such as financing methods, durability, operating costs, maintainability, competitive fuel costs, and escalation rates. Also, this cost measure does not consider the trends of improving system efficiency. Because no reasonable method exists to correct the data for this concern, it is important to be aware of it.

With these concerns in mind it is generally acceptable to use the total solar system cost expressed as a cost per unit of collector area if a broad mix of individual systems is drawn on so as to average out the many variables. No other useful standard for general comparison exists. This cost measure is as stable a platform as can be found. Therefore in the later parts of this chapter, the tables and figures describing cost trends are always presented in dollars per square foot (or square meter) of collector area when sufficient data are available from the literature.

11.2.3   Limitations of Data and Conclusions

We determined that the data used in this chapter best represent the U.S. cost averages at points in time for installed systems broken out by system category and market sector where possible. Data sources with such averages defined or implied were few even though over a hundred documents were reviewed. Formal surveys directed at defining such average cost data were not undertaken until 1980; and even the formal surveys of 1980 and 1981 do not meet the most desirable statistical standards. In the earlier years many "averages" were derived informally without being verified. Therefore over the 1972 to 1982 period only a general picture of the cost trends can be presented, spotted with specific examples. For any particular average figure presented, a range is always implied if not stated.

11.3   Cost Data and Data Sources

Cost data in this chapter are, in general, arranged and presented by operating temperature categories as follows: low temperature, $80°-120°F$ $(27°-50°C)$; medium temperature, $100°-200°F$ $(40°-90°C)$; high temperature, $200°-750°F$ $(90°-400°C)$; very high temperature, $750°-2,500°F$ $(400°-1,370°C)$; passive, not specified.

Most of the useful historical data of a comparative nature are on low- and medium-temperature systems, those systems used for pool, space, and service hot water heating. Historical data on space cooling systems are not numerous enough to indicate trends.

High-temperature cost data are also sparse, primarily because high-temperature systems constitute a small portion of the total market; for example, according to the U.S. Department of Energy (DOE, 1983), less than 3% of total collector shipments in 1982 were concentrator or high-temperature types, such as parabolic trough collectors. An additional 0.5% fell into the evacuated tube classification, some of which operate in the low end of the high-temperature range. However, some high-temperature data are available and are presented.

Very-high-temperature systems, such as parabolic dish concentrators and heliostat central receiver systems, have not passed the prototype stage; data for these systems are therefore limited. Heliostats have been the subject of considerable study, however, and some cost data are available and are presented.

Passive system cost data are available but, for reasons explained later, are extremely difficult to work with so far as cost trends are concerned. Viable cost ranges for passive can be analyzed, however, and are presented.

The sources of cost data for these solar systems include twenty-three selected studies and reports, cost and economic analyses, projections, and surveys. Each of the major sources are discussed in the following sections. With few exceptions, only government-sponsored sources are used. The year related to each cost figure is either in the paragraph or immediately adjacent the figure in parentheses. Conversions to constant-year dollars employ the gross national product (GNP) implicit price deflator. Values of the deflator index are given in chapter 1 of this volume. Although the figures are not rounded, this does not imply that the level of accuracy is significant to two decimals.

11.3.1   Low- and Medium-Temperature Cost Data

Early Baseline Work   Government policy makers and a few researchers considering the potential of investments in solar research and development had undertaken studies as early as 1969 to establish a baseline for solar technology costs. The National Science Foundation (NSF) and a small network of individuals conducted these early studies. In 1973, with the advent of the energy crisis, NSF initiated the Phase Zero studies

(Cohen, 1974) to get unbiased estimates of the expected cost and technical performance of solar technology. These early studies provide excellent premarket cost data for medium-temperature systems.

Studies and workshops during 1970–1974 relied primarily on data derived from expert and engineering estimates and a few models. In a 1971 NSF report (MIT Environmental Laboratory, 1971) a range of $1.30–$4.00/ft^2 ($13.99–$43.04/m^2) (1967 dollars) flat-plate collectors including their "associated gadgetry" was used as the basis for residential applications. Furthermore, a cost comparison was presented that showed optimized solar systems to be cost competitive with electricity in all but the northwestern areas of the United States. Delivered costs in terms of dollars per million Btu for solar systems were projected at $1.10–$4.64 (1967 dollars) based on flat-plate collector costs of $2.00–$4.00/ft^2 ($21.52–$43.04/m^2) (1967 dollars). It was noted that "present prices, based on hot-water heaters purchasable in Florida, Israel, and Australia, are not much above [the $4.00] figure." These figures are assumed to be for total installed systems.

Several NSF-sponsored reports and workshops addressed the solar cost question from 1973 to 1974. Engineering, research, and development of designs required such cost data to optimize system size and performance. However, data remained quite vague. In 1973 one NSF-sponsored presentation used a $6.00/ft^2 ($64.56/m^2) installed flat-plate collector cost with an additional comment that "target costs are between $2.00 and $3.00 in high production quantities" (Schlesinger, 1974). These figures are assumed to be 1973 dollars.

By 1974 NSF had three large contractors at work to verify both cost and performance data. These Phase Zero studies constitute one of the first organized sources of baseline solar cost data. According to presentations made by these contractors in 1974, the cost range of $4.70–$5.80/ft^2 ($50.57–$62.41/m^2) (1974 dollars) of installed flat-plate collector area and a total installed solar energy system cost of $13.00–$20.00/ft^2 ($139.88–$215.20/m^2) (1974 dollars) of collector were considered realistic based on optimized simulations and engineering estimates for both space heating and hot water systems. A typical cost breakdown from one of these 1974 presentations, shown in table 11.1, provides a baseline and what turned out to be an accurate estimate aimed at 1980 installations. Further price reductions were considered feasible, and a case for a 1980 range of $10.00–$15.00/ft^2 ($107.60–$161.40/m^2) (1974 dollars) at the

Table 11.1
Typical projected cost breakdown for space heating and hot water heating solar system for a single-family residence[a]

Component/item	Total cost (1974 $)	1974 $/ft^{2b}
Collector cost, installed	2,780	6.95
Storage tank, installed with glycol, insulation, fittings	1,190	2.97
Auxiliary heating system, heat exchangers, pumps, valves, plumbing, controls	2,470	6.18
Subtotal contractor's cost	6,440	16.10
Contractor's overhead, profit	1,932	4.83
Brokerage or developer's fee	502	1.25
Selling price to customer	8,874 = 13,770 (1980 dollars) = 17,948 (1985 dollars)	22.19 = 34.44 (1980 dollars) = 44.89 (1985 dollars)

Source: Cohen (1974).
a. Home installed in Denver, Colorado, in 1980. Collector area = 400 ft^2 (37 m^2).
b. $/ft^2 × 10.76 = $/m^2.

total system bottom line was also presented (Cohen, 1974). These last figures convert to a range of $20.22–$30.34/ft^2 ($217.62–$326/m^2) in 1985 constant dollars.

1976–1979 Studies and Reports In a 1976 report the Energy Research and Development Administration (ERDA, 1976) used total system costs of $10.00, $15.00, and $20.00/ft^2 ($107.64, $161.40, and $215.20/m^2, respectively) of collector in their scenario analyses (1976 dollars); however, ERDA did not consider the $10.00 and $15.00/ft^2 ($107.64 and $161.40/ m^2, respectively) figures as "current" but rather as possibilities "through incentives or as a result of mass production and/or improved designs."

In 1976 the Federal Energy Administration (FEA) publication *Buying Solar* (Dawson, 1976) suggested a range of $15.00–$18.00/ft^2 ($161.40–$193.68/m^2) (1976 dollars) of collector as a reasonable total system cost for residential space heating and/or hot water systems.

A detailed NSF-sponsored study on solar water heating conducted at the University of Delaware during 1975–1976 concluded that a realistic cost of collector for a residential solar hot water system would be between approximately $25.00 and $32.00/ft^2 ($269.00 and $344.32/m^2) (1976 dollars), depending on the specific type of system installed and whether the installation was new or retrofit (Scott, 1977).

Table 11.2
Typical cost breakdown for a residential domestic hot water solar system, installed[a]

Component/item	Total cost (1977 $)	1977 $/ft[2][b]
Collector	660	12.00
Storage tank	283	5.15
Accumulator	32	0.40
Pumps	89	1.62
Plumbing	55	1.00
Controls	86	1.56
Heat exchangers, other	111	2.02
Installation labor	165	3.00
Subtotal contractor's Cost	1,471	26.75
Contractor's overhead, profit	294	5.35
Selling price to customer	1,765 = 2,929 (1985 dollars)	32.10 = 53.28 (1985 dollars)

Source: Hewitt and Spewak (1977) and calculations by author.
a. Collector area = 55 ft^2 (5.1 m^2).
b. $/ft^2 × 10.76 = $/m^2.

In 1977, testimony before the Joint Economic Committee of Congress (Joint Economics Committee, 1977) noted that "the cost of installation for solar space heating on new tract homes in 1977 was assumed to be $1,350 (1977 dollars) as the fixed cost component *plus* a cost of $11.70 per square foot of collector installed." For hot water systems the fixed cost was estimated at $370.00. These figures imply a total system cost of collector of about $17.00/ft^2 ($183/m^2) for space heating and about $18.00/ft^2 ($194/m^2) for hot water (1977 dollars).

One of the most thorough studies—and sources—of cost data on medium-temperature solar systems for the residential and commercial sectors was completed in 1977 by MITRE Corporation (Hewitt and Spewak, 1977). This study defined five generic solar systems and included direct contractor and supplier inputs as well as analyses of solar projects completed through early 1977. The cost structure for a typical residential solar hot water system is derived from this study and presented in table 11.2. This study provides cost profiles for complete systems installed, for the full range of buildings, and for residential and commercial applications. The data in table 11.3 are displayed for reference purposes for combined space and hot water applications. These data provide a profile of the total fixed and variable cost ranges for these types of applications.

A 1978 DOE solar status report (DOE, 1978c) used a total system cost range of $32.00–$50.00/ft^2 ($344.32–$538.00/m^2) (1978 dollars) of col-

Table 11.3
Cost profiles for solar space and hot water heating applications, installed; liquid working fluid[a]

Building type and cost center	Area-related costs (1977 $/ft^2)[b]		Fixed costs in 1977 $ (1985 $ in parentheses)
	Subject to experience	Not experience related	
One and two family			2,630–4,340 (4,365–7,203)
Equipment	8.00	14.30	
Installation	4.73	3.85	
Design	1.92	1.08	
Low-rise residences			37,100–76,300 (61,576–126,638)
Equipment	11.60	11.50	
Installation	5.82	3.06	
Design	2.26	0.74	
Auditoriums			58,700–139,000 (97,426–230,703)
Equipment	11.80	11.00	
Installation	5.86	2.92	
Design	2.31	0.69	
Stores, clinics			19,300–50,300 (32,033–83,485)
Equipment	11.60	11.70	
Installation	5.83	3.11	
Design	2.24	0.76	
Educational buildings			88,600–248,000 (147,052–411,614)
Equipment	11.80	10.80	
Installation	5.85	2.88	
Design	2.82	0.68	
Hospitals			89,700–224,000 (148,878–371,780)
Equipment	11.80	10.80	
Installation	5.85	2.88	
Design	2.32	0.68	
Malls			510,000–1,850,000 (846,464–3,070,500)
Equipment	11.80	10.80	
Installation	5.85	2.88	
Design	2.32	0.68	
Motels			42,200–88,800 (70,040–147,380)
Equipment	11.60	11.50	
Installation	5.82	3.06	
Design	2.26	0.74	
Warehouses			112,000–337,000 (185,890–559,330)
Equipment	11.80	10.80	
Installation	5.85	2.88	
Design	2.32	0.74	

Source: Hewitt and Spewak (1977).
a. Type of collector: double glazed, selective surface, liquid. Type of storage: fiberglass hot water tank, 1 gal/ft^2. Type of backup: electric, gas, or oil. System O&M cost fraction: 0.007.
b. $/ft^2 × 10.76 = $/m^2.

lector for residential hot water and $27.00–$43.00/ft² ($290.52–$462.68/ m²) of collector for combined hot water and space heating. For industrial and agricultural applications a total system cost of $20.00–$60.00/ft² ($215.20–$645.60/m²) of collector was used. These figures were apparently derived from the experience on selected demonstrations and privately funded systems.

According to a DOE report (DOE, 1979), system costs for 177 non-federally funded residential projects yielded averages of $22.00/ft² ($236.72/m²) of solar collector for residential space heat and $18.00/ft² ($193.68/m²) for residential space heat and domestic hot water. These figures are assumed to be in 1978 dollars. As compared to other data, these costs appear to be toward the lower end of the cost spectrum. However, this same report used a $31.00/ft² ($333.56/m²) total system cost (1978 dollars) for a commercial space heating system example.

The Solar Energy Research Institute (SERI) study on solar energy and conservation (SERI, 1981) used 1978 cost data as a baseline for analysis and projections. Drawing on data from both estimates and actual installations, the total installed cost for residential solar domestic hot water systems ranged from $31.65 to $46.60/ft² ($340.55–$501.42/m²) (1978 dollars) of collector. For combined space heating and hot water systems the costs presented in the study ranged from $22.00 to $47.00/ft² ($236.72–$505.72/m²) (1978 dollars) of collector. The authors of the study noted that several other studies and surveys supported these cost ranges.

A 1979 report by the National Academy of Science (NAS, 1979) noted that "there is considerable uncertainty about [costs of installed solar systems] even at present." Costs cited in this study for a working example were $20.00/ft² ($215.20/m²) (1979 dollars) of collector installed for a residential space heating system, although the range noted was from $10.00 to $43.00/ft² ($107.60–$462.68/m²). This range was based on "assessments from several sources" as follows:

1. ERDA/MITRE. An installed system cost of $20.00/ft² ($215.20/m²), (1976 dollars), of collector is assumed (ERDA, 1976).

2. DOE/CS. Installed system costs range from $25.00 to $40.00/ft² ($269.00–$430.40/m²) (1977 dollars) of collector (DOE, 1977).

3. SERI. Surveyed costs of actual installed systems are $39.00–$43.00/ft² ($419.64–$462.68/m²) (1978 dollars) of collector (SERI, 1978; Ward 1979).

Table 11.4
Solar system, installed cost history, federal demonstration program 1974–1978;
commercial sector[a]

Application	1974 NSF[b] ($/ft²)[c]	1975 PONI[d] ($/ft²)	1976 PONII[e] ($/ft²)	1977–1978 HWI[f] ($/ft²)
Service hot water	58	46	—	37
Space heating	114	61	41	—
Space heating and hot water	122	48	46	—
Space heating and cooling	216	105	69	—
Space heating, cooling, and hot water	127	114	79	—
Weighted Average				
Current dollars	89	72	56	37
1985 dollars	184	135	99	59

Source: DOE (1978b).
a. Amounts in current dollars.
b. National Science Foundation.
c. $/ft² × 10.76 = $/m².
d. Program Opportunity Notice I.
e. Program Opportunity Notice II.
f. Hot Water Initiative.

4. Lovins. Intelligently designed and installed systems in today's market are $10.00–$15.00/ft² ($107.60–$161.40/m²) (1978 dollars) of collector (Lovins, 1978).

The Demonstration Programs The residential, commercial, and industrial demonstration programs conducted by NSF and DOE included the objective of gathering economic and cost data. Cost data from the residential program have been incorporated into several of the reports cited in this section. No single comprehensive analysis on costs for the residential demonstration program was performed and published.

For the commercial demonstration projects the cost data that are available display a definite downward trend. Data from the 1978 DOE Solar Update Conference proceedings (DOE, 1978b) show that, from the initial NSF-sponsored demonstrations of 1974 through the DOE-sponsored demonstrations of 1978, the total installed cost of 182 solar systems decreased from an average of $89.00/ft² to $37.00/ft² in current dollars. Data from this report are summarized in table 11.4. These data are primarily for medium-temperature systems.

Table 11.5
Solar system, installed cost history: Federal Demonstration Program, industrial sector[a]

System type	$/ft² (1979 $)	$/ft² (1985 $)
Low temperature	24.60	34.96
Medium temperature	59.50	84.56

Source: Brown (1980).
a. $/ft² × 10.76 = $/m².

In the industrial sector DOE had sponsored, as of 1980, eighteen industrial process heat (IPH) projects. SERI analyzed the available cost data for fourteen of these projects and concluded that the total installed system cost of collector area ranged from $24.60/ft² for a low-temperature application to $87.10/ft² for a medium-temperature application. Figures are assumed to be in 1979 dollars. The cost in terms of capital productivity ranged from $108.00 to $536.00/10⁶ Btu/yr ($/10⁶ Btu/yr × 0.948 = $/GJ/yr). The average, exclusive of the highest and lowest cost, was $276.00/10⁶ Btu/yr (Brown, 1980). Additional analyses based on actual performance of six of these systems yielded an average capital productivity of $962.00/10⁶ Btu/yr. According to Brown (1980), proposed systems could bring these numbers within a range of $100.00 to $219.00/10⁶ Btu/yr. Conventional fuels such as oil or gas displaced by such systems would have to cost between $50.00 and $75.00/10⁶ Btu— about ten times the 1981–1982 average—for these solar systems to be economical. The data from the report are summarized in table 11.5. The temperature ranges are assumed to be based on collector area and annual energy output.

DOE/Energy Information Administration Surveys   The official published statistics from DOE did not address costs of solar systems until the 1980 and 1981 active solar installations surveys (DOE, 1981, 1982). These reports provide the most thorough and accurate source of capital cost data available for single-family, multifamily, and commercial applications of low- and medium-temperature solar systems for the 1980–1981 period. Table 11.6 summarizes the key data from the 1981 survey, formatted for comparison with other data in this section.

As mentioned previously, one of the major problems encountered in analyzing cost trends of solar systems is the significant variation in costs

Table 11.6
U.S. average solar system costs for 1981 installations[a]

Application	Residential single family in 1981 $/ft² (1985 $ in parentheses)	Residential multifamily 1981 $/ft² (1985 $ in parentheses)	Commercial 1981 $/ft² (1985 $ in parentheses)
Low temperature[b]			
Pool	9.19 (10.92)	9.17 (10.90)	12.39 (14.72)
Medium temperature[b]			
Water heating	53.03 (63.01)	29.18 (34.67)	25.57 (30.38)
Space and water heating	43.14 (51.26)	36.65 (43.55)	29.85 (35.47)
Space heating	32.17 (38.23)	20.95 (24.89)	33.01 (39.23)
Space cooling (alone or in combination)	46.13 (54.82)	64.00 (76.05)	63.30 (75.22)

Source: DOE (1982) and calculations by author.
a. Data are derived from weighted average cost and weighted average collector area. $/ft² × 10.76 = $/m².
b. It is assumed that pool heating systems are all low temperature and all others are medium temperature. There are obviously exceptions; however, this is the only practical approach to the data.

because of type of system, differences in installation, and the location for any particular application. Figures 11.1 and 11.2 from the *1980 Active Solar Installations Survey* (DOE, 1981) show an example of this variation in the United States for residential (single-family) hot water systems. Total cost per square foot and total system installed cost are shown, as well as the national averages. Figures are in 1980 dollars. These types of variation are encountered in all solar applications. Over 84% of these 48,908 systems were retrofit installations. Residential retrofit systems averaged about $1.00/ft² ($10.76/m²) of collector more than new installations. [In a separate study performed for DOE covering earlier *commercial* sector projects (King et al., 1979), the difference between new and retrofit systems was over $20.00/ft² ($215.20/m²) of collector.]

Other Sources and Surveys In addition to the government-sponsored reports and surveys already identified, there has been considerable informal tracking of cost data by members of the solar industry and the technical and professional media. In 1982 a summary of the total installed cost trends for low- and medium-temperature systems from 1974 through 1981 was published based on these informal sources. Selected data from Resource and Technology Management (RTM) Corp. (1982)

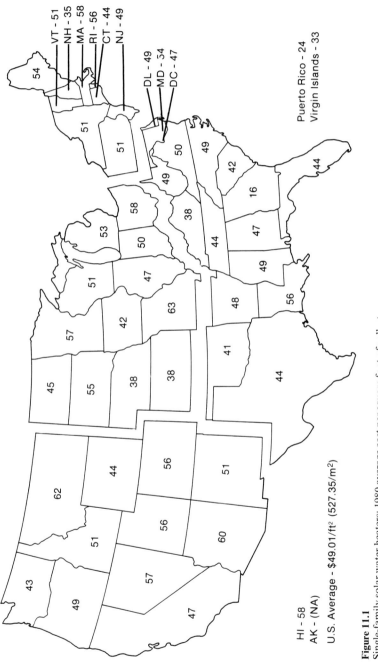

VT - 51
NH - 35
MA - 58
RI - 56
CT - 44
NJ - 49

DL - 49
MD - 54
DC - 47

Puerto Rico - 24
Virgin Islands - 33

HI - 58
AK - (NA)
U.S. Average - $49.01/ft² (527.35/m²)

Figure 11.1
Single-family solar water heaters: 1980 average cost per square foot of collector area (weighted average cost divided by weighted average size, $/ft², multiply by 1.303 to convert to 1985 dollars and by 10.76 to convert to $/m²). From DOE (1981).

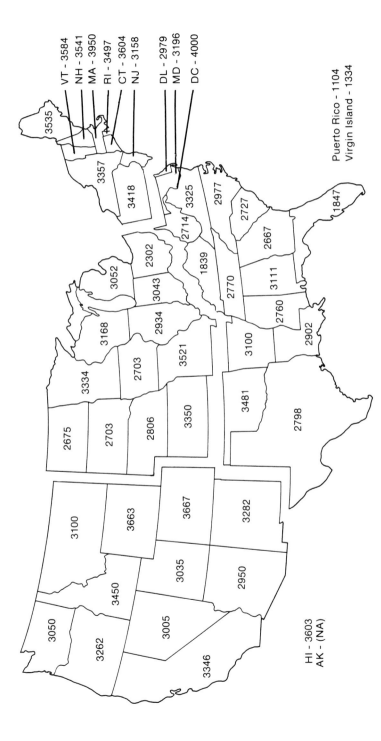

Figure 11.2
Single-family solar water heaters: weighted average cost of 1980 installations in $/ft²
(multiply by 1.303 to convert to 1985 dollars and by 10.76 to convert to $/m²). From DOE
(1981).

are drawn on to fill voids in the sponsored research and data. For low-temperature systems this source shows a 1974 average total system installed cost of $6.00/ft^2 ($64.56/m^2) for the residential sector and $6.50/ft^2 ($69.94/m^2) for the commercial sector (1974 dollars). In 1977 these costs reached $7.80 and $8.50/ft^2 ($83.92 and $91.46/m^2), respectively (1977 dollars).

11.3.2 High-Temperature Cost Data

Few government-sponsored studies have addressed costs of high-temperature collectors or systems. Parabolic trough or other line-focusing collector systems have been in use since the mid-1970s but on a limited scale. These collectors may generate temperatures up to 750°F (400°C).

In an ERDA-sponsored report (Davis et al., 1976), a cost range of $25.00 to $35.00/ft^2 ($269.00–$376.60/m^2) for installed "medium concentration collectors" was used (1976 dollars). These figures did not include other system components.

In another report (SERI, 1980) high-temperature system total costs range from $40.00 to $87.00/ft^2 ($430.40–$936.12/m^2) of collector (1980 dollars). These figures were based on actual costs of installed systems and estimates for systems in the planning and design stages.

An informal survey of professional and technical media and industry professionals published by RTM Corp. (1982) indicates that these types of collector systems averaged about $40.00/ft^2 ($430.40/m^2) in 1975 (1975 dollars) and about $65.00/ft^2 ($699.40/m^2) in 1981 (1981 dollars), inclusive of all system components and installation.

The preceding cost figures are, however, somewhat misleading. The reason is that the efficiency of high-temperature systems has been greatly improved (as much as 400%) over this period compared to that of low- and medium-temperature systems. A data summary by DOE (1984) indicates that in 1978 the average total system cost was about $3,930/kW output; by 1979 the cost was averaging about $2,480/kW$_t$; and by 1980 costs were approaching the $1,030/kW$_t$ figure (all figures in 1985 dollars).

11.3.3 Very-High-Temperature Cost Data

Data sources for very-high-temperature focusing or central receiver solar systems designed primarily for electric power or very-high-temperature industrial applications include pilot projects, prototypes, and simu-

Table 11.7
Percentage of cost profile for heliostat-central receiver system

System component	% of total cost[a]
Heliostat field	52
Electric power system	18
Storage	10
Receiver	5
Tower	3
Piping/heat exchanger	8
Other	4
Total	100

Source: CFEE (1981).
a. For a mature system, according to the source.

lations. These applications have followed the simpler flat-plate medium- and low-temperature systems by several years. There are three collector system categories: line-focusing collectors, dish collectors, and heliostat-central receiver systems. The line-focusing or trough systems are discussed under the high-temperature category (see section 11.3.2), although some of the applications for these systems may reach into the very-high-temperature ranges. There is no adequate source of historical cost data on parabolic dish systems.

In 1975 ERDA envisioned heliostat-central receiver systems to be economically viable with a cost of $1,650–$3,100 per installed kilowatt of capacity (1985 dollars) for commercial plants approaching the 100-MW$_e$– 300-MW$_e$ scale. The first 5-MW$_e$ and 10-MW$_e$ pilot or prototype systems averaged over $12,000 per installed kilowatt (1985 dollars).

Table 11.7 shows a general profile of the system cost for a central receiver electric plant. The high percentage of cost (over 50%) attributable to the heliostat field and the potential savings from new heliostat technology have resulted in concentration on this area as the cost reduction target. Early prototype production of 222 heliostats for the Central Receiver Test Facility in 1975 yielded installed costs of $79.93/ft^2 ($860.00/ m^2) (1980 dollars). The 10-MW$_e$ Barstow pilot plant used 1,818 heliostats that cost $39.03/ft^2 ($420.00/m^2) (1980 dollars) installed in 1980 (Brandt, 1980). In 1980 it was estimated that continuous low-volume-production rates of 2,500 heliostats per year could bring costs down to $18.59/ft^2 ($200.00/m^2) installed (1980 dollars) (Pacific Northwest Laboratory, 1980).

11.3.4   Passive Cost Data

Sources of cost data for passive solar systems include presentations made at conferences and a few special studies. Passive solar systems are difficult to define and difficult to cost. Cost per unit of energy displaced or delivered is the only comparable measure because many passive features replace conventional construction and therefore cannot be isolated in terms of cost. A Trombe wall, for example, may replace a conventional load-bearing wall and thereby serve two purposes: it is part of the basic structure as well as part of the energy system, but only a portion of its cost can be attributed to the energy role. The cost measure used in most studies is to take the additional costs of solar and divide this cost by the estimated or actual energy savings per year. This measure (dollars per 10^6 Btu per year $= 0.948$ GJ/yr) is also called the capital productivity.

In a report issued by DOE (Morse and Maybaum, 1979), a general capital productivity figure of $\$117.00/10^6$ Btu yr (1978 dollars) was used as the midpoint in a range from $\$70.00$ to $\$180.00/10^6$ Btu yr depending on the type of passive solar system, location, and system size.

By 1980 SERI had accumulated considerable cost data on passive systems and published comprehensive cost guidelines. Table 11.8 shows ranges of cost from one of these publications (SERI, 1980) expressed in terms of capital productivity of the additional solar investment for the various types of passive solar systems.

11.3.5   Summary of Data and Sources

The preceding discussion of sources and data indicates the type of cost data available. From this diverse collection of reports, presentations, and analyses we can draw conclusions about general historical cost trends for the various technologies as they have occurred in each major market sector. Table 11.9 summarizes these data. The diversity of systems and costs at any given point in time makes such generalizations risky. The following general cost trends are only indicative of the real world and are constrained by the data available, as discussed at the beginning of this chapter. Data points from the preceding sources are shown in each figure as appropriate in current or nominal dollars and in 1985 constant dollars. Costs are for total system, installed. The GNP implicit price deflator has been used for conversion. Data points are *averages*, representing the midpoints of ranges.

Table 11.8
Costs of passive solar systems; range of additional construction costs; dollars per annual energy delivered

Passive system type	1980 $/10⁶ Btu yr (1985 $ in parentheses)[a]
Direct gain	
Simulated	32–538 (42–701)
Actual (monitored)	30–139 (39–181)
Actual (unmonitored)	33–127 (43–165)
Indirect gain	
Thermal storage wall	
Simulated	85–887 (111–1156)
Actual (monitored)	36–224 (47–292)
Actual (unmonitored)	89–156 (116–203)
Thermal storage roof	
Simulated	179–1092 (233–1423)
Actual (monitored)	83 (108)
Isolated gain	
Attached sunspace	
Simulated	33–58 (43–76)
Actual (unmonitored)	12–120 (16–156)
Connective loop	
Simulated	64–147 (83–192)
Hybrid	
Actual (monitored)	
Residential	26–168 (34–219)
Commercial	278 (362)
Actual (unmonitored)	
Residential	120 (156)
Commercial (institutional)	308 (401)

Source: SERI (1980).
a. $/10⁶ Btu yr × 0.948 = $/GJ yr.

Table 11.9
Cost data for figures with references to source, sector, year; and conversions to 1985 constant dollars[a]

Figure	Technology	Sector	Source	Year of data	Cost/area[b]	1985 $ cost/area[b]
Figure 3	Low temp HW[c]	Residential	RTM (1982)	1974	6.00	12.41
			RTM (1982)	1977	7.80	12.95
			DOE (1982)	1981	9.19	10.92
	Low temp HW	Commercial	RTM (1982)	1974	6.50	13.45
			RTM (1982)	1977	8.50	14.11
			DOE (1982)	1981	12.39	14.72
Figure 4	Med temp HW	Residential	Schlesinger (1974)	1973	6.00	13.54
			Cohen (1974)	1974	16.50	34.13
			ERDA (1976)	1976	20.00	35.40
			Dawson (1976)	1976	16.50	29.21
			Scott (1977)	1976	28.50	50.45
			JEC (1977)	1977	18.00	29.88
			Hewitt (1977)	1977	32.10	53.28
			DOE (1978c)	1978	41.00	63.43
			SERI (1981)	1978	39.12	60.52
			DOE (1981)	1980	49.01	63.88
			DOE (1982)	1981	53.03	63.02
	Med temp HW	Commercial	DOE (1978b)	1974	58.00	119.97
			DOE (1978b)	1975	46.00	86.65
			DOE (1978b)	1977	37.00	61.41
			DOE (1981)	1980	35.43	46.18
			DOE (1982)	1981	25.57	30.38

Figure					
Figure 5	Med temp SH&HW[d]	Residential	Cohen (1974)	1974	16.50	34.13
			ERDA (1976)	1976	20.00	35.40
			Dawson (1976)	1976	16.50	29.21
			DOE (1978c)	1978	35.00	54.15
			DOE (1979)	1978	18.00	27.85
			SERI (1981)	1978	34.50	53.37
			DOE (1981)	1980	20.60	26.85
			DOE (1982)	1981	43.14	51.26
	Med temp SH&HW	Commercial	DOE (1978b)	1974	122.00	252.36
			DOE (1978b)	1975	48.00	90.41
			DOE (1978b)	1976	46.00	81.43
			DOE (1981)	1980	40.43	52.70
			DOE (1982)	1981	29.85	35.47
Figure 6	High temp	All sectors	RTM (1982)	1975	40.00	75.35
			Brown (1980)	1980	63.50	82.76
			RTM (1982)	1981	65.00	77.24
Figure 7	Heliostats	Utility	Brandt (1980)	1975	$1215/m^2$	$2289/m^2$
			Brandt (1980)	1980	$420/m^2$	$547/m^2$

a. Conversion to constant dollars made by using GNP implicit price deflator. All figures are averages.
b. Total system cost installed expressed as $/ft^2$ ($\times 10.76 = \$/m^2$) collector area except for heliostats, which are expressed in $\$/m^2$.
c. HW = hot water.
d. SH&HW = solar heating and hot water.

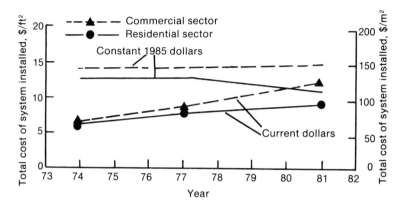

Figure 11.3
Central cost trend for low-temperature systems. See table 11.9 for values and sources.

Low-Temperature Systems   Figure 11.3 shows the general historical cost trends for low-temperature systems. These systems use collectors of plastic or simple metal construction and generate temperatures up to 120°F (50°C). They are used primarily for pool heating. Data from section 11.3.1 are presented for the residential and the commercial sectors. Only the midpoints or averages of the cost ranges presented in the original sources are shown. All figures should be considered as averages for total cost of systems installed.

The cost of low-temperature systems has gradually increased, in nominal or current dollars, over the 1974 to 1981 period for both the residential and commercial sectors. For the residential sector, in terms of current dollars, cost increased from $6.00 to $9.19/ft² ($64.56–$98.88/m²) of collector; for the commercial sector cost increased from $6.50 to $12.39/ft² ($69.94–$133.32/m²). Converted to 1985 constant dollars, however, the average residential application cost about $1.50 less per square foot ($16.14/m²) of collector area in 1981 than in 1974. Constant 1985 dollar costs fell from $12.41 to $10.92/ft² ($133.53 to $117.50/m²) in this sector. For the commercial sector, however, constant 1985 dollar costs increased from $13.45 to $14.72/ft² ($144.72 to $158.39/m²) of collector between 1974 and 1981. In general, commercial applications are more expensive because of greater reliability, maintainability, and durability requirements and the larger size of such basic components as pumps and piping. Commercial systems also usually require engineering design costs that are

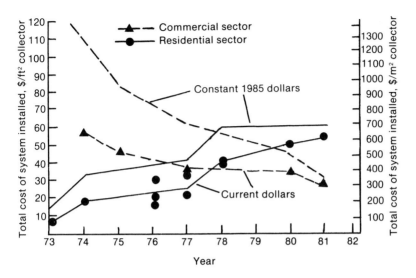

Figure 11.4
General cost trends for medium-temperature hot water systems. See table 11.9 for values
and sources.

seldom encountered in the typical low-temperature residential system.
The primary cause of the real cost increase in the commercial low-
temperature systems over the period is probably the result of technical
improvements in design and materials. The primary cause of the decrease
in real cost for residential applications is most probably a result of increasing competition in this sector.

Medium-Temperature Hot Water Systems   Figure 11.4 shows the general
historical cost trends for medium-temperature hot water systems. These
systems are used for domestic and service hot water only. They typically
generate temperatures of 180°–200°F (82°–93°C). Data are shown for the
residential and commercial sectors from the sources identified in section
11.3.1. Only midpoints or averages of the cost ranges presented in the
original sources are shown. All figures should be considered as averages
for total cost of systems installed.

The costs for medium-temperature hot water systems in the residential
sector increased substantially in terms of current or nominal dollars from
1973 to 1981. However, for commercial sector applications costs have decreased substantially. During this time the average residential system cost

in current dollars increased from about $6.00 to $53.03/ft² ($64.56 to $570.60/m²) of collector area. But for commercial applications the costs fell from about $58.00/ft² ($624.08/m²) in 1974 to $25.57/ft² ($275.13/m²) of collector in 1981. Converted to 1985 constant dollars, these changes are from $13.54 to $63.02/ft² ($145.69 to $678.09/m²) for the residential sector and from $119.97 to $30.38/ft² ($1,291 to $326.90/m²) for the commercial sector. For mature systems commercial applications may be expected to cost somewhat less because of the larger collector area over which fixed costs may be allocated. How much of the 1981 cost difference between the sectors is a result of this factor is not possible to determine, though. Except for the 1981 points, the costs for the two sectors have tended to become more similar.

The dramatic increase in cost, either real or nominal, in the residential sector can be attributed to several factors. First, residential solar systems have become more complex as their applications have been extended from the tropical climates to the temperate zones of the central and northern United States. Second, both collectors and freeze protection subsystems have required expensive improvements to increase durability and reliability from the 1973 system profiles. Third, the costs of competitive energy sources such as electricity have escalated at high rates, especially since 1979, thereby allowing price growth in alternative energy technologies. A fourth factor may be the tax and other incentives made available to buyers in the residential sector since 1977. Even if costs have not been directly adjusted upward to take advantage of such inducements, at the least the inducements encourage purchase of higher-quality and higher-cost systems. In addition, significant increases in system efficiency over this period enabled collector areas to be reduced while the same or greater energy output and system cost were maintained.

The decrease in the cost of commercial medium-temperature hot water systems can be attributed in part to learning and experience and possibly to the use of competitive bidding procedures common to this sector. Before 1974 few if any solar systems were installed in the commercial sector. These larger systems were therefore subject to cost reductions from gained experience. And as will be noted in the behavior of commercial sector costs both for hot water and space heating applications (see the next section), there appear to be greater pressures for cost reductions in this sector. The procedure of closed bidding against rigid specifications commonly used in the commercial sector may be a factor.

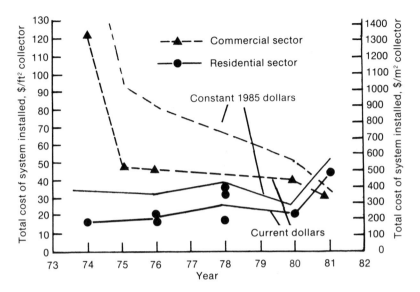

Figure 11.5
General cost trends for medium-temperature space heating and hot water (combined)
systems. See table 11.9 for values and sources.

The difference in total system cost per square foot between the residential and commercial sectors as systems have reached maturity may be
partially explained by closed bidding and specification procedures, but
there appears to be a greater influence at work. This influence is the much
greater collector area in the commercial sector systems over which fixed
costs for system design, components, and installation can be spread. This
is probably the most significant reason for commercial sector cost being
below residential sector cost for the medium-temperature hot water systems at maturity. A similar behavior of costs is indicated for the later
maturing combined space heating and hot water systems, as shown in
figure 11.5. However, the behavior is much more pronounced in the case
of hot-water-only systems, as shown in figure 11.4. This behavior would
be expected because of the earlier maturity of residential systems and the
much greater difference in areas between residential and commercial hot
water systems. For example, the 1981 data (DOE, 1982) gave the average
collector area for a single-family residential medium-temperature hot-
water system as 61 ft^2 (5.7 m^2), whereas for commercial systems it was
517 ft^2 (48 m^2). This difference in average system area, a factor of 8.4, is

considerably greater than the 5.0 factor difference between sectors for combined systems as calculated from 1981 data. The fixed costs are then consistently spread over a larger collector area for commercial sector systems. As system efficiencies increase, this difference becomes even more pronounced. This tends to drive the total system cost per unit of collector area for the commercial sector to values below that for the residential sector.

Medium-Temperature Space Heating and Hot Water Systems   The general historical cost trend for medium-temperature space heating and hot water systems is shown in figure 11.5. Data are presented for both residential and commercial sectors from the sources identified in section 11.3.1. These data are restricted to systems that provide space heating and hot water in combination. These systems typically operate in a temperature range of $100°$–$200°F$ ($38°$–$93°C$). Only midpoints or averages of the cost ranges in the original sources are shown. All figures should be considered to be averages for total costs of systems installed.

Medium-temperature space heating and hot water combined systems follow cost trends similar to hot water systems discussed in the immediately preceding section. For residential applications costs have increased over the 1974 to 1981 period; for commercial applications costs have decreased substantially over the same period. In terms of nominal or current dollars the average cost of collector for residential applications increased from $16.50 to $43.14/ft² ($177.54 to $464.19/m²) for combined space and hot water heating systems between 1974 and 1981, although costs were almost level through 1980. For the commercial sector, however, average cost of collector area decreased from $122.00 to $29.85/ft² ($1,312.72 to $321.19/m²) over the 1974 to 1981 period. In terms of constant 1985 dollars the average residential cost increased from $34.13 to $51.26/ft² ($367.24 to $551.56/m²), and the average commercial sector cost decreased from $252.36 to $35.47/ft² ($2,715.39 to $381.66/m²) over the period. The great cost difference between the two sectors in the early years, with commercial sector costs greatly exceeding costs for the residential sector, is probably the result of lack of both experience and technologies able to accommodate larger commercial applications. At system maturity, apparently reached in about 1978, the sector cost difference in current dollars decreases to between $10.00 and $20.00/ft² ($107.60 and $215.20/m²), and in 1980–1981 a "crossover" occurs with commercial

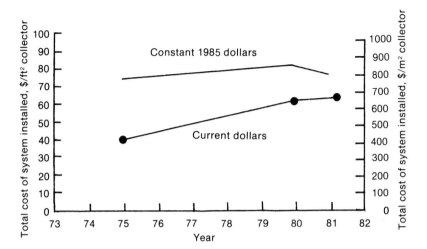

Figure 11.6
General cost trend for high-temperature systems. See table 11.9 for values and sources.

costs falling more than $10.00/\text{ft}^2$ ($107.60/\text{m}^2$) below residential costs. The cost in current dollars for the 1981 data shows commercial applications costing $29.85/\text{ft}^2$ ($321.19/\text{m}^2$) as compared to $43.14/\text{ft}^2$ ($464.19/\text{m}^2$) for the residential sector. Note that the sector data behave in a cross-over fashion for both space heating and hot water applications, as shown in figures 11.4 and 11.5. As observed earlier, greater competitive pressures in the commercial sector may be showing their influence, and there would appear to be basic differences for mature systems, such as the greater collector area over which to spread fixed costs for commercial applications. Again, however, one notices the tendency of the costs of the two sectors to become more similar.

High-Temperature Systems Figure 11.6 shows the general historical cost trend for high-temperature systems. These systems usually employ parabolic trough or other line-focusing devices as collectors. Data are from the sources discussed in section 11.3.2. Temperature ranges typically run from 200° to 750°F (90°–400°C). Only midpoints of ranges presented in the original sources are shown. All figures should be considered to be averages for costs of systems installed.

The cost trend for high-temperature systems in nominal or current

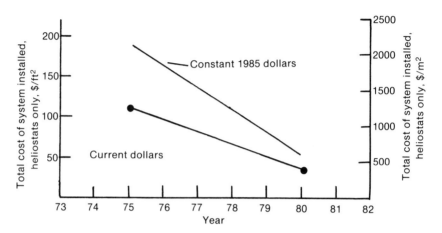

Figure 11.7
Cost trend for heliostats installed. See table 11.9 for values and sources.

dollars increased over the 1975 to 1981 period from between $30.00 and $40.00/ft² ($322.80 and $430.40/m²) of collector to about $65.00/ft² ($699.40/m²). In terms of 1985 constant dollars the increase is from $75.35/ft² ($810.77/m²) to about $77.24/ft² ($831.10/m²) of collector. There are not adequate data to enable the sectors to be separately displayed.

Very-High-Temperature Heliostat Systems Figure 11.7 shows the general historical cost trend for heliostats. These costs represent only the installed heliostat, not the balance of the central receiver systems of which they are a part. Heliostat systems may generate temperatures at the central receiver up to 2,500°F (1,370°C). There are only two specific data points displayed, and these data are from two pilot installations as defined in section 11.3.3. Cost for heliostats for the first pilot installation in 1975 was approximately $112.92/ft² ($1,215.00/m²) in current dollars or about $212.73/ft² ($2,289.00/m²) in 1985 constant dollars. For the 1980 pilot installation, cost in current dollars was $39.03/ft² ($420.00/m²) of heliostat or $50.84/ft² ($547.00/m²) in constant 1985 dollars. A substantial cost reduction is apparent. The principal reasons for such a reduction are most probably the experience gained from the first pilot installation as well as manufacturing and related economies resulting from the larger production quantity for the second installation.

11.4   Experience and Learning

Long-term cost reduction of solar systems has been expected in part because of the experience with furnaces and other manufactured equipment similar in many respects to the collectors, controls, and heat exchangers of the typical solar system. Several long- and short-term variables enter into such cost reductions (or increases). Nonmanufacturing areas such as marketing and distribution can play as important a role as the more hardware-oriented areas of research, development, manufacturing, systems design (conforming to a specific structure), and installation. Cost changes resulting from changing marketing conditions may, for example, outweigh cost changes that are normally expected as a result of repetitive activities and accumulating knowledge. It is also important to separate the aspects of system design and installation from hardware manufacturing when considering potential hardware cost reduction. The experience and learning of the designer and installing contractor may be sporadic, as compared to manufacturing, yet significant for any particular job. Because of the emphasis on installed costs in this study, all these types of learning or experience are included in the figures but are difficult to isolate for analysis.

In a detailed study of the costs of solar systems, Hewitt and Spewak (1977) found that between 40% and 55% of area-related variable costs for a typical solar system may be "subject to experience." No restrictions were placed on the specific functions—manufacturing or nonmanufacturing—that might be included. Reduction in costs from learning and experiences in manufacturing and installation, which tend to be more consistent (in theory), appear to have been outweighed in several sectors by cost increases in marketing and distribution. There is little question that the cost reductions, such as the reduction in the federally sponsored commercial demonstration projects from $89.00 to $37.00/ft^2 ($957.64 to $398.12/m^2) of collector in current dollars from 1974 to 1978, are attributable primarily to hardware manufacturing and installation. However, the cost increases noted in the general trends in the residential sector are attributable primarily to rising costs in the nonhardware, marketing, and distribution areas.

11.4.1   Experience and Learning in Nonmanufacturing Areas

For nonmanufacturing hardware-related areas early prototype and demonstration projects had many cost problems that have been eliminated or greatly reduced as a result of experience and learning. Planners, designers, and contractors have learned to reduce piping and duct run lengths, to use nonpressurized tanks, to limit the number of control modes, and to centralize and modularize components of solar systems. All these factors have tended to reduce the costs of later generation solar systems. Other cost reductions and/or performance improvements have occurred as a result of the experience acquired by planners, designers, and contractors. Such practices as having a few spare collectors on site in case of damage is an example of a cost-saving approach learned from experience. This behavior is to be expected; however, it is a difficult task to determine the cost reductions caused directly by such experience and learning.

In the nonhardware areas early manufacturers and contractors were faced with the problem of introducing an unproven and relatively expensive technology into traditional, well-established markets. Between 1972 and 1982 the traditional market for heating, cooling, and power generation became more difficult, not less, in terms of marketing and distributing solar technologies. Early enthusiasm and public/government support for solar succumbed to the reality that these new technologies could not offer *substantial* cost or technical advantages in most market sectors. Rather than enjoying a cost reduction because of increased learning and experience in the marketing and distribution areas, solar has been subject to increasing cost requirements in these areas, although the exact extent or proportion of these costs as a part of the total systems installed cost is not possible to isolate from currently available data.

11.4.2 Learning and Experience in Manufacturing

Reducing costs through experience and learning in the manufacturing process is a separate matter. In the production process learning and experience are gradually incorporated into the "production function," and consistent, long-term cost reductions are expected. Such learning and experience may take practically any form. People learn. New techniques or new machinery may contribute. The improvement is normally measured as a percentage reduction as production doubles from any given volume. Data for forced warm air, nonelectric furnaces for the years 1950 through 1973, as displayed in figure 11.8, are typical of expectations. Level of investment in production technologies, the degree of competition, and the size of the market as well as the general consistency of the technology all contribute to this type of cost reduction. Figure 11.9 (also discussed in

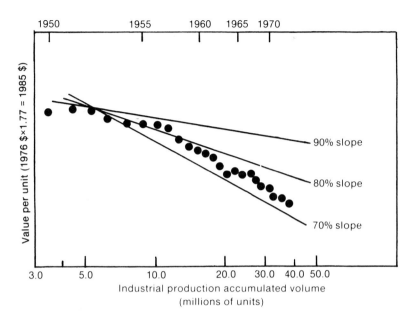

Figure 11.8
Experience curve for nonelectric, forced warm air furnaces, 1950–1973 (MITRE
Corporation, 1978, p. 12).

the following section) is an example of these expectations applied to solar
systems in California.

Manufacturers of solar systems and components most certainly experience cost reductions as a result of learning and experience. However,
measuring such reductions requires consistent data over a long period;
and such data have not yet been isolated; nor has the industry been operating at production volumes high enough and long enough. There is, no
doubt, some such reduction in actual manufacturing costs contained in
the overall trends of installed solar system costs. The reduction in installed heliostat costs from $1,075 to $525 (1984 dollars) as production
increased from a lot of 222 to a lot of 1,818 contains some manufacturing
cost reductions attributable to learning. Most other systems do not have
such obvious data to rely on, and even in this case many other nonmanufacturing factors were involved in this cost reduction.

In conclusion, the actual impact on cost of learning and experience in
the solar technologies, although apparent in some instances, cannot be
isolated or measured in detail because of a lack of data.

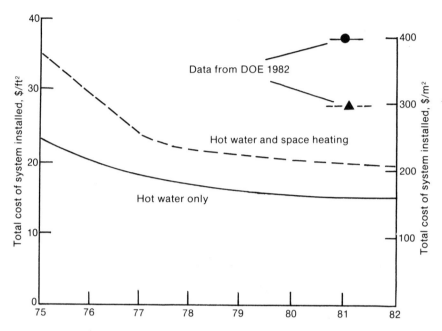

Figure 11.9
Examples of (expected) area dependent cost decreases in California (1977 dollars × 1.660 = 1985 dollars). From Rebibo et al. (1977, pp. 14–20).

11.5   Comparison with Early/Historical Projections

The early cost projections for solar systems as presented in government-sponsored reports and studies appear to have been somewhat optimistic—but not unreasonable. Many early projections were in essence goals directed at competition with 1974–1978 conventional energy prices. Although these years brought steady price escalations, the 1979 oil price shock lifted all energy prices substantially above the expectations of government and industry planners. Inflation from 1974 through 1981 also created unforeseen impacts on costs that few projections could account for in advance. Even with these difficulties most early projections do not deviate significantly from actual data. Projections are converted to 1985 constant dollars in table 11.10 for comparison purposes.

Between 1970 and 1973 several informal cost projections for solar flat-plate (medium-temperature) collectors were made by researchers under

Table 11.10
Cost projections summary with references to sources, sector, year; and conversions to 1985 constant dollars[a]

Technology[b]	Sector	Source	Year of cost data in projection	Cost/area[c]	Target year(s)	1985 cost/area[c]
Collectors	Residential	MIT (1971)	1967	$2.00–$4.00	Open	$6.22–$12.45
Collectors	Residential	Schlesinger (1974)	1973	$2.00–$3.00	Open	$4.50–$6.75
System	Residential	Cohen (1974)	1974	$22.19	1980	$44.89
System	Residential	Cohen (1974)	1974	¢10.00–$15.00	1980	$20.68–$31.03
System	Residential	ERDA (1976)	1976	$10.00–$15.00	Open	$17.70–$26.55
HW[d] system	Residential	Witwer (1978)	1975	$20.00	1980	$37.67
Collectors	Residential	Witwer (1978)	1975	$10.00	1980	$18.84
HW system	Residential	Rebibo (1977)	1977	$16.00	1982	$26.55
SH&HW[e]	Residential	Rebibo (1977)	1977	$20.00	1982	$33.19
Collectors	Res or com	JEC (1977)	1977	$10.00	Open	$16.60
HW system	Residential	DOE (1978c)	1978	$20.00–$30.00	1985	$30.94–$46.41
HW system	Res or com	DOE (1978a)	1978	$30.00–$39.00	1982	$46.41–$60.34

a. Conversion to constant 1985 dollars made by using the GNP implicit price deflator.
b. All medium temperature.
c. Total system cost, except as otherwise noted, expressed as $/ft^2 collector area. Multiply by 10.76 to convert to $/m^2.
d. HW = hot water.
e. SH&HW = space heating and hot water.

NSF sponsorship. During this period figures of \$2.00–\$4.00/ft² (\$21.52–\$43.04/m²) (1973 dollars) of collector installed, excluding other major system components, were projected as target costs for high production quantities—quantities that had not been achieved as of 1985.

The earliest formal projections of significance for medium-temperature systems were generated in 1974 during the Phase Zero studies. In the selected example, shown in table 11.1, the projected cost for a 1980 installation of a residential solar space and water heating system, converted to 1980 dollars, was \$34.44/ft² (\$370.57/m²) of collector area. The 1980 *Active Solar Installations Survey* (DOE, 1981) yielded a national average of \$20.60/ft² (\$221.66/m²) (1980 dollars); the 1981 survey (DOE, 1982) yielded an average of \$43.14/ft² (\$464.19/m²) (1981 dollars). The average of these two figures, about \$30.00/ft² (\$322.80/m²) (1980 dollars) is close to the \$34.25/ft² (\$368.53/m²) Phase Zero projection made in 1974. ERDA (1976) indicated that a total system cost of \$10.00–\$15.00/ft² (\$107.60–\$161.40/m²) (1976 dollars) of collector area was possible. This was, however, not a projection of a formal nature. These costs in 1980 dollars equate to \$13.40–\$20.00/ft² (\$144.18–\$215.20/m²), slightly below the \$20.60/ft² (\$221.66/m²) national average for space and water heating systems in the *1980 Active Solar Installations Survey*, although considerably below the 1981 survey average of \$39.70/ft² (\$427.17/m²) (1980 dollars).

In a major study by Witwer et al. (1978) a figure of approximately \$20.00/ft² (\$215.00/m²) (1975 dollars) of collector area for the total cost of a residential solar hot water system was projected for 1980 [assuming a 50-ft² (5-m²) system size]. The actual national average for 1980 (DOE, 1981) was \$49.00/ft² (\$527.24/m²) (1980 dollars). In this case the authors had projected that installed collector costs would fall from \$15.00 to \$10.00 between 1975 and 1980 (in 1975 dollars). This type of reduction was also anticipated and projected in other studies. The Phase Zero study discussed included a "feasible" 1980 case for a range of \$10.00–\$15.00/ft² (\$107.60–\$161.40/m²) (1974 dollars) of collector area for total system costs—down from the \$13.00–\$20.00/ft² (\$139.88–\$215.20/m²) figure used as the baseline (1974 dollars). Rapid cost reductions such as those noted have been anticipated primarily because of substantial quantity increases projected for production of solar systems. In a 1977 study (Rebibo et al., 1977) rapidly falling costs were anticipated from 1975 to 1982. Figure 11.9 displays a projection from this study that brings total solar system costs of collector (combined space and hot water) from a 1975 range of $35.00–$37.00/ft² ($376.60–$398.12/m²) (1977 dollars) to a 1982 figure of approximately $20.00/ft² ($215.20/m²) (1977 dollars). The 1981 comparative data from the 1981 survey (DOE, 1982) for California shows that these projections, which were based primarily on anticipated learning and experience improvements, were optimistic. In testimony before the Joint Economics Committee (1977) it was observed,

Now that some experience has been gained both in manufacturing and installation of solar systems, it is apparent that the cost of solar collectors installed, but excluding fixed costs, will be about $10 per square foot (1977 dollars). Although this may seem high in comparison to estimates as low as $3 per square foot used in previous studies, we find that feasibility does occur for solar water and space heating systems between now and 1990 if either decontrolled prices of traditional energy sources are used as the basis of comparison or where curtailment of natural gas occurs.

Later in the testimony it was stated that "it is apparent that the cost of solar collectors installed, but excluding fixed costs, will be about $10 per square foot (1977 dollars) with little real hope of long run cost reduction given materials and labor requirements."

As late as June 1978 U.S. DOE (1978c) projected total system costs to fall from a range of $32.00–$50.00/ft² ($344.32–$538.00/m²) (1978 dollars) of collector in 1978 to a range of $20.00–$30.00/ft² ($215.20–$322.80/m²) (1978 dollars) by 1985 for residential hot water systems. Similar reductions were projected for residential space heating and agricultural and industrial applications. These projections were based in part on reaching such national goals as 2.5 million solar homes by 1985. Such cost reductions are not supported by actual data as of 1982, when the average cost for the system specified was about $40.00 (1978 dollars).

DOE (1978a) projected a total system cost range of $30.00–$39.00/ft² ($322.80–$419.64/m²) (1978 dollars) of collector area for the year 1982. This range was considered relevant for both residential and commercial sectors. The estimated costs in terms of delivered energy from a typical solar hot water system ranged from $8.57 to $17.90/10⁶ Btu (1978 dollars) for the 1982 projections. Adjusted to 1982 dollars the cost of collector equates to $41.00–$54.00/ft² ($441.16–$581.04/m²), and the delivered energy adjusted to $11.75–$24.50/10⁶ Btu. The higher projected costs per square foot for solar water heating systems compares reasonably well with the DOE (1982) national average of $56.40/ft² ($606.86/m²) (1982 dollars) for the residential sector, but the $27.20/ft² (292.67/m²) actual figure for commercial sector applications is much below the projected range.

Projections for higher-temperature collection and heliostat systems normally extend beyond 1982 and therefore cannot be compared to actual figures. The broad ranges of passive solar costs and absence of data make most passive projections meaningless for trend analysis or comparison purposes. These technologies are therefore not considered here.

11.6   Economies of Scale and Other Considerations

The ultimate cost of solar technologies is partially dependent on the scale of the total solar industry, the degree of standardization, the amount of competition, and the impacts of government subsidies, programs, and legal constraints. Although these subjects have been discussed in many studies, with few exceptions their direct effect on costs as defined in this chapter have not been analyzed.

11.6.1   Manufacturing Economies of Scale

Over the 1972 to 1982 period some general manufacturing cost reduction of solar thermal systems resulted from increasing economies of scale for manufacturers. In 1974 (the first year for which data are available) collector manufacturers averaged about 28,000 ft² (2,600 m²) of production for the year; but by 1982 they were averaging almost 70,000 ft²/yr (6,500 m2/yr) (DOE, 1983). These averages imply that manufacturers found cost advantages or greater total profitability by increasing the scale of their production over this period. This conclusion is general because the profile of individual manufacturers cannot be presented and specific cost reductions per unit output cannot be calculated; however, it is apparent that increasing economies of scale have influenced the overall cost equation of solar systems. Cost reductions attributable to such economies of scale have not necessarily been visible at the final cost level because of impacts of other factors.

11.6.2   Market and Distribution Economies of Scale

As a manufacturer's production volume increases, the cost per unit for marketing materials, sales force (or sales commissions), and distribution infrastructure will usually decrease. This relationship may not be true after market saturation occurs and competition becomes greater. For a growth industry, such as solar, the marketing and distribution costs per unit most probably decreased as the average manufacturer's production volume increased during the early years of the industry. However, because competition in the market has always been high for solar technologies, it is doubtful that manufacturers enjoyed such cost decreases for long. No detailed DOE reports or other known published sources provide verification of increasing economies of scale in marketing or distribution.

11.6.3 Standardization Impacts

The impact of materials, installation, and performance standards on the cost of solar systems is not possible to measure—but it is certain to be a factor in the cost equation. In general, such standards may tend to increase short-term costs but decrease longer-term costs. For example, the collector performance testing by the National Bureau of Standards against a performance standard for stagnation resulted in some increases in costs for those collectors that needed improvements. But such improvements doubtlessly decreased the life-cycle costs for such improved collectors because they had greater durability. These tests and standards may also decrease the average cost of manufacturing by reducing the high cost of replacing collectors that fail while under warranty. The ultimate impact of standards on cost, however, cannot be verified by existing data. It would appear that a cost reduction is involved when a manufacturer, designer, or installer can rely on established standards for performance, tolerances, and materials such as absorptive coatings, absorber plates, cover plates, sealants, insulation, heat exchangers, and working fluids.

11.6.4 Competition and Government Subsidy Impacts

Both private industry and the federal and state governments have been actively involved in the developing solar industry of the 1970s and early 1980s. The normal competitive nature of the open marketplace has been significantly influenced by federal and state tax laws, direct subsidy of solar purchases, and considerable indirect subsidy through government RD&D programs inclusive of market development activities.

The government-sponsored RD&D programs of the state and federal agencies have most certainly reduced the ultimate costs of solar systems. An indeterminately large amount would have been expended by the private sector and ultimately attached to the cost of solar products had not the government provided this support. As in many other areas the actual impact on costs is not possible to measure. It is a fact, however, that the basic thrust of the majority of government RD&D activities has been to increase the commercial viability of the solar industry primarily by reducing the economic risk for manufacturers, contractors, and buyers. In the competitive market, this risk reduction equates to cost reduction—somewhere in the total cost equation.

Taxes have a significant impact on the life-cycle costs of ownership even though the immediate capital cost (total cash payments made to all suppliers and contractors for all components and installations) may not be directly affected by property and income taxes. Relief from property taxes, or income tax credits, such as those adopted by many cities, states, and the federal government, has had a major effect on the life-cycle costs of solar system ownership and has also affected the installed cost of systems—possibly by increasing rather than reducing the capital cost. As stated in an analysis conducted by the California Foundation on the Environment and Economy (CFEE, 1981), "It is also possible that one effect of the California solar tax credit will be to increase or at least lessen downward pressures on the retail price of solar systems." Special mortgage rates or loan programs also influence the life-cycle ownership costs of solar systems and may add some pressure to increase prices. However, no studies are known that address these concerns, and historical cost data, as summarized in this chapter, do not display significant changes in trend at the point of the major national tax credit program initiation in 1977. Although an increase in the average cost per square foot for residential hot water systems may be observed in the 1977–1979 period in figure 11.4, there are not sufficient data to relate this change or any other change in the figures to the tax subsidy or any other single factor.

11.7   Conclusion

The historical cost trends for solar thermal systems vary considerably depending on the specific technology and the market sector. We have presented data according to generic systems categories and general markets at a national level. We have had to use a general definition of cost—installed capital cost—rather than the delivered cost of energy, which is ultimately the measure of competitiveness in the marketplace. These data, then, provide only a general overview of cost trends from 1972 to

1982 that reflect the type of U.S. average data available and its sources. We have also referenced some of the projections made in the early and late 1970s and found that they coincide rather well with the actual trends in the medium-temperature technologies. A cursory review of broader industry factors, such as economies of scale, competitive forces, and government involvement, has not provided any significant clues to the causes of the cost trends, although these factors certainly have been influential. The actual cost trends do demonstrate cost behavior of systems that have matured significantly in technical terms since the early 1970s; however, they do not necessarily represent the potential cost trends should solar energy technologies achieve the much greater production quantities that could result when and if solar replaces conventional energy sources on a larger scale—an event that still appears imminent.

References

Brandt, L. D. 1980. *Strategy for Heliostat Commercialization*. SAND-80-8239. Albuquerque, NM: Sandia National Laboratory.

Brown, K. 1980. *The Use of Solar Energy to Produce Process Heat for Industry*. SERI/TP-731-626. Golden, CO: Solar Energy Research Institute.

CFEE. 1981. *Solar Energy in the 1980s*. San Francisco, CA: California Foundation on the Environment and the Economy.

Cohen, A. D. 1974. "Solar heating and cooling of buildings Phase Zero feasibility and planning study: A report by General Electric Company," in *Proceedings of Workshop on Solar Heating and Cooling of Buildings*. Washington, DC: National Science Foundation, 4–21.

Davis, J. P., D. Fink, R. J. Raymond, and C. C. Wang. 1976. *Technical and Economic Feasibility of Solar Augmented Process Steam Generation*. Washington, DC: Energy Research and Development Administration, Division of Solar Energy.

Dawson, J. 1976. *Buying Solar*. FEA/G-76/154. Washington, DC: Federal Energy Administration, esp. 40–43.

ERDA. 1976. *An Economic Analysis of Solar Water and Space Heating*. DSE-2322. Washington, DC: Division of Solar Energy, esp. 17.

Hewitt, R., and P. Spewak. 1977. *Systems Descriptions and Engineering Costs for Solar Related Technologies. Volume 2, Solar Heating and Cooling of Buildings (SHACOB)*. MTR 7485. McLean, VA: MITRE Corporation.

Joint Economics Committee. 1977. *The Economics of Solar Home Heating*. Washington, DC: Government Printing Office, 8.

King, T. A., J. G. Shingleton, P. A. Sabatiuk, and J. B. Carlock, III. 1979. "Cost effectiveness: An assessment based on commercial demonstration projects," in *Proceedings of the Second Annual Solar Heating and Cooling Systems Operational Results Conference*. SERI/TP-245-430. Golden, CO: Solar Energy Research Institute, 253–273.

Lovins, A. 1978. "Soft energy technologies," in *Annual Review of Energy*, Jack Hollander, ed. Palo Alto, CA: Annual Reviews, Inc., Vol. 3, 477–517.

MIT Environmental Laboratory. 1971. *Energy Research Needs.* Cambridge, MA: MIT Environmental Laboratory.

MITRE Corporation. 1978. *Solar Energy: A Comparative Analysis to the Year 2000.* MTR-7579. McLean, VA: MITRE Corporation, METREK Division, esp. 12.

Morse, F., and M. W. Maybaum. 1979. *Commercialization Strategy Report for Passive Solar Heating (Draft).* Washington, DC: U.S. Department of Energy.

NAS. 1979. *Energy in Transition 1985–2010.* Final Report of the Committee on Nuclear and Alternative Energy Systems, National Research Council. San Francisco, CA: W. H. Freeman, esp. 355.

Pacific Northwest Laboratory. 1980. *The Cost of Heliostats in Low Volume Production.* SERI/TR-8043-2. Golden, CO: Solar Energy Research Institute.

Rebibo, K., G. Bennington, P. Curto, P. Spewak, and R. Vitray. 1977. *A System for Projecting the Utilization of Renewable Resources: SPURR Methodology.* MTR-7570. McLean, VA: MITRE Corporation.

RTM Corp. 1982. *Alternative Energy Data Summary for the United States,* vol. 4. Arlington, VA: Resource and Technology Management Corporation.

Schlesinger, R. J. 1974. "Sensitivity of solar collector design to solar input" and "Discussion," in *Proceedings Solar Energy Data Workshop.* Washington, DC: National Science Foundation, esp. 118.

Scott, J. E. 1977. *Solar Water Heating Economic Feasibility, Capture Potential, and Incentives.* PB-279 855. Washington, DC: National Science Foundation, esp. 17.

SERI. 1978. *Economic Feasibility and Market Readiness of Eight Solar Technologies.* SERI-34. Golden, CO: Solar Energy Research Institute.

SERI. 1980. *Passive Solar Progress: A Simplified Guide to the 3rd National Passive Solar Conference.* DOE/CS/30046-01. Washington, DC: U.S. Department of Energy.

SERI. 1981. *A New Prosperity: Building a Sustainable Energy Future.* Andover, MA: Brick House Publishing, esp. 90–96.

U.S. Department of Energy. 1977. *An Analysis of the Current Economic Feasibility of Solar Water and Space Heating.* Washington, DC: U.S. Department of Energy.

U.S. Department of Energy. 1978a. *Commercialization Readiness Assessment: Solar Water Heating Systems.* Concept Paper. Washington, DC: U.S. Department of Energy, System Development Branch, Conservation and Solar Applications.

U.S. Department of Energy. 1978b. *The Department of Energy's Solar Update, Proceedings: Atlanta, San Francisco, Chicago, Boston.* Washington, DC: U.S. Department of Energy.

U.S. Department of Energy. 1978c. *Solar Energy: A Status Report.* DOE/ET-0062. Washington, DC: U.S. Department of Energy, esp. 15.

U.S. Department of Energy. 1979. *Commercial Readiness Assessment: Active Solar Space Heating,* draft. Washington, DC: U.S. Department of Energy, Conservation and Solar Applications.

U.S. Department of Energy. 1981. *1980 Active Solar Installations Survey: Final Report.* Washington, DC: U.S. Department of Energy, Energy Information Administration.

U.S. Department of Energy. 1982. *1981 Active Solar Installations Survey.* DOE/EIA-0360(81). Washington, DC: U.S. Department of Energy, Energy Information Administration.

U.S. Department of Energy. 1983. *Solar Collector Manufacturing Activity, 1982.* DOE/EIA-0174(82). Washington, DC: U.S. Department of Energy, esp. 20.

U.S. Department of Energy. 1984. *National Solar Thermal Technology Program*. Washington, DC: U.S. Department of Energy, Solar Thermal Technologies Division.

Ward, D. S. 1979. *Solar Heating and Cooling Systems Operational Results Conference*. SERI/TP-49-209, Summary Report. Golden, CO: Solar Energy Research Institute.

Witwer, J. G., J. A. Alich, S. M. Kohan, M. D. Levine, P. C. Meagher, E. E. Pickering, F. A. Schooley, A. J. Slemmons, and T. E. Thompson. 1978. *A Comparative Evaluation of Solar Alternatives: Implications for Federal RD&D*, vols. 1 and 2. TID-28533/1 and TID-28533/2. Menlo Park, CA: SRI International.

Contributors

Gerald E. Bennington

Gerald E. Bennington is President of X*Press Information Services, a partnership of McGraw-Hill Publishing and Tele-Communications, Inc. From 1976 to 1984 Bennington managed the Renewable Energy Systems Department of the MITRE Corporation and later formed his own energy consulting firm, Bennington Enterprises, Limited. He led a team that developed the SPURR model and published several of the major solar energy studies during this period. He received a PhD in operations research from John Hopkins University and has taught at North Carolina State University, the University of British Columbia, and George Washington University.

Kenneth C. Brown

Kenneth C. Brown is Vice President for Planning and Development at Inspiration Resources Corporation, a diversified producer of coal, metals, and agricultural products in the United States and Canada. From 1977 to 1981, Mr. Brown was employed at the Solar Energy Research Institute, where he was instrumental in the development of application analysis methodologies for industrial solar energy systems. He has spent over ten years addressing both the technical and economic aspects of energy development for firms involved in nuclear power, synthetic fuels, and solar energy technology. Brown has a BS in mechanical engineering from Cornell University and an MA in engineering science and economics from Oxford University, where he was a Rhodes Scholar.

Ronald Edelstein

Ronald Edelstein is the Director of Planning and Appraisal for the Gas Research Institute. His perspectives in solar thermal derive from having been employed at the Solar Energy Research Institute, serving as chair of the Solar Thermal Cost Goals Committee, and being a consultant for the U.S. Department of Energy's Solar Thermal Program. Edelstein has been involved in planning energy R&D programs for the past eleven years and was an engineer in gas turbine analysis for eight years before that. He holds master's degrees in solid mechanics and environmental engineering and a bachelor's degree in aerospace engineering.

Charles E. Hansen

Charles E. Hansen is an economic consultant who specialized in solar and other alternative energy technologies from 1976 to 1984. He has directed several solar energy projects for the Department of Energy. While President of Resource and Technology Management Corporation, he directed one of the first comprehensive national energy balance and cost/price studies that incorporated the solar technologies: the Alternative Data Summary for the United States. He received a BS in business with a minor in engineering from the University of Colorado and an MBA from Wright State University.

Charles R. Hauer

Charles R. Hauer is a consultant in Washington, D.C., specializing in energy economics and science policy-related problems. His experience ranges from the management of research and development efforts in the private sector to the management of research programs at the National Science Foundation and the Department of Energy. As Vice President at Ion Physics Corporation, a division of High Voltage Engineering, he managed the development of nuclear weapons effects simulators and radiation-hardened silicon solar cells. At the National Science Foundation he directed the early solar energy proof-of-concept experiments. He holds degrees in chemical and nuclear reactor engineering.

Robert A. Herendeen

Robert A. Herendeen is Associate Professional Scientist at the Illinois Natural History Survey, Champaign, where his interests are the behavior of ecosystems under resource constraints and the parallels between ecology and economics. For the past fifteen years he has worked in energy and resource analysis at the Oak Ridge National Laboratory, the University of Illinois, and the Norwegian Institute of Technology. He holds a PhD in physics from Cornell University.

Frank Kreith

Frank Kreith is a Senior Research Fellow at the Solar Energy Research Institute. From 1977 to 1984 Kreith was the Chief of Solar Thermal Research at the institute, where he built up the laboratory for solar heating and cooling, direct contact heat transfer, thermal storage, ocean thermal energy conversion, double diffusion convection, and high-temperature solar thermal power generation. Kreith has been on the faculties of the University of Colorado, Lehigh University, the University of California at Berkeley, UCLA, the University of California at Santa Barbara, and the University of Utah. He was also on the staff of the Jet Propulsion Laboratory, where he established a heat transfer laboratory. Kreith received a doctorate in engineering science from the University of Paris. He has an MS in engineering from UCLA and a BSME from the University of California, Berkeley.

Rosalie T. Ruegg

Rosalie T. Ruegg is a Senior Economist at the National Bureau of Standards. She has served as a consultant to government and industry on methods and procedures of economic analysis, energy pricing policy, building design criteria, government cost-sharing programs, life-safety technologies, government research programs, and finance and investment decisions. She received an MBA in finance from the American University, an MA in economics from the University of Maryland, where she was a Woodrow Wilson Fellow, and a BA from the University of North Carolina, where she was elected to Phi Beta Kappa.

G. Thomas Sav

G. Thomas Sav is Professor and Chair, Department of Economics, Wright State University. Previously, he was a Senior Economist at the Center for Economics Research, Research Triangle Institute, and a Visiting Associate Professor at North Carolina State University. He has also held positions as an Economist at the National Bureau of Standards and the U.S. Nuclear Regulatory Commission and Associate Professor and Chairman of Economics at West Georgia College. Sav received a PhD in economics from George Washington University.

Walter Short

Walter Short is a member of the Thermal Systems Research Branch at the Solar Energy Research Institute. He joined the institute in 1980 and works on end-use modeling and economic analysis methods for the evaluation of solar energy systems. Short has also worked at SRI International. He holds an MS in operations research from Stanford University.

Peter C. Spewak

Peter C. Spewak is a Vice President with Vanguard Technologies Corporation. He recently completed a solar market penetration analysis for the Virginia Electric Power Company using a modified and updated version of the SPURR model. He has been involved in the economic analysis of solar energy since 1974, specializing in the economic analysis of national policy. He was one of the primary architects of the MITRE SPURR model and played an active part in the preparation of the solar energy sections of the Project Independence Blueprint. Through the years Spewak has participated in many major national solar/renewable policy analyses and plans. He has served as Head of Policy Planning and Evaluation for the Southern Solar Energy Center and as Vice President of Bennington Enterprises.

Wesley L. Tennant

Wesley L. Tennant is a consultant who has specialized in technological and social change studies. He has worked on several Department of Energy projects involving market development and monitoring of alternative energy technologies. He has held research and project management positions with the Stanford Research Institute, George Washington University, and the Denver Research Institute. Tennant was Vice President for Information Research for Resource and Technology Management Corporation. He is a graduate of Drake University.

Mashuri L. Warren

Mashuri L. Warren is presently Product Manager for ASI Controls, a small California company developing computer technology for improved comfort and energy performance in buildings. Previously he was Staff Scientist in the Building Energy Systems Program in the Applied Science Division of Lawrence Berkeley Laboratory. From 1978 to 1988 he headed the Systems Simulation and Analysis Group for the Active Solar Cooling Project at LBL and actively participated in the DOE Active Program Research Assessment. From 1978 to 1980 Warren was responsible for the Solar Controls Test Facility at LBL. Before joining the LBL staff, Warren taught at California State University, Hayward, and at San Francisco State University. He received a PhD in physics from the University of California, Berkeley, and is a professional engineer registered in California.

Ronald E. West

Ronald E. West is Professor of Chemical Engineering at the University of Colorado, Boulder. West first worked on solar energy in the early 1960s and returned to the subject in 1978 as a consultant and sometime visiting professional at the Solar Energy Research Institute. Efficient utilization of energy and materials continue to be his chief professional interests. He holds a PhD in chemical engineering from the University of Michigan.

Name Index

Subject Index

Page numbers in *italics* refer to figures.

Absolute energy ratio. *See* AER
Absorption cooling systems, 287, *290*, *293*, *301*, 302, 305
Active Program Research Requirements (APRR) assessment, 294, 298
Active solar heating and cooling systems
  absorption cooling and, 298–305
  collector array costs of, 295–297
  cost requirements for, 274–308
  economic goals of, 279, 281
  electrical energy consumption of, 299
  electrical performance improvement of, 297–298
  establishing cost goals for, 288–291
  future, *296*
  incremental cost goals for, *289*
  market potential for, 275–278
  net present value vs. simple payback, 284–286
  payback and real ROI goals of, *282*
  payback period of, *281*
  ROI goals for, 279–284
  ROI of, *284*, *285*
  subsystem cost and performance goals of, 294–295
  system cost goals of, 286
  test locations for, 287
  thermal performance analysis of, 287–288
  thermal performance improvement of, 297
Adjusted internal rate of return (AIRR), 23
Advanced Concentrator Research Program, 364
Advanced Technology Subprogram, 212
AEC (Atomic Energy Commission), 132
AER (absolute energy ratio), 257–271
Aerospace Corp., 210, 218, 219–220, 224
Agricultural Extension Service, 195
AIA Research Corp., 312
Alcohol fuels, 212
American Society of Heating, Refrigeration and Air-Conditioning Engineers. *See* ASHRAE
Analytic methods, 40
Applications analysis, 205–254
  and end-use matching, 234–240
  federal research in, 209–223
  industrial case studies in, 244–247
  and industrial energy requirements analysis, 240–244
  for large power systems, 223–225
  major program elements of, 223–247
  and market analysis compared, 207

  programs, 211
  repowering programs, 225–234
  scope of, 207–209
  of solar energy, 3
Applications mapping, 247
Architectural Aluminum Manufacturers Association, 315
Area dependent cost decreases in California, *404*
ARKLA Corp., 288
Arthur D. Little, Inc., 39, 128, 149, *162*, 163
ASHRAE (American Society of Heating, Refrigeration and Air-Conditioning Engineers), 116, 288, 313, 332
Assessing market potential, 163–165
Assistant Secretary for Conservation and Renewable Energy, 220, 223
Assistant Secretary for Energy Technology, 210, 212
Atomic Energy Commission (AEC). *See* AEC
Attached sunspace, 310
Attainability-base goals, 345–347

Backup energy systems, 85
Base case, defined, 20
Battelle Memorial Institute, 173
Battelle Pacific Northwest Laboratories, 43, 210, 219, 225, 242, 312
Benefit/cost ratio (BCR or B/C) method, 24–25
Biomass energy, 134, 267, 268
BLAST, 34, 106, 110, 208, 313
BNL (Brookhaven National Laboratory), 128, 149, *150*, 296–297
Bonneville Power Administration (TVA), 200. *See also* TVA
Break-even (B–E) method, 26
Brookhaven National Laboratory. *See* BNL
Brookhaven Time-Stepped Energy System Optimization Model. *See* TESOM
Building Energy Performance Standards (BEPS), 199–200
Building load coefficient (BLC), 313

CALPAS 3, 321
Census of Manufacturers, 224, 243
Central receiver technology, 233
Coal incentives, *189*
Coal industry, 189